Formal Verification of Floating-Point Hardware Design

David M. Russinoff

Formal Verification of Floating-Point Hardware Design

A Mathematical Approach

Second Edition

 Springer

David M. Russinoff
Arm, Inc.
Austin, TX, USA

ISBN 978-3-030-87183-3   ISBN 978-3-030-87181-9   (eBook)
https://doi.org/10.1007/978-3-030-87181-9

This Springer imprint is published by the registered company Springer Nature Switzerland AG
The registered company address is: Gewerbestrasse 11, 6330 Cham, Switzerland

To Lin, Joshua, and Solomon

Foreword

Stone masons were building bridges—and using mathematics—long before 1773, when Coulomb published his groundbreaking mathematical analysis of some fundamental problems in civil engineering: the bending of beams, the failure of columns, and the determination of abutment thrusts imposed by arches. While Coulomb's work was appreciated by the mathematicians and physicists of the day, it was largely irrelevant to the bridge builders who were guided by experience, tradition, and intuition. However, over the next several centuries attitudes changed; statics is now a standard part of the civil engineer's training. One reason is that as materials and requirements changed, and as mathematics and computation further developed, what was once impractical became practical and then, perhaps, automatic. Mathematical tools facilitated the design of safe, cost-effective, reliable structures that could not have been built by earlier techniques.

But acceptance took time and, sometimes, dreadful experience. For example, a catastrophic bridge failure occurred a full century after Coulomb's work. The bridge carrying the Edinburgh to Dundee train over the estuary of the River Tay in Scotland collapsed in 1879, killing 75 people. A commission conducted a rigorous investigation that identified a number of contributing factors, including failure of the bridge designer to allow for wind loading and shoddy quality control over the manufacturing of the ironwork. The commission's recommendations, based on mathematical models and empirical studies of wind speeds and pressures, were immediately adopted by engineers designing a new rail bridge over the Forth estuary near Edinburgh. That bridge, the longest cantilever bridge in the world when completed in 1890 and still the second longest, has stood for over 120 years and carries about 200 trains daily. The lesson is clear: while mathematics alone will not guarantee your bridge will stand, mathematics, if properly applied, can guarantee the bridge will stand if its construction, environment, and use are as modeled in the design.

That mathematics can guide and reassure the engineer is nothing new (though the fact that symbolic mathematics can model the physical world is really awe-inspiring if one just stops taking it for granted). Engineers have been using mathematics this way for millennia. But as our mathematical and computational tools grow more sophisticated, they find more use in engineering.

I have witnessed this firsthand in hardware and software verification. My own specialty is the construction and use of mechanical theorem provers, i.e., software that attempts to prove a formula by deriving it from a small set of axioms using a small set of inference rules like chaining together previously proved results, substitution of equals for equals, and induction. If the axioms are all valid ("always true") and if the rules of inference preserve validity, then any formula so proved must be valid. Thus, proof is a way to establish truth. Mathematicians have been using the axiomatic method since Euclid.

A *theorem* is a formula that has been proved. A *lemma* is just another name for a theorem, but the label *lemma* is generally used only for formulas whose primary role is in the proofs of more interesting formulas. If Lemma 1 tells us that when P is true then Q is true, and Lemma 2 tells us that when Q is true, R is true, then we can chain them together to prove the theorem that when P is true, R is true.

But how do we know that a "proof" is a proof? Traditionally, when a mathematician publicizes a proof, other mathematicians interested in that result scrutinize it, frequently finding flaws, typos, forgotten cases, unstated assumptions, etc. Over time these flaws are fixed—if possible—and eventually the mathematical community accepts the formula as proved.

Mechanical theorem provers are designed to circumvent this "social process" to some extent. If a result has been proved by a trusted mechanical prover, then one can rest assured that the "proof" is a proof and that the formula is "always true". The social process need only inspect the formula itself and its underlying definitions and decide whether the formula actually means what the author intended to say.

An important application of mechanical theorem provers is to prove properties of computer hardware and software. Exhaustive testing is impractical for modern computing artifacts: there are just too many cases to consider. Proof is the obvious way to establish the truth of properties. But the mathematical social process is not well suited to these proofs: the proofs are often long; there are myriad cases to consider because the artifacts, their specifications, and the underlying logical concepts are often complicated; the various logical concepts are related in a wide variety of ways by many lemmas; and the artifacts and their properties may be proprietary. Just keeping track of all the assumptions and known relationships can be an almost impossible challenge. Using a machine to check proofs is an ideal solution.

I said above that I have witnessed firsthand the slow acceptance of "new" mathematical methods by the engineering community. Here is a part of the story:

In 1987, after sixteen years of working on mechanical theorem provers for hardware and software verification, my colleagues and I at the University of Texas at Austin started a company whose mission was to spread the technology to industry. One of those colleagues was David Russinoff, who had joined our theorem proving group in 1983. When we learned that David had a PhD in number theory from the Courant Institute, we challenged him to prove Wilson's Theorem with the prover and offered him a Master's degree for doing so. He succeeded and went on to prove mechanically many other results in number theory, including Gauss's Law of Quadratic Reciprocity. His was the first mechanically checked proof of that theorem and, as he writes in his online bibliography, "The primary significance of this last result, of course, was that it finally put to rest any suspicion that the 197 previously published proofs of this theorem were all flawed." His humor obscures the fact that the result was of sufficient interest that 197 proofs were worthy of publication!

The company that the UT Austin group spun off in 1987 was called Computational Logic, Inc. (CLI). There we continued to develop our tools, and in 1989, together with my colleagues Bob Boyer and Matt Kaufmann, I started work on a new prover that we called ACL2: A Computational Logic for Applicative Common Lisp. The characteristics of ACL2 are unimportant except for three things. First, ACL2's logic is a subset of a standard programming language and is thus well suited to modeling computational artifacts like hardware designs, programming languages, algorithms, etc. Second, the more complicated a proof is, the more help the ACL2 user has to provide. Think of the prover as trying to find a path from point A to point B. The more intermediate milestones the user provides, the more successful the prover will be. Those milestones are the key lemmas in the proof. Third, the existence of prior ACL2 work in a domain can be a great aid to the user because it may not be necessary to formalize and prove so many lemmas.

In 1995, Advanced Micro Devices (AMD) hired the company to prove the mathematical correctness of the microcode for its floating-point division operation on the soon-to-be-fabricated AMD K5 microprocessor. This was especially important because the Intel Pentium FDIV bug was just beginning to be publicized and AMD had recently made major changes in its floating-point unit, requiring the design team to discard previously tested hardware for division and re-implement it in microcode using floating-point addition, subtraction, and multiplication.

I temporarily joined the AMD floating-point design team to lead the CLI proof effort. The goal was to prove that the FDIV algorithm complied with IEEE Floating-Point Standard 754. Roughly speaking, the tasks could be thought of as follows: formalize the relevant part of the standard, formalize the algorithm, formalize the desired properties, and then lead the mechanical theorem prover to a proof of the properties by discovering, formalizing, and piecing together lemmas about floating-point operations and concepts such as rounding. ACL2 had never been used to do proofs about floating-point, so there was no "bookshelf" of previously proved lemmas to draw upon. And while I had been programming for decades and thus had a passing familiarity with floating-point, I was basically ignorant of the technical details. The designers in the group, many with a decade or more of experience in floating-point design, were the experts. But their explanations of the various steps taken by the division algorithm often involved examples, graphs plotting trends in error reduction, assertions that this or that technique was well known and had been reliably used in earlier products, etc. My real job was to distill that informal "shop talk" down into formulas and prove them.

Once in a weekly meeting, when asked what I had accomplished that week, I displayed a lemma I had proved only to be met by the universal response "we knew that". However, there is a big difference between "knowing" something and writing it down precisely so that it is always true even when interpreted by the most malicious of readers! The lemma under discussion was a version of Lemma 6.80 concerning what was called "sticky" rounding. I strongly suspect that while everybody in the room could state some of the necessary hypotheses, I was the only person in the room who could state them all and do so with confidence.

Working with Matt Kaufmann and the lead designer, Tom Lynch, we eventually got ACL2 to prove the theorem in question. The project took a total of 9 weeks and the K5 was fabricated on time.

AMD was interested in proving other properties of components of its floating-point unit, but it was uninterested in hiring CLI to do so: sharing AMD proprietary designs with outsiders was unusual, to say the least. Aside from offering AMD assurance that its FDIV operation was correct, AMD considered the CLI project a test to see if our technology was useful. We passed the test, but they wanted an AMD employee to drive the prover. The person they hired away from CLI was David Russinoff.

One might have hoped that upon arrival at AMD, David could simply build on the library of lemmas developed for the FDIV proof. But that library was just what you would expect from a 9-week crash effort: an ad hoc collection of inelegant formulas. Each was, of course, valid—that is the beauty of mechanized proof—but they were far from a useful theory of floating-point. So David started over and followed a disciplined approach to developing such a theory in ACL2.

Here he presents a carefully considered collection of the key properties of floating-point operations of use when proving a wide variety of theorems about many different kinds of algorithms. The properties are stated accurately and in complete detail, there are no hidden assumptions, and each lemma and theorem is valid. I consider this work a tour de force in formal reasoning about floating-point designs. It truly provides a formal mathematical basis for the analysis of floating-point.

The definitions and theorems in this book are shown in conventional mathematical notation, not the rigid formal syntax of ACL2. But underlying this volume is a large collection of definitions and theorems in ACL2's syntax, all formally processed and certified by ACL2. That collection—built by David over twenty years of industrial application of ACL2 to floating-point designs—is a powerful aid to formal verification. Many familiar companies have used David's library. This book is exquisite documentation of that library.

If you trust ACL2 and care only about validity, the proofs shown here are unimportant, since the formulas have been mechanically verified. But in a much deeper sense, following the traditions of mathematics, the proofs are everything because *they explain why these formulas are valid*. Each proof also illustrates how the previous lemmas can be used—a point driven home mainly by Part V of the book, where David formally analyzes a commercially interesting floating-point unit.

Returning to the general theme of this foreword—the role of mathematics in the construction of useful artifacts and the adoption of techniques that were impractical just decades earlier—it is not unusual today to see logic-based tools, such as mechanized theorem provers, in use in design and verification groups in the microprocessor industry. Companies such as AMD, Arm, Centaur, Intel, Oracle, Samsung, and Rockwell Collins have used mechanical theorem provers to prove important theorems about their products.

That last sentence would have been unthinkable just decades ago. Formal logic, once studied only by philosophers and logicians, now is a branch of applied mathematics. Mechanized provers are used in the design of amazing but now common everyday objects found in everything from high-end servers to mobile devices, from medical instruments to aircraft avionics. Indeed, many modern microprocessor designs simply cannot be built with confidence without such tools. The work presented in this book makes it easier to do the once unimaginable.

Edinburgh, Scotland   J Strother Moore
April 2018

Preface to the Second Edition

During the three years since the publication of the first edition, my work in arithmetic circuit design verification has continued at Arm, Inc. It is unsurprising that the underlying mathematical theory, which had been developed through investigations of industrial hardware designs over the preceding two decades, has continued to evolve.

As always, increasingly sophisticated algorithms and implementation techniques have presented new verification challenges, requiring new proofs and extensions of the theory. Consequently, the final chapters, consisting of correctness proofs of Arm floating-point designs, have been replaced with new chapters, reflecting the next generation of processors. In particular, the radix-4 divider described in the first edition has been replaced with a more efficient radix-8 version, and the underlying theory presented in Chap. 10 has been further developed accordingly. Another new development is the introduction to the Arm architecture of "Intel compatibility mode" of floating-point underflow detection. This requires implementations of both the traditional Arm detection method, based on the unrounded result, and the x86 method, which involves an additional rounding. We have designed an efficient solution to this problem that requires very little additional hardware as described in a new section of Chap. 6.

As of 2018, my work in formal verification had been restricted to the higher-precision floating-point modules of so-called big core CPUs. These are generally the most complex and error-prone arithmetic designs and are typically resistant to the more traditional methods of verification by equivalence checking. They are also the focus of the Arm Austin arithmetic group. As the member of that group who was solely responsible for this task, I had little time for other investigations. The following year, however, I was joined by Cuong Chau, a recently graduated Ph.D. from the University of Texas with experience and expertise in interactive theorem proving with ACL2. Chau's collaboration has allowed me to explore different flavors of arithmetic design at various Arm sites, which has led to a few surprises.

While integer operations are generally simpler than floating-point, the two's complement encoding scheme introduces a number of complications that are not present with a sign and magnitude representation. For example, my encounter with an integer multiplier resulted in an extension of Chap. 9 to provide for multiplication of signed integers. A new chapter treats integer division as implemented by the new divider along with the floating-point operations. A novel feature of this design is fast detection of a power-of-two divisor, allowing early exit in this case. For a floating-point number, this is a simple matter of noting a zero mantissa, but for a negative two's complement integer, it is more difficult. The design employs an efficient new algorithm to solve this problem, the verification of which involves a challenging inductive proof.

Similarly, the arithmetic implementations found in smaller cores and graphics processors, for which execution speed is less important than conservation of power and silicon area, face a different set of problems. Another new chapter describes a small core radix-2 SRT divider, a design that is unlikely to be found in a big core, for which the speed provided by a higher radix (at the expense of power and area) is a necessity. Although a simpler implementation than the higher-radix dividers, it was found to be too complex for verification by equivalence checking. One interesting twist is a feature that addresses a problem unique to maximally redundant SRT dividers (a necessary property of a radix-2 divider) pertaining to the accuracy of the partial remainder approximation, as described in Sect. 10.1. This may be easily solved by widening the adder used in the approximation, but in this case, in order to avoid that cost in area, a more sophisticated method of "remainder compression" is employed. The proof of correctness of this method is an interesting aspect of that chapter.

The final chapter of this edition, which is also new, describes a fused multiply-add unit of a GPU. Since this module is replicated many times in a single core, its area is a high priority, which influenced the design of the integer adder. The designer of a big core FPU usually enjoys the luxury of leaving this task to a synthesis tool, which generally selects a fast addition algorithm with little regard for area. In this case, a hand-coded solution was preferable. This led to a new section of Chap. 8 on integer adders (specifically, parallel prefix adders), a subject that was largely ignored in the first edition.

The same GPU module also features an interesting new leading zero anticipator. The first LZA of Sect. 8.3, which is used in the big core adder of Chap. 17, is both faster and simpler but must be duplicated in hardware because its design is based on an assumed ordering of the operands, which is known only after the addition is performed. The new LZA, which is described in the same section, does not rely on this assumption and is consequently more complex, as is its correctness proof.

These additions to the book serve as illustrations of the benefits of a rigorous approach to arithmetic circuit verification. In particular, the topics of remainder compression, parallel prefix adders, and leading zero anticipation mentioned above are all commonly addressed throughout the engineering literature, but to my knowledge, never before in a sufficiently rigorous manner to support formal verification.

The first six chapters, which cover the basic theory of register-transfer logic and floating-point arithmetic, form the foundation of the book and are intended to be read in order. The remaining chapters are less strictly ordered; thus, (1) Chap. 7 may be omitted by the reader who is not interested in the problem of formally specifying the behavior of the square root operation in a logic restricted to rational arithmetic; (2) the chapters of Part III are unordered and largely independent of Part II; (3) Part IV is independent of Part III; and (4) there is little interdependence among the final seven chapters, each of which is independent of all but the one relevant chapter of Part III.

For each of the applications of those final chapters, the corresponding RAC model, its pseudocode version, and the ACL2 proof script are available in the ACL2 directory books/projects/arm/second/. ACL2 remains freely available at www.cs.utexas.edu/users/moore/acl2.

Perhaps the most important contribution of this edition is the correction of numerous errors, typographical and otherwise, that plagued the first. I appreciate the assistance received from readers, especially Cuong Chau, Chris Larsen, and Mayank Manjrekar, but I am certain that we have not found them all, and moreover that I have introduced new ones. I intend to maintain a Web site associated with the book, including a list of errata, at https://go.sn.pub/fvfphd.

I am grateful to the entire Arm Austin arithmetic group for their contributions to my understanding of computer arithmetic and circuit design: Javier Bruguera, Remi Chaintreuil, Cuong Chau, Thomas Elmer, David Lutz, Nicholas Pfister, Thomas Tarridec, and Harsha Valsaraju. I am also indebted to the engineers with whom I have worked at various other Arm sites, notably Neil Burgess, Luka Dejanovic, David Gilday, Michael Kennedy, Jørn Nystad, and Harsh Rungta. Nystad is responsible for the FMA of Chapter 22; the principal designer of the divider of Chapter 21 was Fergus MacGarry. It must also be acknowledged that much of this work would not have been possible without the managerial guidance of Ram Narayan and Vaibhav Agrawal.

Block Island, RI, USA   D.M.R.
July 2021

Preface to the First Edition

It is not the purpose of this book to expound the principles of computer arithmetic algorithms, nor does it presume to offer instruction in the art of arithmetic circuit design. A variety of publications spanning these subjects are readily available, including general texts as well as more specialized treatments, covering all areas of functionality and aspects of implementation. There is one relevant issue, however, that remains to be adequately addressed: the problem of eliminating human error from arithmetic hardware designs and establishing their ultimate correctness.

As in all areas of computer architecture, the designer of arithmetic circuitry is preoccupied with efficiency. His objective is the rapid development of logic that optimizes resource utilization and maximizes execution speed, guided by established practices and intuition. Subtle conceptual errors and miscalculations are accepted as inevitable, with the expectation that they will be eliminated through a separate validation effort.

As implementations grow in complexity through the use of increasingly sophisticated techniques, errors become more difficult to detect. It is generally acknowledged that testing alone is insufficient to provide a satisfactory level of confidence in the functional correctness of a state-of-the-art floating-point unit; formal verification methods are now in widespread use. A common practice is the use of an automated sequential logic equivalence checker [21, 36] to compare a proposed register-transfer logic (RTL) design either to an older trusted design or to a high-level C++ model. One deficiency of this approach is that the so-called "golden model", whether coded in Verilog or C++, has typically never been formally verified itself and thus cannot be guaranteed to be free of errors. Another is the inherent complexity limitations of such tools [35], which have been found to render them inadequate for the comprehensive verification of complex high-precision floating-point modules.

A variety of projects have attempted to address these issues by combining such automatic methods with the power of interactive mechanical theorem proving [22, 27]. This book is an outgrowth of one such effort in the formal verification of commercial floating-point units, conducted over the course of two decades during which I was employed by Advanced Micro Devices, Inc. (1996–2011), Intel

Corp. (2012–2016), and Arm, Inc. (2016–present). My theorem prover of choice is ACL2 [11–13], a heuristic prover based on first order logic, list processing, rational arithmetic, recursive functions, and mathematical induction. ACL2 is a freely available software system, developed and maintained at the University of Texas by Matt Kaufmann and J Moore.

The principal advantages of theorem proving over equivalence and model checking are greater flexibility, derived from a more expressive underlying logical notation, and scalability to more complex designs. The main drawback is the requirement of more control and expertise on the part of the user. In the domain of computer arithmetic, effective use of an interactive prover entails a thorough understanding and a detailed mathematical exposition of the design of interest. Thus, the success of our project requires an uncommonly scrupulous approach to the design and analysis of arithmetic circuits. Loose concepts, intuition, and arguments by example must be replaced by formal development, explicit theorems, and rigorous proofs.

One problem to be addressed is the semantic gap between abstract behavioral specifications and concrete hardware models. While the design of a circuit is modeled for most purposes at the bit level, its prescribed behavior is naturally expressed in terms of high-level arithmetic concepts and algorithms. It is often easier to prove the correctness of an algorithm than to demonstrate that it has been implemented accurately.

As a simple illustration, consider the addition of two numbers, x and y, as diagrammed below in binary notation. Suppose the sum $z = x + y$ is to be truncated at the dotted line and that the precision of x is such that its least significant nonzero bit lies to the left of the line.

$$
\begin{array}{r}
\texttt{1xxxxx.xxxxxxxxx00} \cdots \\
+ \quad \texttt{1yyy.yyyyyyyyy\,yy} \cdots \\
\hline
\texttt{1zzzzzz.zzzzzzzzz\,zz} \cdots
\end{array}
$$

Instead of computing the exact value of z and then extracting the truncated result, an implementation may choose to perform the truncation on y instead (at the same dotted line) before adding it to x. The designer, in order to convince himself of the equivalence of these two approaches, might resort to a diagram like the one above.

The next logical step would be to formulate the underlying principle in precise terms (see Lemma 6.14 in Sect. 6.1), explicitly identifying the necessary conditions for equivalence and establishing its correctness by rigorous mathematical proof. The result could then be integrated into an evolving theory of computer arithmetic and thus become available for reuse and subject to extension and generalization (e.g., Lemma 6.90 in Sect. 6.5) as appropriate for new applications.

Such results are nowhere to be found in the existing literature, which is more concerned with advanced techniques and optimizations than with their theoretical underpinnings. What prevents the organization of the essential properties of the basic data objects and operations of this domain into a theory suitable for systematic application? The foregoing example, however trivial, serves to illustrate one of the main obstacles to this objective: the modeling of data at different levels of abstraction. Numbers are naturally represented as bit vectors, which, for some purposes, may be viewed simply as strings of boolean values. Truncation, for example, is conveniently described as the extraction of an initial segment of bits. For the purpose of analyzing arithmetic operations, on the other hand, the same data must be interpreted as abstract numbers. Although the correspondence between a number and its binary representation is straightforward, a rigorous proof derived from this correspondence requires more effort than an appeal to intuition based on a simple diagram. Consequently, the essential properties of bit vector arithmetic have never been formalized and compiled into a well-founded comprehensive theory. In the absence of such a theory, the designer must rediscover the basics as needed, relying on examples and intuition rather than theorems and proofs.

Nearly two centuries ago, the Norwegian mathematician Niels Abel complained about a similar state of affairs in another area of mathematical endeavor [14]:

> It lacks so completely all plan and system that it is peculiar that so many men could have studied it. The worst is, it has never been treated stringently. There are very few theorems ... which have been demonstrated in a logically tenable manner. Everywhere one finds this miserable way of concluding from the specific to the general ...

The subject in question was the calculus, a major mathematical development with a profound impact on the sciences, but lacking a solid logical foundation. Various attempts to base it on geometry or on intuition derived from other areas of mathematics proved inadequate as the discipline grew in complexity. The result was a climate of uncertainty, controversy, and stagnation. Abel and others resolved to restore order by rebuilding the theory of calculus solely on the basis of arithmetical concepts, thereby laying the groundwork for modern mathematical analysis.

While the contemporary hardware engineer may not be susceptible to the same philosophical qualms that motivated nineteenth-century mathematicians, he is certainly concerned with the "bugs" that inevitably attend undisciplined reasoning Even if the analogy overreaches the present problem, it charts a course for its solution and sets the direction of this investigation.

Contents and Structure of the Book

Our initial objective is a unified mathematical theory, derived from the first principles of arithmetic, encompassing two distinct domains of discourse. The first of these, which is the subject of Part I, is the realm of register-transfer logic (RTL), comprising the primitive data types and operations on which microprocessor designs are built: bit vectors and logical operations. A critical first step is the careful formulation of these primitives in a manner consistent with our goals, which sometimes requires resistance to intuition. Thus, notwithstanding its name, we define a bit vector of width k to be a natural number less than 2^k rather than a sequence of k boolean values. This decision will seem unnatural to those who are accustomed to dealing with these objects in the context of hardware description languages, but it is a critical step in the master plan of an arithmetic-based theory. As a consequence, which may also be disturbing to some, the logical operations are defined recursively rather than bit-wise.

Part II addresses the second domain of interest, the more abstract world of rational arithmetic, focusing on floating-point representations of rational numbers as bit vectors. The benefits of a rigorous approach are most evident in the chapter on floating-point rounding, which includes a variety of results that would otherwise be difficult to state or prove, especially those that relate abstract arithmetic operations on rationals to lower-level properties of bit vectors. All of the architectural rounding modes prescribed by IEEE Standard 754 [9] are thoroughly analyzed. Moreover, since hardware implementation is of central interest, further attention is given to several other modes that are commonly used for internal computations but are not normally covered in treatises on floating-point arithmetic.

In Part III, the theory is extended to the analysis of several well-known algorithms and techniques used in the implementation of elementary arithmetic operations. The purpose here is not to present a comprehensive survey of the field but merely to demonstrate a methodology for proving the correctness of implementations, providing guidance to those who are interested in applying it further. There is a chapter on addition, including a discussion of leading zero anticipation, and another on multiplication, describing several versions of Booth encoding. Two division algorithms are analyzed: a subtractive SRT algorithm, which is also applied to square root extraction, and a multiplicative algorithm based on a fused multiplication-addition operation.

Although IEEE 754 is routinely cited as a specification of correctness of floating-point implementations, it contains a number of ambiguities and leaves many aspects of behavior unspecified, as reflected in the divergent behaviors exhibited by various "compliant" architectures, especially in the treatment of exceptional conditions. In particular, the two primary floating-point instruction sets of the x86 architecture, known as *SSE* and *x87*, employ distinct exception-handling procedures that have been implicitly established across the microprocessor industry. Another important floating-point instruction set, with its own variations of exception handling, is provided by the Arm architecture. For every new implementation of any of these architectures, backward compatibility is a strict requirement. Unfortunately, no existing published reference is adequate for this purpose. When a verification engineer seeks clarification of an architectural detail, he is likely to consult an established expert, who may refer to a trusted RTL module or even a comment embedded in a microcode file, but rarely a published programming manual and never, in my experience, an IEEE standard. We address this problem in Part IV, presenting comprehensive behavioral specifications for the elementary arithmetic instructions of these three instruction sets—SSE, x87, and Arm—which were compiled and tested over the course of more than twenty years through simulation and analysis of commercial RTL models.

Part V describes and illustrates our verification methodology. In Chap. 15, we present a functional programming language, essentially a primitive subset of C augmented by C++ class templates that implement integer and fixed-point registers. This language has proved suitable for abstract modeling of floating-point RTL designs and is susceptible to mechanical translation to the ACL2 logic. The objective in coding a design in the language is a model that is sufficiently faithful to the RTL to allow efficient equivalence checking by a standard commercial tool, but as abstract as possible in order to facilitate formal analysis. The ACL2 translation of such a model may be verified to comply with the appropriate behavioral specification of Part IV as an application of the results of Parts I–III, thereby establishing correctness of the original RTL. Each of the final chapters contains a correctness proof of a module of a state-of-the-art floating-point unit that has been formally verified through this process.

Formalization: The Role of ACL2

All definitions, lemmas, and theorems presented in this exposition have been formalized in the logical language of ACL2 and mechanically checked with the ACL2 prover. The results of Parts I–IV have been collected in an evolving library— a component of the standard ACL2 release—which has been used in the formal verification of a variety of arithmetic RTL designs [25, 29–34]. These include the modules presented in Part V, the proof scripts for which also reside in the ACL2 repository and are thus available to the ACL2 user, who may wish to "replay" and experiment with these proofs on his own machine.

The role of ACL2 in the development of the theory is evidenced in various ways throughout the exposition, including its emphasis on recursion and induction, but mainly in the level of rigor that is inherent in the logic and enforced by the prover. Any vague arguments, miscalculations, or missing hypotheses may be assumed to have been corrected through the mechanization process. With regard to style of presentation, the result will appeal to those readers whose thought processes most closely resemble the workings of a mechanical theorem prover; others may find it pedantic.

The above claim regarding the correspondence between the results presented here and their ACL2 formalizations requires a caveat. Since the ACL2 logic is limited to rational arithmetic,[1] properties that hold generally for real numbers can be stated formally only as properties of rationals. In the chapters relevant to the theory of rounding—Chaps. 1, 4, and 6—all results, with the single exception of Lemma 4.13,

---

[1] It should be noted that an extension of ACL2 supporting the reals through nonstandard analysis has been implemented by Ruben Gamboa [6].

are stated and proved more generally, with real variables appearing in place of the rational variables found in the corresponding ACL2 code.

This limitation of the logic is especially relevant to the square root operator, which cannot be defined as an ACL2 function. This presents a challenge in the formalization of IEEE compliance (i.e., correctly rounded results) of a square root implementation, which is addressed in Chap. 7. Here we introduce an equivalent formulation of the IEEE requirement that is confined to the domain of rational arithmetic and thus provides for the ACL2 formalization of the architectural specifications presented in Part IV.

On the other hand, the ACL2 connection may be safely ignored by those uninterested in the formal aspect of the theory. Outside of Sect. 15.6, no familiarity with ACL2, LISP, or formal logic is required of (or even useful to) the reader. Several results depend on computations that have been executed with ACL2 and cannot reasonably be expected to be carried out by hand,[2] but in each case, the computation is explicitly specified and may be readily confirmed in any suitable programming language. Otherwise, the entire exposition is surveyable and self-contained, adhering to the most basic conventions of mathematical notation, supplemented only by several RTL constructs common to hardware description languages, all of which are explicitly defined upon first use. Nor is any uncommon knowledge of mathematics presupposed. The entire content should be accessible to a competent high school student who has been exposed to the algebra of real numbers and the principle of mathematical induction. It must be conceded, however, that repeated attempts to substantiate this claim have been consistently unsuccessful.

The book has been written with several purposes in mind. For the ACL2 user interested in applying or extending the associated RTL library, it may be read as a user's manual. The theory might also be used to guide other verification efforts, either without mechanical support or encoded in the formalism of another theorem prover. A more ambitious goal is a rigorous approach to arithmetic circuit design that is accessible and useful to architects and RTL writers.

Obtaining the Associated ACL2 Code

The ACL2 distribution [13] includes a books directory, consisting of libraries contributed and maintained by members of the ACL2 user community. Three of its subdirectories are related to this project:

- books/rtl/: The formalization of the theory presented in Parts I–IV;
- books/projects/rac/: The parser and ACL2 translator for the language described in Chap. 15;

---

[2] These pertain to table-based reciprocal computation (Lemma 11.1), FMA-based division (Lemmas 11.7 and 11.9), and SRT division and square root (Lemmas 10.7, 10.8, and 10.15).

- `books/projects/arm`: The scripts for the proofs presented in the final chapters.

A more up-to-date version of the `books` directory is available more directly through the GitHub hosting service at https://github.com/acl2/acl2/tree/master/books/.

Acknowledgments

In 1995, in the wake of the Intel FDIV affair, Moore and Kaufmann were contracted by AMD to verify the correctness of the division instruction of the K5 processor—AMD's answer to the Pentium—with their new theorem prover. Soon thereafter, I was assigned the task of extending their work to the K5 square root operation. I am indebted to them for creating a delightful prover and demonstrating its effectiveness as a floating-point verification tool, thus laying a foundation for this venture.

Whatever I learned about arithmetic circuits over the intervening twenty-two years was explained to me by the designers with whom I have been privileged to collaborate: Mike Achenbach, Javier Bruguera, Michael Dibrino, David Dean, Steve Espy, Warren Ferguson, Kelvin Goveas, Carl Lemonds, Dave Lutz, Tom Lynch, Stuart Oberman, Simon Rubanovich, and Peter Seidel.

A critical component of the AMD effort was a Verilog-ACL2 translator originally implemented by Art Flatau and me and later refined by Kaufmann, Hanbing Liu, and Rob Sumners. The ACL2 formalization of some of the early proofs was done by Kaufmann, Liu, and Eric Smith.

The development and application of the verification methodology described in Part V began at Intel and has continued at Arm. The initial design of the modeling language was a collaboration with John O'Leary and benefited from suggestions by Rubanovich. Bruguera and Lutz are the architects of the Arm FPU on which the final chapters are based.

I am also grateful to Matthew Bottkol, Warren Ferguson, Cooky Goldblatt, David Hardin, Linda Ness, and Coke Smith for their suggestions for improvement of earlier versions of the manuscript, and to Glenn Downing for telling me that it was time to publish the book. Finally, I cannot deny myself the pleasure of thanking Lin Russinoff, who designed the cover graphics and is my continual source of inspiration.

Austin, Tex, USA   D.M.R.
April 2018

Contents

Part I
Register-Transfer Logic

An arithmetic circuit design may be modeled at various levels of abstraction, ranging from a *netlist*, which specifies an interconnection of electronic components, to a high-level numerical algorithm. Although the former is required for implementation, modern synthesis tools can readily create such a model from a somewhat higher-level *register-transfer logic* (RTL) design coded in a hardware description language (HDL), such as Verilog. There also exist high-level synthesis tools that can convert numerical algorithms coded in C++ to Verilog RTL, but these are not nearly mature enough to achieve the level of efficiency required of today's commercial floating-point units. Consequently, most of the design and verification effort is conducted at the RTL level, i.e., in terms of the flow of data between hardware registers and the logical operations performed on them.

The following three chapters, on which the entire book is based, may be viewed as an exposition of basic HDL semantics. The fundamental data type is the *bit vector*, which is implemented as an ordered set of two-state devices. A possible starting point for our theory, therefore, is the definition of a bit vector as a sequence of boolean values. An advantage of this approach is that it allows straightforward definitions of primitive operations, such as bit extraction and concatenation. In the realm of computer arithmetic, however, a bit vector is viewed more fruitfully as the binary expansion of an integer. In our formalization, we shall identify a bit vector with the number that it so represents, i.e., we shall define a bit vector simply to be an integer. This naturally leads to the definition of the primitive RTL operations as arithmetic functions. While the consequences of this decision may sometimes seem cumbersome, its benefits will become clear in later chapters as we further explore the arithmetic of bit vectors.

In Chap. 2, we formalize and examine the properties of bit vectors and the primitive operations of bit slice, bit extraction, and concatenation. The bit-wise logical operations are discussed in Chap. 3. All of these operations are defined in terms of the basic arithmetic functions *floor* and *modulus*, which are the subject of Chap. 1.

Chapter 1
Basic Arithmetic Functions

This chapter examines the properties of the *floor*, *ceiling*, and *modulus* functions, which are central to our formulation of the RTL primitives as well as the floating-point rounding modes. Thus, their definitions and properties are prerequisite to a reading of the subsequent chapters. We also define and investigate the properties of a function that truncates to a specified number of fractional bits, which is related to the floor and is relevant to the analysis of fixed-point encodings, as discussed in Sect. 2.5, among other applications.

The reader will find many of the lemmas of this chapter to be self-evident and may question the need to include them all, but it will prove convenient to have these results collected for later reference.

Notation The symbols $\mathbb{R}$, $\mathbb{Q}$, $\mathbb{Z}$, $\mathbb{N}$, and $\mathbb{Z}^+$ will denote the sets of all real numbers, rational numbers, integers, natural numbers (i.e., nonnegative integers), and positive integers, respectively.

1.1 Floor and Ceiling

The functions $\lfloor x \rfloor$ and $\lceil x \rceil$, known as the *floor* and *ceiling*, are approximations of reals by integers. The floor is also known as the *greatest integer* function, because the value of $\lfloor x \rfloor$ may be characterized as the greatest integer not exceeding x:

Definition 1.1 For each $x \in \mathbb{R}$, $\lfloor x \rfloor$ is the unique integer that satisfies

$$\lfloor x \rfloor \leq x < \lfloor x \rfloor + 1.$$

We list several obvious consequences of the definition:

Lemma 1.1 *Let* $x \in \mathbb{R}$, $y \in \mathbb{R}$, *and* $n \in \mathbb{Z}$.

© The Author(s), under exclusive license to Springer Nature Switzerland AG 2022
D. M. Russinoff, *Formal Verification of Floating-Point Hardware Design*,
https://doi.org/10.1007/978-3-030-87181-9_1

(a) $\lfloor x \rfloor = x \Leftrightarrow x \in \mathbb{Z}$;
(b) $x \le y \Rightarrow \lfloor x \rfloor \le \lfloor y \rfloor$;
(c) $n \le x \Rightarrow n \le \lfloor x \rfloor$;
(d) $\lfloor x + n \rfloor = \lfloor x \rfloor + n$.

Lemma 1.2 *If $x \in \mathbb{R}$ and $n \in \mathbb{Z}^+$, then $\lfloor \lfloor x \rfloor / n \rfloor = \lfloor x/n \rfloor$.*

Proof Since $\lfloor x \rfloor \le x$, $\lfloor x \rfloor / n \le x/n$, and, by monotonicity, $\lfloor \lfloor x \rfloor / n \rfloor \le \lfloor x/n \rfloor$. To derive the reverse inequality, note that since $x/n \ge \lfloor x/n \rfloor$, we have $x \ge n \lfloor x/n \rfloor$. It follows that $\lfloor x \rfloor \ge n \lfloor x/n \rfloor$, and hence $\lfloor x \rfloor / n \ge \lfloor x/n \rfloor$, which implies $\lfloor \lfloor x \rfloor / n \rfloor \ge \lfloor x/n \rfloor$. □

The next result is used in a variety of inductive proofs pertaining to bit vectors. (See, for example, the proof of Lemma 2.22 and the discussion following Definition 3.1.)

Lemma 1.3 *If $n \in \mathbb{Z}$, then $|\lfloor n/2 \rfloor| \le |n|$, and if $n \notin \{0, -1\}$, then $|\lfloor n/2 \rfloor| < |n|$.*

Proof It is clear that equality holds for $n = 0$ or $n = -1$. If $n \ge 1$, then

$$|\lfloor n/2 \rfloor| = \lfloor n/2 \rfloor \le n/2 < n = |n|.$$

If $n \le -2$, then since $\lfloor n/2 \rfloor > n/2 - 1$ and $n/2 \le -1$,

$$|\lfloor n/2 \rfloor| = -\lfloor n/2 \rfloor < -(n/2 - 1) = n/2 + 1 - n \le -n = |n|.$$ □

The floor commutes with negation only for integer arguments:

Lemma 1.4 *For all $x \in \mathbb{R}$,*

$$\lfloor -x \rfloor = \begin{cases} -\lfloor x \rfloor & \text{if } x \in \mathbb{Z} \\ -\lfloor x \rfloor - 1 & \text{if } x \notin \mathbb{Z}. \end{cases}$$

Proof If $x \in \mathbb{Z}$, then Lemma 1.1 implies

$$\lfloor -x \rfloor = -x = -\lfloor x \rfloor.$$

Otherwise,

$$\lfloor x \rfloor < x < \lfloor x \rfloor + 1,$$

which implies

$$-\lfloor x \rfloor - 1 < -x < -\lfloor x \rfloor,$$

and by Definition 1.1,

$$\lfloor -x \rfloor = -\lfloor x \rfloor - 1.$$

□

When x is expressed as a ratio of integers, we also have the following unconditional expression for $\lfloor -x \rfloor$.

Lemma 1.5 *If $m \in \mathbb{Z}$, $n \in \mathbb{Z}$, and $n > 0$, then*

$$\lfloor -m/n \rfloor = -\lfloor (m-1)/n \rfloor - 1.$$

Proof Suppose first that $m/n \in \mathbb{Z}$. Then

$$\lfloor (m-1)/n \rfloor = \lfloor m/n - 1/n \rfloor = m/n + \lfloor -1/n \rfloor = m/n - 1$$

and

$$-\lfloor (m-1)/n \rfloor - 1 = -m/n = \lfloor -m/n \rfloor.$$

Now suppose $m/n \notin \mathbb{Z}$. Then $m/n > \lfloor m/n \rfloor$, which implies $m > \lfloor m/n \rfloor n$, and hence $m \geq \lfloor m/n \rfloor n + 1$. Thus,

$$\lfloor m/n \rfloor \leq (m-1)/n < m/n < \lfloor m/n \rfloor + 1,$$

and by Definition 1.1, $\lfloor (m-1)/n \rfloor = \lfloor m/n \rfloor$. Finally, by Lemma 1.4,

$$\lfloor -m/n \rfloor = -\lfloor m/n \rfloor - 1 = -\lfloor (m-1)/n \rfloor - 1. \qquad \qquad □$$

Examples

$$\left\lfloor -\frac{6}{5} \right\rfloor = -\left\lfloor \frac{6-1}{5} \right\rfloor - 1 = -\lfloor 1 \rfloor - 1 = -1 - 1 = -2$$

$$\lfloor -1 \rfloor = \left\lfloor -\frac{5}{5} \right\rfloor = -\left\lfloor \frac{5-1}{5} \right\rfloor - 1 = -\left\lfloor \frac{4}{5} \right\rfloor - 1 = 0 - 1 = -1$$

$$\left\lfloor -\frac{4}{5} \right\rfloor = -\left\lfloor \frac{4-1}{5} \right\rfloor - 1 = -\left\lfloor \frac{3}{5} \right\rfloor - 1 = 0 - 1 = -1$$

The ceiling is defined most conveniently using the floor:

Definition 1.2 For all $x \in \mathbb{R}$, $\lceil x \rceil = -\lfloor -x \rfloor$.

We have an alternative characterization of $\lceil x \rceil$, analogous to Definition 1.1, as the least integer not exceeded by x:

Lemma 1.6 *For all $x \in \mathbb{R}$, $\lceil x \rceil \in \mathbb{Z}$ and $\lceil x \rceil \geq x > \lceil x \rceil - 1$.*

Proof By Definition 1.1, $\lfloor -x \rfloor \leq -x < \lfloor -x \rfloor + 1$, which leads to $-\lfloor -x \rfloor \geq x > -\lfloor -x \rfloor - 1$. The lemma now follows from Definition 1.2. □

We also have analogs of Lemmas 1.1 and 1.2:

Lemma 1.7 *Let $x \in \mathbb{R}$, $y \in \mathbb{R}$, and $n \in \mathbb{Z}$.*

(a) $\lceil x \rceil = x \Leftrightarrow x \in \mathbb{Z}$;
(b) $x \le y \Rightarrow \lceil x \rceil \le \lceil y \rceil$;
(c) $n \ge x \Rightarrow n \ge \lceil x \rceil$;
(d) $\lceil x + n \rceil = \lceil x \rceil + n$.

Lemma 1.8 *If $x \in \mathbb{R}$ and $n \in \mathbb{Z}^+$, then $\lceil \lceil x \rceil / n \rceil = \lceil x / n \rceil$.*

Proof By Definition 1.2 and Lemma 1.2,

$$\lceil \lceil x \rceil / n \rceil = -\lfloor -\lceil x \rceil / n \rfloor = -\lfloor \lfloor -x \rfloor / n \rfloor = -\lfloor -x / n \rfloor = \lceil x / n \rceil. \qquad \square$$

The floor and the ceiling are related as follows.

Lemma 1.9 *For all $x \in \mathbb{R}$,*

$$\lceil x \rceil = \begin{cases} \lfloor x \rfloor & \text{if } x \in \mathbb{Z} \\ \lfloor x \rfloor + 1 & \text{if } x \notin \mathbb{Z}. \end{cases}$$

Proof If $x \in \mathbb{Z}$, then of course, $\lceil x \rceil = \lfloor x \rfloor = x$. Otherwise, by Lemma 1.1, $x \ne \lfloor x \rfloor$, and hence Definition 1.1 yields $\lfloor x \rfloor < x < \lfloor x \rfloor + 1$. Rearranging these inequalities, we have $\lfloor x \rfloor + 1 > x > (\lfloor x \rfloor + 1) - 1$. By Lemma 1.6, $\lceil x \rceil = \lfloor x \rfloor + 1$.
$\square$

1.2 Modulus

The integer quotient of x and y may be defined as $\lfloor x/y \rfloor$. This formulation leads to the following characterization of the modulus function:

Definition 1.3 For all $x \in \mathbb{R}$ and $y \in \mathbb{R}$,

$$x \bmod y = \begin{cases} x - \lfloor x/y \rfloor y & \text{if } y \ne 0 \\ x & \text{if } y = 0. \end{cases}$$

Notation For the purpose of resolving ambiguous expressions, the precedence of this operator is higher than that of addition and lower than that of multiplication.

Although $x \bmod y$ is of interest mainly when $x \in \mathbb{Z}$ and $y \in \mathbb{Z}^+$, the definition is less restrictive, and arbitrary real arguments must be considered. We note the following closure properties, which follow from Definitions 1.3 and 1.1.

Lemma 1.10 *Let $m \in \mathbb{Z}$ and $n \in \mathbb{Z}$.*

(a) $m \bmod n \in \mathbb{Z}$;
(b) If $n > 0$, then $m \bmod n \in \mathbb{N}$.

We have the following upper bounds in the integer case:

Lemma 1.11 *Let $m \in \mathbb{Z}$ and $n \in \mathbb{Z}$.*

(a) If $n > 0$, then $m \bmod n < n$;
(b) If $m \geq 0$, then $m \bmod n \leq m$;
(c) If $n > m \geq 0$, then $m \bmod n = m$.

Proof

(a) By Definitions 1.3 and 1.1,

$$m \bmod n = m - \lfloor m/n \rfloor n < m - ((m/n) - 1)n = n.$$

(b) By Definition 1.3, $m \bmod n = m - \lfloor m/n \rfloor n$. If $n > 0$, then $m/n > 0$, $\lfloor m/n \rfloor \geq 0$ by Lemma 1.1, and $\lfloor m/n \rfloor n \geq 0$. If $n < 0$, then $\lfloor m/n \rfloor \leq m/n \leq 0$, and again, $\lfloor m/n \rfloor n \geq 0$.
(c) Since $0 \leq m/n < 1$, $\lfloor m/n \rfloor = 0$ by Definition 1.1. Now by Definition 1.3,

$$m = \lfloor m/n \rfloor n + m \bmod n = m \bmod n. \qquad \square$$

Definition 1.4 For $a \in \mathbb{R}$, $b \in \mathbb{R}$, and $n \in \mathbb{R}$, a is congruent to b modulo n, or

$$a \equiv b \pmod{n},$$

if $a \bmod n = b \bmod n$.

Lemma 1.12 *If $a \in \mathbb{Z}$, $b \in \mathbb{Z}$, and $n \in \mathbb{Z}^{+}$, then*

(a) $a \equiv b \pmod{n} \Leftrightarrow (a - b)/n \in \mathbb{Z}$;
(b) $a \bmod n = 0 \Leftrightarrow a/n \in \mathbb{Z}$.

Proof (a) By Definition 1.3,

$$(a - b)/n = \lfloor a/n \rfloor - \lfloor b/n \rfloor + ((a \bmod n) - (b \bmod n))/n$$

and

$$(a - b)/n \in \mathbb{Z} \Leftrightarrow ((a \bmod n) - (b \bmod n))/n \in \mathbb{Z}.$$

By Lemmas 1.10 and 1.11, $0 \leq a \bmod n < n$ and $0 \leq b \bmod n < n$, and hence,

$$((a \bmod n) - (b \bmod n))/n \in \mathbb{Z} \Leftrightarrow a \bmod n = b \bmod n.$$

(b) By (a) and Lemma 1.11 (c), $a \bmod n = 0 \Leftrightarrow a \equiv 0 \pmod{n} \Leftrightarrow a/n \in \mathbb{Z}$. $\quad\square$

Lemma 1.13 *For all $a \in \mathbb{Z}$, $m \in \mathbb{Z}$, and $n \in \mathbb{Z}$,*

$$m + an \equiv m \pmod{n}.$$

Proof Since $((m + an) - m)/n = a \in \mathbb{Z}$, this follows from Lemma 1.12.   □

Lemma 1.14 *If $m \in \mathbb{Z}$ and $n \in \mathbb{Z}$, then $m \bmod n \equiv m \pmod{n}$.*

Proof By Definition 1.3, $(m - m \bmod n)/n = \lfloor m/n \rfloor \in \mathbb{Z}$.   □

Lemma 1.15 *For all $a \in \mathbb{Z}$, $b \in \mathbb{Z}$, and $n \in \mathbb{Z}$,*

(a) $a + b \bmod n \equiv a + b \pmod{n}$;
(b) $a - b \bmod n \equiv a - b \pmod{n}$;
(c) $a(b \bmod n) \equiv ab \pmod{n}$.

Proof These identities all follow from Lemmas 1.12 and 1.14.   □

Lemma 1.16 *For all $m \in \mathbb{Z}$, $k \in \mathbb{Z}$, and $n \in \mathbb{Z}$,*

$$m \bmod kn \equiv m \pmod{n}.$$

Proof This also follows from Lemmas 1.12 and 1.14: $(m \bmod kn) - m$ is divisible by kn and therefore by n.   □

Lemma 1.16 is used most frequently with power-of-two moduli.

Corollary 1.17 *For all $a \in \mathbb{Z}$, $b \in \mathbb{Z}$, and $m \in \mathbb{Z}$, if $a \geq b \geq 0$, then*

$$m \bmod 2^a \equiv m \pmod{2^b}.$$

Proof This is the case of Lemma 1.16 with $n = 2^b$ and $k = 2^{a-b}$.   □

Lemma 1.18 *If $n \in \mathbb{Z}^+$ and $m \in \mathbb{Z}$, then*

$$(-m) \bmod n = n - 1 - (m - 1) \bmod n.$$

Proof Let $p = n - 1 - (m - 1) \bmod n$. Since $0 \leq (m - 1) \bmod n < n$, $-1 < (m-1) \bmod n \leq n-1$, which implies $0 \leq p < n$. By Lemmas 1.11, 1.15, and 1.13,

$$p = p \bmod n$$
$$= (n - 1 - (m - 1) \bmod n) \bmod n$$
$$= (n - 1 - (m - 1)) \bmod n$$
$$= (n - m) \bmod n$$
$$= (-m) \bmod n.$$ □

Lemma 1.19 *Let $a \in \mathbb{Z}$, $m \in \mathbb{Z}$, $n \in \mathbb{Z}$, and $r \in \mathbb{Z}$.*

(a) *If* $an \leq m < (a + 1)n$, *then* $m \bmod n = m - an$;

(b) *If* $an \leq m < an + r$, *then* $m \bmod n < r$.

Proof By Lemmas 1.13 and 1.11, $an \leq m < (a + 1)n$ implies

$$m \bmod n = (m - an) \bmod n = m - an,$$

and if $an \leq m < an + r$, then

$$m \bmod n = (m - an) \bmod n \leq m - an < r. \qquad \square$$

Lemma 1.20 *For all* $k \in \mathbb{Z}$, $m \in \mathbb{Z}$, *and* $n \in \mathbb{Z}$,

$$km \bmod kn = k(m \bmod n).$$

Proof By Definition 1.3,

$$
\begin{aligned}
km \bmod kn &= km - \left\lfloor \frac{km}{kn} \right\rfloor kn \\
&= k(m - \lfloor m/n \rfloor n) \\
&= k(m \bmod n). \qquad \square
\end{aligned}
$$

As another consequence of Lemmas 1.10 and 1.11, *mod* is an idempotent operator in the sense that for $n > 0$,

$$(m \bmod n) \bmod n = m \bmod n.$$

This observation may be generalized as follows:

Lemma 1.21 *Let* $a \in \mathbb{Z}$, $m \in \mathbb{Z}^+$, *and* $n \in \mathbb{Z}^+$. *Then*

$$\left\lfloor \frac{a \bmod mn}{n} \right\rfloor = \left\lfloor \frac{a}{n} \right\rfloor \bmod m.$$

Proof By Lemmas 1.3, 1.1, and 1.2,

$$
\begin{aligned}
\left\lfloor \frac{a \bmod mn}{n} \right\rfloor &= \left\lfloor \frac{a - mn \left\lfloor \frac{a}{mn} \right\rfloor}{n} \right\rfloor \\
&= \left\lfloor \frac{a}{n} \right\rfloor - m \left\lfloor \frac{a}{mn} \right\rfloor \\
&= \left\lfloor \frac{a}{n} \right\rfloor - m \left\lfloor \frac{\left\lfloor \frac{a}{n} \right\rfloor}{m} \right\rfloor \\
&= \left\lfloor \frac{a}{n} \right\rfloor \bmod m. \qquad \square
\end{aligned}
$$

1.3 Truncation

The following function truncates a real number to a specified number of fractional bits:

Definition 1.5 For $x \in \mathbb{R}$ and $k \in \mathbb{Z}$,

$$x^{(k)} = \frac{\lfloor 2^k x \rfloor}{2^k}.$$

Note that according to Definition 1.3, an equivalent definition is

$$x^{(k)} = x - x \bmod 2^{-k}.$$

Example Let

$$x = \frac{51}{8} = 6 + \frac{3}{8}.$$

Then

$$x^{(2)} = \frac{\lfloor 2^2 x \rfloor}{2^2} = \frac{\left\lfloor \frac{51}{2} \right\rfloor}{4} = \frac{25}{4} = 6 + \frac{1}{4}.$$

———————

Lemma 1.22 *If $x \in \mathbb{R}$ and $k \in \mathbb{Z}$, then*

$$x - 2^{-k} < x^{(k)} \le x.$$

Proof $x - 2^{-k} = 2^{-k}(2^k x - 1) < 2^{-k}\lfloor 2^k x \rfloor \le 2^{-k} 2^k x = x.$ □

Lemma 1.23 *If $x \in \mathbb{R}$, $m \in \mathbb{N}$, and $n \in \mathbb{Z}$, then $n \le x \Leftrightarrow n \le x^{(m)}$.*

Proof $n \le x \Leftrightarrow 2^m n \le 2^m x \Leftrightarrow 2^m n \le \lfloor 2^m x \rfloor \Leftrightarrow n \le \frac{\lfloor 2^m x \rfloor}{2^m}.$ □

Lemma 1.24 *If $x \in \mathbb{R}$, $m \in \mathbb{Z}$, $k \in \mathbb{Z}$, and $k \le m$, then*

$$\left(x^{(k)}\right)^{(m)} = \left(x^{(m)}\right)^{(k)} = x^{(k)} \le x^{(m)}.$$

Proof

$$\left(x^{(k)}\right)^{(m)} = \frac{\left\lfloor 2^m \cdot \frac{\lfloor 2^k x \rfloor}{2^k} \right\rfloor}{2^m} = \frac{\left\lfloor 2^{m-k} \lfloor 2^k x \rfloor \right\rfloor}{2^m} = \frac{2^{m-k} \lfloor 2^k x \rfloor}{2^m} = \frac{\lfloor 2^k x \rfloor}{2^k} = x^{(k)}$$

and by Lemma 1.2,

$$\left(x^{(m)}\right)^{(k)} = \frac{\left\lfloor 2^k \cdot \frac{\lfloor 2^m x\rfloor}{2^m}\right\rfloor}{2^k} = \frac{\left\lfloor \frac{\lfloor 2^m x\rfloor}{2^{m-k}}\right\rfloor}{2^k} = \frac{\left\lfloor \frac{2^m x}{2^{m-k}}\right\rfloor}{2^k} = \frac{\lfloor 2^k x\rfloor}{2^k} = x^{(k)}.$$

The inequality follows from Lemma 1.22, with $x^{(m)}$ substituted for x.   □

Lemma 1.25 *If $x \in \mathbb{R}$, $y \in \mathbb{R}$, and $k \in \mathbb{Z}$, then*

(a) $x^{(k)} + y^{(k)} = \left(x + y^{(k)}\right)^{(k)} = \left(x^{(k)} + y^{(k)}\right)^{(k)}$;

(b) $x^{(k)} - y^{(k)} = \left(x - y^{(k)}\right)^{(k)} = \left(x^{(k)} - y^{(k)}\right)^{(k)}$.

Proof Since $2^k y^{(k)} = \lfloor 2^k y\rfloor \in \mathbb{Z}$,

$$\lfloor 2^k x\rfloor + \lfloor 2^k y\rfloor = \lfloor 2^k x\rfloor + 2^k y^{(k)} = \lfloor 2^k x + 2^k y^{(k)}\rfloor = \lfloor 2^k (x + y^{(k)})\rfloor$$

and

$$x^{(k)} + y^{(k)} = \frac{\lfloor 2^k x\rfloor + \lfloor 2^k y\rfloor}{2^k} = \frac{\lfloor 2^k (x + y^{(k)})\rfloor}{2^k} = \left(x + y^{(k)}\right)^{(k)}.$$

Substituting $x^{(k)}$ for x in this result and applying Lemma 1.24, we have

$$x^{(k)} + y^{(k)} = \left(x^{(k)}\right)^{(k)} + y^{(k)} = \left(x^{(k)} + y^{(k)}\right)^{(k)}.$$

The proof of (b) is similar.   □

Lemma 1.26 *If $x \in \mathbb{R}$, $k \in \mathbb{Z}$, and $m \in \mathbb{Z}$, then*

$$(2^k x)^{(m)} = 2^k x^{(k+m)}.$$

Proof $(2^k x)^{(m)} = 2^{-m}\lfloor 2^m (2^k x)\rfloor = 2^k (2^{-(k+m)}\lfloor 2^{k+m} x\rfloor) = 2^k x^{(k+m)}$.   □

Lemma 1.27 *If $x \in \mathbb{R}$, $m \in \mathbb{Z}$, and $-2^{-m} \le x < 2^{-m}$, then*

$$x^{(m)} = \begin{cases} 0 & \text{if } x \ge 0 \\ -2^{-m} & \text{if } x < 0. \end{cases}$$

Proof Since $-1 \le 2^m x < 1$, $-1 \le \lfloor 2^m x\rfloor < 1$, which implies

$$\lfloor 2^m x\rfloor = \begin{cases} 0 \text{ if } x \ge 0 \\ -1 \text{ if } x < 0 \end{cases}$$

and

$$x^{(m)} = 2^{-m} \lfloor 2^m x \rfloor = \begin{cases} 0 & \text{if } x \geq 0 \\ -2^{-m} & \text{if } x < 0. \end{cases}$$ □

If $k > 0$, then $x^{(-k)} = 2^k \lfloor 2^{-k} x \rfloor$ is the largest multiple of 2^k that does not exceed x.

Example Let

$$x = \frac{51}{8} = 6 + \frac{3}{8}.$$

Then

$$x^{(-2)} = 2^2 \lfloor 2^{-2} x \rfloor = 4 \left\lfloor \frac{51}{32} \right\rfloor = 4.$$

The following two lemmas were discovered in the analysis of the FMA design of Chap. 17. (See the proof of Lemma 17.10.)

Lemma 1.28 *If $k \in \mathbb{N}$, $n \in \mathbb{N}$, and $x \in \mathbb{R}$, then*

(a) $\lfloor \frac{x}{2^n} \rfloor^{(-k)} \leq \frac{x^{(-k)}}{2^n}$;

(b) $\frac{x^{(-k)} + 2^k}{2^n} \leq \lfloor \frac{x}{2^n} \rfloor^{(-k)} + 2^k$.

Proof

(a) By Lemma 1.2,

$$\left\lfloor \frac{x}{2^n} \right\rfloor^{(-k)} = 2^k \left\lfloor \frac{\lfloor \frac{x}{2^n} \rfloor}{2^k} \right\rfloor = 2^k \left\lfloor \frac{x}{2^{k+n}} \right\rfloor = 2^k \left\lfloor \frac{\lfloor \frac{x}{2^k} \rfloor}{2^n} \right\rfloor \leq 2^k \frac{\lfloor \frac{x}{2^k} \rfloor}{2^n} = \frac{x^{(-k)}}{2^n}.$$

(b) By Lemmas 1.3, 1.11, and 1.2,

$$\frac{x^{(-k)} + 2^k}{2^n} = \frac{2^k \left(\lfloor \frac{x}{2^k} \rfloor + 1 \right)}{2^n}$$

$$= 2^{k-n} \left(\left\lfloor \frac{x}{2^k} \right\rfloor + 1 \right)$$

$$= 2^{k-n} \left(2^n \left\lfloor \frac{\lfloor \frac{x}{2^k} \rfloor}{2^n} \right\rfloor + \left\lfloor \frac{x}{2^k} \right\rfloor \bmod 2^n + 1 \right)$$

$$\leq 2^{k-n} \left(2^n \left\lfloor \frac{\lfloor \frac{x}{2^k} \rfloor}{2^n} \right\rfloor + (2^n - 1) + 1 \right)$$

$$= 2^k \left(\left\lfloor \frac{\left\lfloor \frac{x}{2^n} \right\rfloor}{2^k} \right\rfloor + 1 \right)$$

$$= \left\lfloor \frac{x}{2^n} \right\rfloor^{(-k)} + 2^k. \qquad \qquad \square$$

Lemma 1.29 *Let* $k \in \mathbb{N}$, $n \in \mathbb{N}$, $x \in \mathbb{R}$, *and* $y \in \mathbb{R}$. *If* $\lfloor 2^{-k} x \rfloor = \lfloor 2^{-k} y \rfloor$ *and* $2^{-k} x \notin \mathbb{Z}$, *then*

$$\left(-\left\lfloor \frac{y}{2^n} \right\rfloor - 1 \right)^{(-k)} = \left(-\frac{x}{2^n} \right)^{(-k)}.$$

Proof We simplify the expression on the left using Lemmas 1.2 and 1.5:

$$\left(-\left\lfloor \frac{y}{2^n} \right\rfloor - 1 \right)^{(-k)} = 2^k \left\lfloor -\frac{\left\lfloor \frac{y}{2^n} \right\rfloor + 1}{2^k} \right\rfloor$$

$$= -2^k \left(\left\lfloor \frac{\left\lfloor \frac{y}{2^n} \right\rfloor}{2^k} \right\rfloor + 1 \right)$$

$$= -2^k \left(\left\lfloor \frac{\left\lfloor \frac{y}{2^k} \right\rfloor}{2^n} \right\rfloor + 1 \right)$$

$$= -2^k \left(\left\lfloor \frac{\left\lfloor \frac{x}{2^k} \right\rfloor}{2^n} \right\rfloor + 1 \right)$$

$$= -2^k \left(\left\lfloor \frac{x}{2^{k+n}} \right\rfloor + 1 \right).$$

For the expression on the right, we apply the same two lemmas and Lemma 1.4:

$$\left(-\frac{x}{2^n} \right)^{(-k)} = 2^k \left\lfloor -\frac{\frac{x}{2^n}}{2^k} \right\rfloor$$

$$= 2^k \left\lfloor \frac{-x}{2^{k+n}} \right\rfloor$$

$$= 2^k \left\lfloor \frac{\left\lfloor \frac{-x}{2^k} \right\rfloor}{2^n} \right\rfloor$$

$$= 2^k \left\lfloor \frac{-\left\lfloor \frac{x}{2^k} \right\rfloor - 1}{2^n} \right\rfloor$$

$$= -2^k \left(\left\lfloor \frac{\left\lfloor \frac{x}{2^k} \right\rfloor}{2^n} \right\rfloor + 1 \right)$$

$$= -2^k \left(\left\lfloor \frac{x}{2^{k+n}} \right\rfloor + 1 \right)$$

$$= \left(-\left\lfloor \frac{y}{2^k} \right\rfloor - 1 \right)^{(-k)}. \qquad \square$$

Chapter 2
Bit Vectors

We shall use the term *bit vector* as a synonym of *integer*. Thus, a bit vector may be positive, negative, or zero. However, only a nonnegative bit vector may be associated with a *width*:

Definition 2.1 If $x \in \mathbb{N}$, $n \in \mathbb{N}$, and $x < 2^n$, then x is a bit vector of width n, or an n-bit vector.

Note that the width of a bit vector is not unique, since an n-bit vector is also an m-bit vector for all $m > n$.

The most fundamental operations on bit vectors are the *bit slice* and *bit extraction* functions. These are discussed in Sects. 2.1 and 2.2, respectively, since the latter is defined as a special case of the former. Section 2.3 deals with the RTL operation of *concatenation*.

Arithmetic hardware employs a variety of encoding schemes to represent integers and rational numbers as bit vectors. Floating-point representations are the subject of Chap. 5. In Sects. 2.4 and 2.5, we address the simpler integer and fixed-point formats.

2.1 Bit Slices

The *bit slice* function is defined as follows:

Definition 2.2 For $x \in \mathbb{Z}$, $i \in \mathbb{Z}$, and $j \in \mathbb{Z}$,

$$x[i : j] = \lfloor (x \bmod 2^{i+1})/2^j \rfloor,$$

Notation For the purpose of resolving ambiguous expressions, this operator takes precedence over the basic arithmetic operators, e.g.,

$$xy[i : j][k : \ell] = x((y[i : j])[k : \ell]).$$

The following properties are immediate consequences of Definition 2.2:

Lemma 2.1 *Let $x \in \mathbb{Z}$, $y \in \mathbb{Z}$, $i \in \mathbb{N}$, and $j \in \mathbb{N}$.*

(a) *If $i + 1 \geq j$, then $x[i : j]$ is an $(i + 1 - j)$-bit vector;*
(b) *If $x \equiv y \pmod{2^{i+1}}$, then $x[i : j] = y[i : j]$;*
(c) *$x[i : 0] = x \bmod 2^{i+1}$;*
(d) *If x is a j-bit vector, then $x[i : j] = 0$.*

Example Let $x = 93 = (1011101)_2$. Then

$$x[4 : 2] = \lfloor (x \bmod 2^5)/2^2 \rfloor = \lfloor (93 \bmod 32)/4 \rfloor = \lfloor 29/4 \rfloor = 7 = (111)_2$$

is a 3-bit vector and

$$x[10 : 7] = \lfloor (93 \bmod 2^{11})/2^7 \rfloor = \lfloor 93/128 \rfloor = 0 = (0000)_2$$

is a 4-bit vector.

In most cases of interest, the index arguments of $x[i : j]$ satisfy $i \geq j \geq 0$. However, the following lemma is worth noting.

Lemma 2.2 *Let $x \in \mathbb{Z}$, $i \in \mathbb{Z}$, and $j \in \mathbb{Z}$. If $i < 0$ or $i < j$, then $x[i : j] = 0$.*

Proof Suppose $i < 0$. Applying Definition 1.3 and Lemma 1.1, we have

$$x \bmod 2^{i+1} = x - \left\lfloor \frac{x}{2^{i+1}} \right\rfloor 2^{i+1} = x - \frac{x}{2^{i+1}} \cdot 2^{i+1} = 0.$$

If $i < j$, then by Lemma 1.11, $x \bmod 2^{i+1} < 2^{i+1} \leq 2^j$, and hence

$$x[i : j] = \lfloor (x \bmod 2^{i+1})/2^j \rfloor = 0. \qquad \square$$

Lemma 2.3 *Let $x \in \mathbb{Z}$ and $i \in \mathbb{N}$. If $-2^{i+1} \leq x < 2^{i+1}$, then*

$$x[i : 0] = \begin{cases} x & \text{if } x \geq 0 \\ x + 2^{i+1} & \text{if } x < 0. \end{cases}$$

Proof If $x \geq 0$, the claim follows from Lemma 2.1 (c). If $-2^{i+1} \leq x < 0$, then by Lemmas 2.1 (c), 1.13, and 1.11,

$$x[i : 0] = x \bmod 2^{i+1} = (x + 2^{i+1}) \bmod 2^{i+1} = x + 2^{i+1}. \qquad \square$$

If $-2^j \leq x < 0$, then $x[i : j]$ is the bit vector of width $i - j + 1$ consisting of all 1s:

Lemma 2.4 Let $x \in \mathbb{Z}$, $i \in \mathbb{N}$, and $j \in \mathbb{N}$. If $i \geq j$ and $-2^j \leq x < 0$, then $x[i : j] = 2^{i-j+1} - 1$.

Proof By Lemmas 2.1 (c) and 1.13, $x \bmod 2^{i+1} = x + 2^{i+1}$. Thus, by Definition 2.2, Lemma 1.1, and Definition 1.1,

$$x[i : j] = \lfloor (x + 2^{i+1})/2^j \rfloor = \lfloor x/2^j + 2^{i-j+1} \rfloor = \lfloor x/2^j \rfloor + 2^{i-j+1} = 2^{i-j+1} - 1.$$

$\square$

Corollary 2.5 If $i \in \mathbb{N}$, $j \in \mathbb{N}$, and $i \geq j$, then $(-1)[i : j] = 2^{i-j+1} - 1$.

Corollary 2.6 Let $x \in \mathbb{N}$, $i \in \mathbb{N}$, and $j \in \mathbb{N}$. If $i \geq j$ and $2^{i+1} - 2^j \leq x < 2^{i+1}$, then $x[i : j] = 2^{i-j+1} - 1$.

Proof Let $x' = x - 2^{i+1}$. By Lemmas 2.1 (b) and 2.4,

$$x[i : j] = x'[i : j] = 2^{i-j+1} - 1.$$

$\square$

By expanding the modulus, we may express a bit slice in terms of the floor alone:

Lemma 2.7 Let $x \in \mathbb{Z}$, $i \in \mathbb{Z}$, and $j \in \mathbb{Z}$. If $i \geq j$, then

$$x[i-1 : j] = \left\lfloor \frac{x}{2^j} \right\rfloor - 2^{i-j} \left\lfloor \frac{x}{2^i} \right\rfloor.$$

Proof By Definitions 2.2 and 1.3,

$$x[i-1:j] = \left\lfloor \frac{x \bmod 2^i}{2^j} \right\rfloor = \left\lfloor \frac{x - \left\lfloor \frac{x}{2^i} \right\rfloor 2^i}{2^j} \right\rfloor = \left\lfloor \frac{x}{2^j} - \left\lfloor \frac{x}{2^i} \right\rfloor 2^{i-j} \right\rfloor,$$

and the claim follows from Lemma 1.1.

$\square$

Lemma 2.8 Let $x \in \mathbb{Z}$, $i \in \mathbb{Z}$, and $j \in \mathbb{Z}$. If $i \geq j$, then

$$x[i-1 : j] = \left\lfloor \frac{x}{2^j} \right\rfloor \bmod 2^{i-j}.$$

Proof This follows from Definition 1.3 and Lemmas 1.2 and 2.7:

$$\left\lfloor \frac{x}{2^j} \right\rfloor \bmod 2^{i-j} = \left\lfloor \frac{x}{2^j} \right\rfloor - 2^{i-j} \left\lfloor \frac{\lfloor 2^{-j} x \rfloor}{2^{i-j}} \right\rfloor = \left\lfloor \frac{x}{2^j} \right\rfloor - 2^{i-j} \left\lfloor \frac{x}{2^i} \right\rfloor = x[i-1 : j]$$

$\square$

We have the following formula for extracting a bit slice of a sum:

Lemma 2.9 *Let $x \in \mathbb{Z}$, $y \in \mathbb{Z}$, $i \in \mathbb{Z}$, and $j \in \mathbb{Z}$, with $i \geq j \geq 0$. Let $g = \lfloor (x[i-1:0] + y[i-1:0])/2^j \rfloor$. Then*

$$(x + y)[i:j] = (x[i:j] + y[i:j] + g) \bmod 2^{i+1-j}.$$

Proof By Definition 1.3 and Lemma 2.1 (c),

$$x + y = 2^j \left(\left\lfloor \frac{x}{2^j} \right\rfloor + \left\lfloor \frac{y}{2^j} \right\rfloor \right) + x[i-1:0] + y[i-1:0],$$

and hence, by Lemma 1.2,

$$\left\lfloor \frac{x+y}{2^j} \right\rfloor = \left\lfloor \left\lfloor \frac{x}{2^j} \right\rfloor + \left\lfloor \frac{y}{2^j} \right\rfloor + \frac{x[i-1:0] + y[i-1:0]}{2^j} \right\rfloor = \left\lfloor \frac{x}{2^j} \right\rfloor + \left\lfloor \frac{y}{2^j} \right\rfloor + g.$$

Thus, by Lemmas 2.8 and 1.15,

$$\begin{aligned}
(x + y)[i:j] &= \left\lfloor \frac{x+y}{2^j} \right\rfloor \bmod 2^{i+1-j} \\
&= \left(\left\lfloor \frac{x}{2^j} \right\rfloor + \left\lfloor \frac{y}{2^j} \right\rfloor + g \right) \bmod 2^{i+1-j} \\
&= \left(\left\lfloor \frac{x}{2^j} \right\rfloor \bmod 2^{i+1-j} + \left\lfloor \frac{y}{2^j} \right\rfloor \bmod 2^{i+1-j} + g \right) \bmod 2^{i+1-j} \\
&= (x[i:j] + y[i:j] + g) \bmod 2^{i+1-j}. \qquad \square
\end{aligned}$$

Here is an important lemma that decomposes a slice into two subslices.

Lemma 2.10 *Let $x \in \mathbb{Z}$, $m \in \mathbb{N}$, $n \in \mathbb{N}$, and $p \in \mathbb{N}$. If $m \leq p \leq n$, then*

$$x[n:m] = 2^{p-m} x[n:p] + x[p-1:m].$$

Proof The proof consists of three applications of Lemma 2.7:

$$2^{p-m} x[n:p] = 2^{p-m} \left(\left\lfloor \frac{x}{2^p} \right\rfloor - 2^{n+1-p} \left\lfloor \frac{x}{2^{n+1}} \right\rfloor \right),$$

$$x[p-1:m] = \left\lfloor \frac{x}{2^m} \right\rfloor - 2^{p-m} \left\lfloor \frac{x}{2^p} \right\rfloor,$$

and hence,

$$2^{p-m} x[n:p] + x[p-1:m] = \left\lfloor \frac{x}{2^m} \right\rfloor - 2^{n+1-m} \left\lfloor \frac{x}{2^{n+1}} \right\rfloor = x[n:m]. \qquad \square$$

The next four lemmas address slices of shifted bit vectors.

Lemma 2.11 *For all $x \in \mathbb{N}$, $i \in \mathbb{N}$, $j \in \mathbb{N}$, and $k \in \mathbb{N}$,*

$$\lfloor x/2^k \rfloor[i : j] = x[i + k : j + k].$$

Proof By Definition 2.2 and Lemmas 2.8 and 1.2,

$$\begin{aligned}
\lfloor x/2^k \rfloor[i : j] &= \left\lfloor \frac{\lfloor x/2^k \rfloor \bmod 2^{i+1}}{2^j} \right\rfloor \\
&= \left\lfloor \frac{x[i + k : k]}{2^j} \right\rfloor \\
&= \left\lfloor \frac{\lfloor (x \bmod 2^{i+k+1})/2^k \rfloor}{2^j} \right\rfloor \\
&= \left\lfloor \frac{x \bmod 2^{i+k+1}}{2^j + k} \right\rfloor \\
&= x[i + k : j + k].
\end{aligned}$$ $\qquad\square$

Lemma 2.12 *For all $x \in \mathbb{N}$, $i \in \mathbb{N}$, and $k \in \mathbb{N}$,*

$$\lfloor x/2^k \rfloor[i : 0] = \lfloor x[i + k : 0]/2^k \rfloor.$$

Proof Applying Lemma 2.11, Definition 2.2, and Lemma 2.1 (c) in succession, we have

$$\lfloor x/2^k \rfloor[i : 0] = x[i + k : k] = \lfloor (x \bmod 2^{i+k+1})/2^k \rfloor = \lfloor x[i + k : 0]/2^k \rfloor. \quad \square$$

Lemma 2.13 *For all $x \in \mathbb{N}$, $i \in \mathbb{N}$, $j \in \mathbb{N}$, and $k \in \mathbb{N}$,*

$$(2^k x)[i : j] = x[i - k : j - k].$$

Proof By Definition 2.2, $(2^k x)[i : j] = \lfloor (2^k x \bmod 2^{i+1})/2^j \rfloor$. If $k \leq i$, then by Lemma 1.20, this reduces to

$$\lfloor 2^k (x \bmod 2^{i-k+1})/2^j \rfloor = \lfloor (x \bmod 2^{i-k+1})/2^{j-k} \rfloor = x[i - k : j - k].$$

On the other hand, if $i < k$, then by Lemma 2.2, the same expression reduces to

$$\lfloor 0/2^j \rfloor = 0 = x[i - k : j - k]. \qquad\square$$

Lemma 2.14 *For all $x \in \mathbb{N}$, $i \in \mathbb{N}$, and $k \in \mathbb{N}$,*

$$2^k x[i : 0] = (2^k x)[i + k : 0].$$

Proof By Lemmas 1.20 and 2.1 (c),

$$(2^k x)[i + k : 0] = 2^k x \bmod 2^{i+k+1} = 2^k (x \bmod 2^{i+1}) = 2^k x[i : 0]. \qquad \square$$

Finally, we have the following formula for composing bit slices:

Lemma 2.15 *For all $x \in \mathbb{N}$, $i \in \mathbb{N}$, $j \in \mathbb{N}$, $k \in \mathbb{N}$, and $\ell \in \mathbb{N}$,*

$$x[i : j][k : l] = \begin{cases} x[k + j : \ell + j] \ \text{if } k \leq i - j \\ x[i : \ell + j] \qquad \text{if } k > i - j. \end{cases}$$

Proof By Lemma 2.8,

$$x[i : j][k : \ell] = (\lfloor x/2^j \rfloor \bmod 2^{i-j+1})[k : \ell]$$
$$= \lfloor ((\lfloor x/2^j \rfloor \bmod 2^{i-j+1}) \bmod 2^{k+1})/2^\ell \rfloor.$$

If $k \leq i - j$, then this reduces, by Corollary 1.17 and Lemma 2.11, to

$$\lfloor (\lfloor x/2^j \rfloor \bmod 2^{k+1})/2^\ell \rfloor = \lfloor x/2^j \rfloor[k : \ell] = x[k + j : \ell + j].$$

On the other hand, if $k > i - j$, then by Lemma 1.11,

$$\lfloor x/2^j \rfloor \bmod 2^{i-j+1} < 2^{i-j+1} < 2^{k+1},$$

and the expression reduces instead to

$$\lfloor (\lfloor x/2^j \rfloor \bmod 2^{i-j+1})/2^\ell \rfloor = \lfloor x/2^j \rfloor[i - j : \ell] = x[i : \ell + j]. \qquad \square$$

2.2   Bit Extraction

Bit extraction is defined as a case of bit slice:

Definition 2.3 For $x \in \mathbb{Z}$ and $i \in \mathbb{Z}$,

$$x[i] = x[i : i].$$

Notation Like the bit slice, this operator takes precedence over the basic arithmetic operators.

A number of important properties of bit extraction are special cases of the results of Sect. 2.1. We list some of them here without proof.

Lemma 2.16 *Let $x \in \mathbb{Z}$, $x \in \mathbb{Z}$, $n \in \mathbb{Z}$, $i \in \mathbb{Z}$, $j \in \mathbb{Z}$, and $k \in \mathbb{Z}$.*

(a) $x[n] \in \{0, 1\}$;

(b) If $n < 0$, then $x[n] = 0$;

(c) If $x \equiv y \pmod{2^n}$ and $k < n$, then $x[k] = y[k]$;

(d) If x is an n-bit vector, then $x[n] = 0$;

(e) If $-2^n \le x < 0$, then $x[n] = 1$;

(f) If $n \ge 0$, then $(-1)[n] = 1$;

(g) If $n \ge 0$, then $x[n] = \begin{cases} x \bmod 2 & \text{if } n = 0 \\ \lfloor x/2 \rfloor [n-1] & \text{if } n > 0; \end{cases}$

(h) $(2^k x)[n + k] = x[n]$;

(i) If $i \ge 0$ and $k \ge 0$, then $\lfloor x/2^k \rfloor [i] = x[i + k]$;

(j) If $0 \le k \le i - j$, then $x[i : j][k] = x[j + k]$;

(k) If $m \le n$, then $x[n : m] = 2^{n-m} x[n] + x[n-1 : m] = x[m] + 2x[n : m+1]$;

(l) $x[n] = \lfloor x/2^n \rfloor \bmod 2$.

We note the following consequences of Lemma 2.16 (l):

Lemma 2.17 Let $n \in \mathbb{N}$, $k \in \mathbb{N}$, and $x \in \mathbb{N}$. If $k < n$ and $2^n - 2^k \le x < 2^n$, then $x[k] = 1$.

Proof The hypothesis implies $2^{n-k} - 1 \le x/2^k < 2^{n-k}$, and by Definition 1.1, $\lfloor x/2^k \rfloor = 2^{n-k} - 1$. By Lemma 2.16 (l), $x[k] = (2^{n-k} - 1) \bmod 2 = 1$. □

Corollary 2.18 For all $n \in \mathbb{Z}$ and $x \in \mathbb{N}$, if $2^n \le x < 2^{n+1}$, then $x[n] = 1$.

Lemma 2.19 For all $n \in \mathbb{N}$ and $i \in \mathbb{Z}$, $(2^n)[i] = 1 \Leftrightarrow i = n$.

Proof By Lemma 2.16 (l),

$$(2^n)[i] = \lfloor 2^n/2^i \rfloor \bmod 2 = \begin{cases} 2^{n-i} \bmod 2 = 0 \text{ if } i < n \\ 1 \bmod 2 = 1 \text{ if } i = n \\ 0 \bmod 2 = 0 \text{ if } i > n. \end{cases}$$ □

Every integer may be expressed as a sum of powers of 2 using bit extraction:

Lemma 2.20 Let $x \in \mathbb{Z}$, $i \in \mathbb{N}$, $j \in \mathbb{N}$, and $n \in \mathbb{Z}^+$.

(a) $x[i : j] = \sum_{k=j}^{i} 2^{k-j} x[k]$;

(b) If $0 \le x < 2^n$, then $x = \sum_{k=0}^{n-1} 2^k x[k]$;

(c) If $-2^n \le x < 0$, then $x = -2^n + \sum_{k=0}^{n-1} 2^k x[k]$;

Proof If $i < j$, then both sides of the equation of (a) reduce to 0 by Lemma 2.2. We proceed by induction. Thus, for $i \ge j$, applying Lemma 2.16 (k), we have

$$\sum_{k=j}^{i} 2^{k-j} x[k] = 2^{i-j} x[i] + \sum_{k=j}^{i-1} 2^{k-j} x[k]$$

$$= 2^{i-j} x[i] + x[i - i : j]$$

$$= x[i : j].$$

(b) and (c) follow from (a) and Lemma 2.3. □

The next lemma allows us to define a bit vector in a natural way as a sequence of bits.

Lemma 2.21 *Let $n \in \mathbb{N}$ and $b_i \in \{0, 1\}$ for $i = 0, \ldots, n - 1$. Let $x = \sum_{i=0}^{n-1} 2^i b_i$ and $y = x - 2^n$.*

(a) For $0 \le k < n$, $x[k] = y[k] = b_k$;
(b) For $k \ge n$, $x[k] = 0$ and $y[k] = 1$.

Proof (a) By Lemma 2.16 (c), $x[k] = y[k]$, and by Lemma 2.16 (l), since

$$\sum_{i=0}^{k-1} 2^i b_i \le \sum_{i=0}^{k-1} 2^i = 2^k - 1 < 2^k,$$

$$x[k] = \left\lfloor \frac{x}{2^k} \right\rfloor \bmod 2$$

$$= \left(\sum_{i=k}^{n-1} 2^{i-k} b_i \right) \bmod 2$$

$$= \left(2 \sum_{i=k+1}^{n-1} 2^{i-k} b_i + b_k \right) \bmod 2$$

$$= b_k.$$

(b) Since $0 \le x < 2^n$ and $-2^n \le y < 0$, this follows from Lemma 2.16 (d) and (e). $\qquad\square$

A bit vector is uniquely determined by its bits:

Lemma 2.22 *Let $x \in \mathbb{Z}$ and $y \in \mathbb{Z}$. If $x[k] = y[k]$ for all $k \in \mathbb{N}$, then $x = y$.*

Proof The proof is by induction on $|x| + |y|$.

Suppose $x \ne y$. We must show that for some $k \in \mathbb{N}$, $x[k] \ne y[k]$. We may assume that $x[0] = y[0]$, and hence $\lfloor x/2 \rfloor \ne \lfloor y/2 \rfloor$, for otherwise

$$x = 2\lfloor x/2 \rfloor + x[0] = 2\lfloor y/2 \rfloor + y[0] = y.$$

Since $x \ne y$ and $x[0] = y[0]$, at least one of x and y must be different from both 0 and -1, and hence, by Lemma 1.3,

$$|\lfloor x/2 \rfloor| + |\lfloor y/2 \rfloor| < |x| + |y|.$$

By induction, there exists $k \in \mathbb{N}$ such that $\lfloor x/2 \rfloor[k] \ne \lfloor y/2 \rfloor[k]$, and consequently, by Lemma 2.16 (i),

$$x[k+1] = \lfloor x/2 \rfloor[k] \ne \lfloor y/2 \rfloor[k] = y[k+1]. \qquad\square$$

Notation For $x \in \mathbb{Z}$, the *binary representation* of x is $(\ldots b_2 b_1 b_0)_2$, where $b_i = x[i]$ for all $i \in \mathbb{N}$. We may omit the parentheses and final subscript when the intention is clear. By Lemma 2.22, distinct integers have distinct binary representations, so that we may identify an integer with its representation:

$$x = \ldots b_2 b_1 b_0.$$

By Lemmas 2.16 (d) and (e) and 2.20, if $0 \leq x < 2^n$, then

$$x = \ldots 000 b_{n-1} \ldots b_0 = \sum_{k=0}^{n-1} 2^k x[k],$$

and if $-2^n \leq x < 0$, then

$$x = \ldots 111 b_{n-1} \ldots b_0 = -2^n + \sum_{k=0}^{n-1} 2^k x[k].$$

In the former case, we may omit the leading zeroes and write

$$x = b_{n-1} \ldots b_0.$$

By Lemma 2.21, every sequence of either of these two forms represents some integer. Thus, in particular, given a sequence of bits $\{b_0, \ldots, b_{n-1}\}$, we may say, without ambiguity, *Let x be the bit vector of width n defined by $x[k] = b_k$ for $k = 0, \ldots, n-1$.*

In the sequel, we shall extend this notation to non-integral floating-point numbers: for $k \in \mathbb{N}$,

$$2^{-k} x = \ldots b_k . b_{k-1} \ldots b_1 b_0.$$

2.3 Concatenation

If $x = (\beta_{m-1} \cdots \beta_0)_2$ and $y = (\gamma_{n-1} \cdots \gamma_0)_2$ are considered as bit vectors of widths m and n, respectively, then the concatenation of x and y is the $(m+n)$-bit vector

$$(\beta_{m-1} \cdots \beta_0 \gamma_{n-1} \cdots \gamma_0)_2.$$

This notion is extended by the following function, which takes a list of bit vectors and widths, coerces each bit vector to its associated width, and concatenates the results:

Definition 2.4 For all $x \in \mathbb{Z}$, $y \in \mathbb{Z}$, $m \in \mathbb{N}$, and $n \in \mathbb{N}$,

$$cat(x, m, y, n) = 2^n x[m-1 : 0] + y[n-1 : 0].$$

If $k > 2$ and $x_i \in \mathbb{Z}$ and $n_i \in \mathbb{N}$ for $i = 1, \ldots, k$, then

$$cat(x_1, n_1, x_2, n_2, \ldots, x_k, n_k) = cat(x_1, n_1, cat(x_2, n_2, \ldots, x_k, n_k), n_2 + \ldots + n_k).$$

Associativity follows immediately:

Lemma 2.23 *For all $x \in \mathbb{Z}$, $y \in \mathbb{Z}$, $z \in \mathbb{Z}$, $m \in \mathbb{N}$, $n \in \mathbb{N}$, and $p \in \mathbb{N}$,*

$$cat(cat(x, m, y, n), z, p) = cat(x, m, y, n, z, p).$$

Lemma 2.24 *For all $x \in \mathbb{Z}$, $y \in \mathbb{Z}$, $m \in \mathbb{N}$, and $n \in \mathbb{N}$, $cat(x, m, y, n)$ is an $(m + n)$-bit vector.*

Proof By Lemma 2.1 (a), $x[m-1 : 0] \leq 2^m - 1$ and $y[n-1 : 0] \leq 2^n - 1$. Thus,

$$\begin{aligned}
cat(x, m, y, n) &= 2^n x[m-1 : 0] + y[n-1 : 0] \\
&\leq 2^n(2^m - 1) + (2^n - 1) \\
&= 2^{n+m} - 1 \\
&< 2^{n+m}.
\end{aligned}$$
□

We note several trivial cases:

Lemma 2.25 *For all $x \in \mathbb{Z}$, $y \in \mathbb{Z}$, $m \in \mathbb{N}$, and $n \in \mathbb{N}$,*

$$cat(x, m, y, 0) = x[m-1 : 0]$$

and

$$cat(x, 0, y, n) = cat(0, m, y, n) = y[n-1 : 0].$$

Proof These are simple consequences of Definition 2.4 and Lemmas 2.2 and 2.1 (d).
□

Notation In standard RTL syntax, the concatenation of 2 bit vectors ϕ and ψ is denoted by $\{\phi, \psi\}$. This notation depends on a characteristic shared by conventional hardware description languages: any expression that represents a bit vector has an associated (explicit or implicit) width. For example, the expression `sig[3:0]` is understood to be of width 4, and the expression `5'b01001` identifies the constant 9 as a bit vector of width 5. We shall incorporate this construct into our informal mathematical notation through an abuse of Verilog syntax, representing $cat(x, m, y, n)$ as

$$\{m' x, n' y\}.$$

The width specifier may be omitted in a context in which it can be inferred by default.

Example If $x \in \{0, 1\}$ and y has been identified as a bit vector of width n, then

$$\{x, y, z[i : j], w[k]\} = cat(x, 1, y, n, z[i : j], i + 1 - j, w[k], 1).$$

The following is a restatement of Lemma 2.10:

Lemma 2.26 *Let $x \in \mathbb{Z}$, $m \in \mathbb{N}$, $n \in \mathbb{N}$, and $p \in \mathbb{N}$. If $m \leq p \leq n$, then*

$$x[n : m] = \{x[n : p], x[p-1 : m]\}.$$

Corollary 2.27 *Let $x \in \mathbb{Z}$, $m \in \mathbb{N}$, and $n \in \mathbb{N}$. If $m \leq n$, then*

$$x[n : m] = \{x[n], x[n-1 : m]\} = \{x[n : m+1], x[m]\}.$$

Lemma 2.28 *Let $z = \{m' x, n' y\}$, where $x \in \mathbb{Z}$, $y \in \mathbb{Z}$, $m \in \mathbb{N}$, and $n \in \mathbb{N}$. Then*

$$z[n-1 : 0] = y[n-1 : 0]$$

and

$$z[n + m-1 : n] = x[m-1 : 0].$$

Proof By Definition 2.4, we have

$$z = 2^n x[m-1 : 0] + y[n-1 : 0],$$

where $0 \leq y[n-1 : 0] < 2^n$ by Lemma 2.1 (a). Thus, by Lemmas 2.1 (c), 1.13, and 1.11,

$$z[n-1 : 0] = z \bmod 2^n = y[n-1 : 0] \bmod 2^n = y[n-1 : 0].$$

Now by Definition 2,

$$z[n + m-1 : n] = \lfloor (z \bmod 2^{n+m})/2^n \rfloor.$$

But Lemma 2.24 yields $z < 2^{n+m}$ and hence, by Lemma 1.11,

$$z[n + m-1 : n] = \lfloor z/2^n \rfloor = \lfloor x[m-1 : 0] + y[n-1 : 0]/2^n \rfloor.$$

Finally, by Lemma 1.1, this reduces to

$$x[m-1:0] + \lfloor y[n-1:0]/2^n \rfloor = x[m-1:0]. \qquad \square$$

Lemma 2.29 *Let* $x \in \mathbb{Z}$, $y \in \mathbb{Z}$, $m \in \mathbb{N}$, $n \in \mathbb{N}$, *and* $z = \{x[m-1:0], y[n-1:0]\}$. *If* $i \in \mathbb{N}$, $j \in \mathbb{N}$, *and* $i \geq j$, *then*

$$z[i:j] = \begin{cases} y[i:j] & \text{if } n > i \\ x[i-n:j-n] & \text{if } m+n > i \geq j \geq n \\ x[m-1:j-n] & \text{if } i \geq m+n \text{ and } j \geq n \\ \{x[i-n:0], y[n-1:j]\} & \text{if } m+n > i \geq n > j \\ \{x[m-1:0], y[n-1:j]\} & \text{if } i \geq n+m \text{ and } n > j. \end{cases}$$

Proof By Lemma 2.28,

$$y[n-1:0] = z[n-1:0]$$

and

$$x[m-1:0] = z[n+m-1:n],$$

and by Lemma 2.24, z is an $(m+n)$-bit vector. We consider five cases as suggested by the lemma statement, each of which involves two or more applications of Lemma 2.15.

Case 1: $n > i$

By Lemma 2.15,

$$z[i:j] = z[n-1:0][i:j] = y[n-1:0][i:j] = y[i:j].$$

Case 2: $m+n > i \geq j \geq n$

By Lemma 2.15,

$$\begin{aligned} z[i:j] &= z[m+n-1:n][i-n:j-n] \\ &= x[m-1:0][i-n:j-n] \\ &= x[i-n:j-n]. \end{aligned}$$

Case 3: $i \geq m+n$ and $j \geq n$

By Lemma 2.26,

$$z[i:j] = \{z[i:m+n], z[m+n-1:j]\}.$$

But $z[i : m + n] = 0$ by Lemma 2.1 (d), and hence

$$z[i : j] = z[m + n - 1 : j]$$

by Lemma 2.25. Now by Lemma 2.15,

$$z[m + n - 1 : j] = z[m + n - 1 : n][m - 1 : j - n]$$
$$= x[m - 1 : 0][m - 1 : j - n]$$
$$= x[m - 1 : j - n].$$

Case 4: $m + n > i \geq n > j$
By Lemma 2.26,

$$z[i : j] = \{z[i : n], z[n - 1 : j]\}.$$

But by Lemma 2.15,

$$z[i : n] = z[m + n - 1 : n][i - n : 0]$$
$$= x[m - 1 : 0][i - n : 0]$$
$$= x[i - n : 0]$$

and

$$z[n - 1 : j] = z[n - 1 : 0][n - 1 : j]$$
$$= y[n - 1 : 0][n - 1 : j]$$
$$= y[n - 1 : j].$$

Case 5: $i \geq n + m$ and $n > j$
By Lemma 2.26,

$$z[i : j] = \{z[i : m + n], z[m + n - 1 : n], z[n - 1 : j]\}.$$

As in Case 4, $z[n - 1 : j] = y[n - 1 : j]$. By Lemma 2.1 (d), $z[i : m + n] = 0$, and hence by Lemma 2.25,

$$z[i : j] = \{z[m + n - 1 : n], z[n - 1 : j]\} = \{x[m - 1 : 0], y[n - 1 : j]\}. \qquad \square$$

Corollary 2.30 *If $x \in \mathbb{Z}$, $y \in \mathbb{Z}$, $m \in \mathbb{N}$, $n \in \mathbb{N}$, and $i \in \mathbb{N}$, then*

$$\{m'x, n'y\}[i] = \begin{cases} y[i] & \text{if } i < n \\ x[i - n] & \text{if } n \leq i < m + n \\ 0 & \text{if } n + m \leq i. \end{cases}$$

Proof The cases listed correspond to the first three cases of Lemma 2.29 with $i = j$. Note that for the third case, the lemma gives $x[m-1 : i - n]$, but since $i > n + m$, i.e., $i - n > m - 1$, this reduces to 0 by Lemma 2.2. $\qquad\square$

2.4   Integer Formats

The simplest of all bit vector encoding schemes is the *unsigned integer* format, whereby the first 2^n natural numbers, i.e., the bit vectors of width n, are represented by themselves under the identity mapping. However trivial, it will be convenient to have an explicit definition of this correspondence:

Definition 2.5 If x is an n-bit vector, where $n \in \mathbb{N}$, then $ui(x, n) = x$.

Somewhat more interesting is the *signed integer* format, which maps the set of 2^n integers x in the range $-2^{n-1} \leq x < 2^{n-1}$ to the set of bit vectors of width n under the bijection

$$x \longmapsto x[n-1 : 0].$$

With respect to this mapping, the most significant bit of the encoding of x,

$$x[n-1 : 0][n-1] = x[n-1],$$

is 0 if $0 \leq x < 2^{n-1}$ (by Lemma 2.16 (d)) and 1 if $-2^{n-1} \leq x < 0$ (by Lemma 2.16 (e)) and is therefore considered the *sign bit* of the encoding.

This scheme is also known as the n-bit *two's complement* encoding, because if $0 < x < 2^{n-1}$, then the encoding of $-x$ is the complement of x with respect to 2^n in the sense that, according to Lemma 2.3,

$$x + (-x)[n-1 : 0] = x + (-x + 2^n) = 2^n.$$

The integer represented by a given encoding is computed by the following function.

Definition 2.6 If x is an n-bit vector, where $n \in \mathbb{N}$, then

$$si(x, n) = \begin{cases} x - 2^n & \text{if } x[n-1] = 1 \\ x & \text{if } x[n-1] = 0. \end{cases}$$

Lemma 2.31 *If x is an n-bit vector, then $-2^{n-1} \leq si(x, n) < 2^{n-1}$.*

Proof If $x[n-1] = 0$, then $0 \leq x < 2^{n-1}$ and $si(x, n) = x$. If $x[n-1] = 1$, then $2^{n-1} \leq x < 2^n$, $-2^{n-1} \leq x - 2^n < 0$, and $si(x, n) = x - 2^n$. $\qquad\square$

Lemma 2.32 *Let $n \in \mathbb{N}$ and $x \in \mathbb{Z}$. If $-2^{n-1} \le x < 2^{n-1}$, then*

$$si(x[n-1:0], n) = x.$$

Proof If $0 \le x < 2^{n-1}$, then $x[n-1:0] = x$ by Lemma 2.3 and $x[n-1] = 0$ by Lemma 2.16 (d). Thus,

$$si(x[n-1:0], n) = si(x, n) = x.$$

If $-2^{n-1} \le x < 0$, then $x[n-1:0] = x+2^n$ by Lemma 2.3 and $(x+2^n)[n-1] = 1$ by Corollary 2.18. Thus,

$$si(x[n-1:0], n) = si(x+2^n, n) = (x+2^n) - 2^n = x. \qquad \square$$

Lemma 2.33 *If $n \in \mathbb{N}$, $x \in \mathbb{N}$, $i \in \mathbb{N}$, and $j \in \mathbb{N}$ with $j \le i < n$, then*

$$si(x, n)[i:j] = x[i:j].$$

Lemma 2.34 *If $n \in \mathbb{N}$, $k \in \mathbb{N}$, and x is an n-bit vector, then*

$$si(2^k x, k+n) = 2^k si(x, n).$$

Proof This follows easily from Definition 2.6 and Lemma 2.16 (h). $\qquad \square$

Lemma 2.35 *Let $n \in \mathbb{N}$ and $m \in \mathbb{N}$ with $n > m$, and let x be an n-bit vector. Then*

$$si(x, n) = 2^m si\left(\left\lfloor \frac{x}{2^m} \right\rfloor, n-m\right) + x \bmod 2^m.$$

Proof If $x < 2^{n-1}$, then

$$\left\lfloor \frac{x}{2^m} \right\rfloor \le \frac{x}{2^m} < 2^{n-m-1},$$

and by Definitions 2.6 and 1.3,

$$2^m si\left(\left\lfloor \frac{x}{2^m} \right\rfloor, n-m\right) + x \bmod 2^m = 2^m \left\lfloor \frac{x}{2^m} \right\rfloor + x \bmod 2^m$$
$$= x$$
$$= si(x, n).$$

On the other hand, if $x \ge 2^{n-1}$, then

$$\frac{x}{2^m} \ge 2^{n-m-1}$$

which implies

$$\left\lfloor \frac{x}{2^m} \right\rfloor \geq 2^{n-m-1},$$

and

$$2^m si\left(\left\lfloor \frac{x}{2^m} \right\rfloor, n - m\right) + x \bmod 2^m = 2^m \left(\left\lfloor \frac{x}{2^m} \right\rfloor - 2^{n-m-1}\right) + x \bmod 2^m$$

$$= \left(2^m \left\lfloor \frac{x}{2^m} \right\rfloor + x \bmod 2^m\right) - 2^{n-1}$$

$$= x - 2^{n-1}$$

$$= si(x, n). \qquad \square$$

An n-bit integer encoding is converted to an m-bit encoding, where $m > n$, by *sign extension*:

Definition 2.7 Let x be an n-bit vector, where $n \in \mathbb{N}$, and let $m \in \mathbb{N}, m \geq n$. Then

$$sextend(m, n, x) = si(x, n)[m-1 : 0].$$

A sign extension of an integer encoding x represents the same value as x:

Lemma 2.36 *Let x be an n-bit vector, where $n \in \mathbb{N}$, and let $m \in \mathbb{N}, m \geq n$.*

(a) $sextend(m, n, x) = \begin{cases} x & \text{if } x[n - 1] = 0 \\ x + 2^m - 2^n & \text{if } x[n - 1] = 1; \end{cases}$

(b) $si(sextend(m, n, x), m) = si(x, n).$

Proof First, suppose $x[n-1] = 0$. Then $si(x, n) = x$ and, by Corollary 2.18, $0 \leq x < 2^{n-1}$. By Lemma 2.3,

$$sextend(m, n, x) = si(x, n)[m-1 : 0] = x[m-1 : 0] = x,$$

and since Lemma 2.16 (d) implies $x[m-1] = 0$,

$$si(sextend(m, n, x), m) = si(x, m) = x = si(x, n).$$

Now suppose $x[n-1] = 1$. Then by Lemma 2.16 (d), $2^{n-1} \leq x < 2^n$. Now $si(x, n) = x - 2^n$, where $-2^{m-1} \leq -2^{n-1} \leq x - 2^n < 0$. By Lemma 2.3,

$$sextend(m, n, x) = si(x, n)[m-1 : 0] = (x - 2^n)[m-1 : 0] = x - 2^n + 2^m.$$

But since $2^{m-1} \leq x - 2^n + 2^m < 2^m$. Corollary 2.18 implies $(x - 2^n + 2^m)[m-1] = 1$, and hence

$$\begin{aligned} si(sextend(m, n, x), m) &= si(x - 2^n + 2^m, m) \\ &= (x - 2^n + 2^m) - 2^m \\ &= x - 2^n \\ &= si(x, n). \end{aligned} \qquad \square$$

As illustrated throughout Part V, the result of a computation may be represented in a *redundant form*, i.e., as a sum or difference of 2 bit vectors, in order to avoid an expensive carry-propagate addition (see Sect. 8.1). An explicit approximation of such a result is typically derived by adding or subtracting narrow leading slices of these vectors. If the result is to be interpreted as a signed integer encoding and its value lies near a boundary of the representable range, then "wrap-around" may occur, resulting in a severe loss of accuracy. Given an approximation Y of an n-bit integer X, the following lemma provides a condition under which the signed integer represented by Y is an equally accurate approximation of the signed integer represented by X. (For an example of the utility of this result, see the proof of Lemma 18.6.)

Lemma 2.37 Let $X \in \mathbb{Z}$, $Y \in \mathbb{Z}$, and $n \in \mathbb{Z}^+$. If

$$|si(X \bmod 2^n, n)| + |X - Y| < 2^{n-1},$$

then

$$si(X \bmod 2^n, n) - si(Y \bmod 2^n, n) = X - Y.$$

Proof Since

$$si(X \bmod 2^n, n) + Y - X \equiv X + Y - X \equiv Y \pmod{2^n}$$

and

$$|si(X \bmod 2^n, n) + Y - X| \leq |si(X \bmod 2^n, n)| + |Y - X| < 2^{n-1},$$

Lemma 2.32 implies

$$si(X \bmod 2^n, n) + Y - X = si(Y \bmod 2^n, n). \qquad \square$$

2.5 Fixed-Point Formats

A fixed-point format may be thought of as derived from an integer format by inserting an implicit binary point following some specified number of leading bits. The rational value represented by an n-bit vector r with respect to an unsigned or signed fixed-point format of width n with m integer bits is computed as follows:

Definition 2.8 Let $n \in \mathbb{Z}^+$ and $m \in \mathbb{Z}$ and let x be a bit vector of width n.

(a) $uf(x, n, m) = 2^{m-n} ui(x) = 2^{m-n} x$;

(b) $sf(x, n, m) = 2^{m-n} si(x, n) = \begin{cases} 2^{m-n} x & \text{if } x < 2^{n-1} \\ 2^{m-n} x - 2^m & \text{if } x \geq 2^{n-1} \end{cases}$.

The number of fractional bits of a fixed-point format of width n and m integer bits is $f = n - m$. Note that while n must be positive, there is no restriction on m. If $m > n$, then the interpreted value is an integer with $m - n$ trailing zeroes and $f < 0$; if $m < 0$, then the interpreted value is a fraction with $-m$ leading zeroes and $f > n$.

We have the following expression for a bit slice of an encoding in terms of the encoded value:

Lemma 2.38 *Let $n \in \mathbb{N}$, $m \in \mathbb{N}$, $i \in \mathbb{N}$, and $j \in \mathbb{N}$ with $m \leq n$ and $j \leq i < n$. Let $f = n - m$. Let x be an n-bit vector, and suppose that either $r = uf(x, n, m)$ or $r = sf(x, n, m)$. Then*

$$x[i : j] = 2^{f-j} \left(r^{(f-j)} - r^{(f-i-1)} \right).$$

Proof If $r = uf(r, n, m)$, then $x = 2^f r$; if $r = sf(x, n, m)$, then either $x = 2^f r$ or $x = 2^f r + 2^n$. In any case, by Lemmas 2.1 (b) and 2.7,

$$x[i : j] = (2^f r)[i : j] = \left\lfloor \frac{2^f r}{2^j} \right\rfloor = \left\lfloor \frac{2^f r}{2^{1+i}} \right\rfloor = 2^{f-j} \left(r^{(f-j)} - r^{(f-i-1)} \right). \quad \square$$

Corollary 2.39 *Let $n \in \mathbb{N}$ and $m \in \mathbb{N}$ with $m \leq n$ and let $f = n - m$. Let $k \in \mathbb{Z}$ with $f - n \leq k < f$. Let x be an n-bit vector, and suppose that either $r = uf(x, n, m)$ or $r = sf(x, n, m)$. Then*

$$r^{(k)} = x \Leftrightarrow x[f - k-1 : 0] = 0.$$

Proof By Lemma 2.38,

$$x[f - k-1 : 0] = 2^f \left(r^{(f)} - r^{(k)} \right) = 2^f \left(r - r^{(k)} \right). \quad \square$$

The following result is useful in determining the value of a fixed-point encoding:

Lemma 2.40 *Let $n \in \mathbb{N}$ and $m \in \mathbb{N}$ with $m \le n$. Let x be an n-bit vector and $r = sf(x, n, m)$. If $y \in \mathbb{Z}$ satisfies $x \equiv y \pmod{2^n}$ and $-2^{n-1} \le y < 2^{n-1}$, then*

$$r = 2^{m-n} y.$$

Proof Since $x \equiv y \pmod{2^n}$ and $0 \le x < 2^n$, $x = y \bmod 2^n = y[n-1 : 0]$. By Lemma 2.32, $y = si(x, n)$, and hence

$$r = sf(x, n, m) = 2^{m-n} si(x, n) = 2^{m-n} y. \qquad \square$$

Chapter 3
Logical Operations

In this chapter, we define and analyze the four basic logical operations: the unary "not", or complement, and the binary "and", "inclusive or", and "exclusive or". These are commonly known as *bit-wise* operations, as each may be computed by performing a certain operation on each bit of its argument (in the unary case) or each pair of corresponding bits of its arguments (for binary operations). For example, the logical "and" x & y of 2 bit vectors may be specified in a bit-wise manner as the bit vector z such that for all $k \in \mathbb{N}$, $z[k] = 1$ iff $x[k] = y[k] = 1$.

In the context of our formalization, however, the logical operations are more naturally defined as arithmetic functions: the complement is constructed as an arithmetic difference, and the binary operations are defined by recursive formulas, which facilitate inductive proofs of their relevant properties. Among these are the bit-wise characterizations, as represented by Lemmas 3.4 (d) and 3.19.

3.1 Binary Operations

Following standard RTL syntax, we denote "and", "inclusive or", and "exclusive or" with the infix symbols &, |, and ^, respectively.

Definition 3.1 For all $x \in \mathbb{Z}$ and $y \in \mathbb{Z}$,

(a) $x \mathbin{\&} y = \begin{cases} 0 & \text{if } x = 0 \text{ or } y = 0 \\ x & \text{if } x = y \\ 2 \cdot (\lfloor x/2 \rfloor \mathbin{\&} \lfloor y/2 \rfloor) + x[0] \mathbin{\&} y[0] & \text{otherwise,} \end{cases}$

(b) $x \mathbin{|} y = \begin{cases} y & \text{if } x = 0 \text{ or } x = y \\ x & \text{if } y = 0 \\ 2 \cdot (\lfloor x/2 \rfloor \mathbin{|} \lfloor y/2 \rfloor) + x[0] \mathbin{|} y[0] & \text{otherwise;} \end{cases}$

© The Author(s), under exclusive license to Springer Nature Switzerland AG 2022
D. M. Russinoff, *Formal Verification of Floating-Point Hardware Design*,
https://doi.org/10.1007/978-3-030-87181-9_3

(c) $x \mathbin{\char`\^} y = \begin{cases} y & \text{if } x = 0 \\ x & \text{if } y = 0 \\ 0 & \text{if } x = y \\ 2 \cdot (\lfloor x/2 \rfloor \mathbin{\char`\^} \lfloor y/2 \rfloor) + x[0] \mathbin{\char`\^} y[0] & \text{otherwise.} \end{cases}$

Notation For the purpose of resolving ambiguous expressions, the precedence of $\char`\^$ is higher than that of $|$ and lower than that of &.

It is not obvious that these are admissible recursive definitions, i.e., that each of them is satisfied by a unique function. To establish this, it suffices to demonstrate the existence of a *measure* function $\mu : \mathbb{Z} \times \mathbb{Z} \to \mathbb{N}$ that strictly decreases on each recursive call. Thus, we define

$$\mu(x, y) = \begin{cases} 0 & \text{if } x = y \\ |xy| & \text{if } x \neq y. \end{cases}$$

For the admissibility of each of the three definitions, we must show that μ satisfies the following two inequalities, corresponding to the two recursive calls, under the restrictions $x \neq 0$, $y \neq 0$, and $x \neq y$:

(1) $\mu(\lfloor x/2 \rfloor, \lfloor y/2 \rfloor) < \mu(x, y)$.
(2) $\mu(x[0], y[0]) < \mu(x, y)$.

Since the restrictions imply that at least one of x and y is neither 0 nor -1, (1) follows from Lemma 1.3. To establish (2), note that one of the following must hold: $x[0] = 0$, $y[0] = 0$, or $x[0] = y[0] = 1$. In each of these three cases,

$$\mu(x[0], y[0]) = 0 < |xy| = \mu(x, y).$$

The proof of the following is a typical inductive argument based on the recursion of Definition 3.1.

Lemma 3.1 *If $x \in \mathbb{N}$ and $y \in \mathbb{Z}$, then $0 \leq x$ & $y \leq x$. In particular, if x is an n-bit vector, then so is x & y.*

Proof We may assume that $x \neq 0$, $y \neq 0$, and $x \neq y$. Thus,

$$x \mathbin{\&} y = 2(\lfloor x/2 \rfloor \mathbin{\&} \lfloor y/2 \rfloor) + x[0] \mathbin{\&} y[0]$$

and by induction,

$$0 \leq x \mathbin{\&} y \leq 2\lfloor x/2 \rfloor + x[0] = x. \qquad \square$$

Lemma 3.2 *If x and y are n-bit vectors, then so are $x \mid y$ and $x \mathbin{\char`\^} y$.*

Proof The same argument applies to both operations. The claim is trivial if $n = 0$, $x = 0$, $y = 0$, or $x = y$. In all other cases, $\lfloor x/2 \rfloor$ and $\lfloor y/2 \rfloor$ are $(n - 1)$-bit vectors and by induction, so is, for example, $\lfloor x/2 \rfloor \mid \lfloor y/2 \rfloor$. Thus,

$$x \mid y = 2(\lfloor x/2 \rfloor \mid \lfloor y/2 \rfloor) + x[0] \mid y[0] \le 2 \cdot (2^{n-1} - 1) + 1 < 2^n. \qquad \square$$

Lemma 3.3 *For $x \in \mathbb{Z}$ and $y \in \mathbb{Z}$, $x \mathbin{\&} y + x \mathbin{\char94} y = x \mid y$.*

Proof The claim is trivial if $x = 0$, $y = 0$, or $x = y$. Otherwise, by induction and a case analysis on $x[0]$ and $y[0]$,

$$x \mathbin{\&} y + x \mathbin{\char94} y = 2(\lfloor x/2 \rfloor \mathbin{\&} \lfloor y/2 \rfloor + \lfloor x/2 \rfloor \mathbin{\char94} \lfloor y/2 \rfloor) + (x[0] \mathbin{\&} y[0] + x[0] \mathbin{\char94} y[0])$$

$$= 2(\lfloor x/2 \rfloor \mid \lfloor y/2 \rfloor) + x[0] \mid y[0]$$

$$= x \mid y. \qquad \square$$

Lemma 3.4 *Let $x \in \mathbb{Z}$, $y \in \mathbb{Z}$, $n \in \mathbb{N}$, $m \in \mathbb{N}$, $i \in \mathbb{N}$, and $j \in \mathbb{N}$. Let "$\circ$" denote any of the binary operations of Definition 3.1.*

(a) $(x \circ y) \bmod 2^n = (x \bmod 2^n) \circ (y \bmod 2^n)$;
(b) $\lfloor (x \circ y)/2^n \rfloor = \lfloor x/2^n \rfloor \circ \lfloor y/2^n \rfloor$;
(c) $(x \circ y)[i : j] = x[i : j] \circ y[i : j]$;
(d) $(x \circ y)[n] = x[n] \circ y[n]$;
(e) $\{m' \, x_1, n' \, y_1\} \circ \{m' \, x_2, n' \, y_2\} = \{m' \, (x_1 \circ x_2), n' \, (y_1 \circ y_2)\}$;
(f) $2^n(x \circ y) = 2^n x \circ 2^n y$.

Proof We present the proofs for $\&$; the other two operations are similar. We may ignore the trivial cases $x = 0$, $y = 0$, and $x = y$, and when relevant, $n = 0$ and $m = 0$.
(a) By Definition 3.1 and Lemma 1.15,

$$(x \mathbin{\&} y) \bmod 2^n$$

$$= (2 \cdot (\lfloor x/2 \rfloor \mathbin{\&} \lfloor y/2 \rfloor) + x[0] \mathbin{\&} y[0]) \bmod 2^n$$

$$= \big((2 \cdot (\lfloor x/2 \rfloor \mathbin{\&} \lfloor y/2 \rfloor)) \bmod 2^n + x[0] \mathbin{\&} y[0]\big) \bmod 2^n.$$

By induction and Lemmas 1.20, 2.1 (c), and 2.12, the first addend may be written as

$$(2 \cdot (\lfloor x/2 \rfloor \mathbin{\&} \lfloor y/2 \rfloor)) \bmod 2^n = 2 \cdot \big((\lfloor x/2 \rfloor \mathbin{\&} \lfloor y/2 \rfloor) \bmod 2^{n-1}\big)$$

$$= 2 \cdot \big((\lfloor x/2 \rfloor \bmod 2^{n-1}) \mathbin{\&} (\lfloor y/2 \rfloor \bmod 2^{n-1})\big)$$

$$= 2 \cdot (\lfloor x/2 \rfloor[n - 2 : 0] \mathbin{\&} \lfloor y/2 \rfloor[n - 2 : 0])$$

$$= 2 \cdot (\lfloor x[n-1 : 0]/2 \rfloor \mathbin{\&} \lfloor y[n-1 : 0]/2 \rfloor),$$

and by Lemma 2.16 (g) and (j), the second addend is

$$x[0] \ \& \ y[0] = x[0] \ \& \ y[0] = x[n-1:0][0] \ \& \ y[n-1:0][0]$$
$$= (x[n-1:0] \bmod 2) \ \& \ (y[n-1:0] \bmod 2).$$

Thus, by Definition 3.1 (a) and Lemmas 3.1 and 2.1 (c),

$$(x \ \& \ y) \bmod 2^n = (x[n-1:0] \ \& \ y[n-1:0]) \bmod 2^n$$
$$= x[n-1:0] \ \& \ y[n-1:0]$$
$$= (x \bmod 2^n) \ \& \ (y \bmod 2^n).$$

(b) By Lemma 1.2, induction, and Definition 3.1 (a),

$$\lfloor (x \ \& \ y)/2^n \rfloor = \left\lfloor \lfloor (x \ \& \ y)/2^{n-1} \rfloor / 2 \right\rfloor$$
$$= \left\lfloor (\lfloor x/2^{n-1} \rfloor \ \& \ \lfloor y/2^{n-1} \rfloor)/2 \right\rfloor$$
$$= \left\lfloor \lfloor x/2^{n-1} \rfloor / 2 \rfloor \ \& \ \lfloor x/2^{n-1} \rfloor / 2 \rfloor \right\rfloor$$
$$= \lfloor x/2^n \rfloor \ \& \ \lfloor y/2^n \rfloor.$$

(c) By Definition 2.2 and (a) and (b),

$$(x \ \& \ y)[i:j] = \left\lfloor ((x \ \& \ y) \bmod 2^{i+1})/2^j \right\rfloor$$
$$= \left\lfloor ((x \bmod 2^{i+1}) \ \& \ (y \bmod 2^{i+1}))/2^j \right\rfloor$$
$$= \lfloor (x \bmod 2^{i+1})/2^j \rfloor \ \& \ \lfloor (y \bmod 2^{i+1})/2^j \rfloor$$
$$= x[i:j] \ \& \ y[i:j].$$

(d) This follows from (c).

(e) Let $C = \{m' x_1, n' y_1\} \ \& \ \{m' x_2, n' y_2\}$. By (a) and Lemma 2.29,

$C \bmod 2^n$

$= \{x_1[m-1:0], y_1[n-1:0]\}[n-1:0] \ \& \ \{x_2[m-1:0], y_2[n-1:0]\}[n-1:0]$

$= y_1[n-1:0] \ \& \ y_2[n-1:0].$

By (b),

$$\lfloor C/2^n \rfloor = \lfloor \{x_1[m-1:0], y_1[n-1:0]\}/2^n \rfloor \ \& \ \lfloor \{x_2[m-1:0], y_2[n-1:0]\}/2^n \rfloor,$$

where, by Definition 2.4 and the properties of the floor,

$$\lfloor \{x_i[m-1:0], y_i[n-1:0]\}/2^n \rfloor = \lfloor (2^n x_i[m-1:0] + y_i[n-1:0])/2^n \rfloor$$
$$= x_i[m-1:0] + \lfloor y_i[n-1:0]/2^n \rfloor$$
$$= x_i[m-1:0].$$

Thus,

$$\lfloor C/2^n \rfloor = x_1[m-1:0] \ \& \ x_2[m-1:0].$$

Finally, by Definitions 1.3 and 2.4,

$$C = \lfloor C/2^n \rfloor 2^n + (C \bmod 2^n))$$
$$= 2^n(x_1[m-1:0] \ \& \ x_2[m-1:0]) + y_1[m-1:0] \ \& \ y_2[m-1:0])$$
$$= \{x_1[m-1:0] \ \& \ x_2[m-1:0], y_1[n-1:0] \ \& \ y_2[n-1:0]\}.$$

(f) This follows from Lemmas 2.16 (h) and 2.22 and (d): for $k \in \mathbb{N}$,

$$(2^n(x \ \& \ y))[k] = (x \ \& \ y)[k-n]$$
$$= x[k-n] \ \& \ y[k-n]$$
$$= (2^n x)[k] \ \& \ (2^n y)[k]$$
$$= (2^n x \ \& \ 2^n y)[k] \qquad \square$$

Lemma 3.4 (d) is particularly useful in proving logical identities, when used in conjunction with Lemma 2.22 as illustrated by the following:

Lemma 3.5 *For all $x \in \mathbb{Z}$,*

(a) $x \ \& \ (-1) = x;$
(b) $x \ | \ (-1) = -1;$

Proof By Lemmas 3.4 (d) and 2.16 (f), for all $k \in \mathbb{N}$,

$$(x \ \& \ (-1))[k] = x[k] \ \& \ (-1)[k] = x[k] \ \& \ 1 = x[k],$$

and (a) follows from Lemma 2.22. A similar argument applies to (b). $\square$

The same technique, combining Lemmas 3.4 (d) and 2.22, may be used to derive the algebraic properties listed in Lemmas 3.6–3.9. The proofs are sufficiently similar to that of Lemma 3.5 that they may be safely omitted.

Lemma 3.6 *For all $x \in \mathbb{Z}$ and $y \in \mathbb{Z}$,*

(a) $x \mid y = 0 \Leftrightarrow x = y = 0$;
(b) $x \,\hat{}\, y = 0 \Leftrightarrow x = y$.

Lemma 3.7 *For all $x \in \mathbb{Z}$ and $y \in \mathbb{Z}$,*

(a) $x \,\&\, y = y \,\&\, x$;
(b) $x \mid y = y \mid x$;
(c) $x \,\hat{}\, y = y \,\hat{}\, x$.

Lemma 3.8 *For all $x \in \mathbb{Z}$, $y \in \mathbb{Z}$, and $z \in \mathbb{Z}$,*

(a) $(x \,\&\, y) \,\&\, z = x \,\&\, (y \,\&\, z)$;
(b) $(x \mid y) \mid z = x \mid (y \mid z)$;
(c) $(x \,\hat{}\, y) \,\hat{}\, z = x \,\hat{}\, (y \,\hat{}\, z)$.

Lemma 3.9 *For all $x \in \mathbb{Z}$, $y \in \mathbb{Z}$, and $z \in \mathbb{Z}$,*

(a) $(x \mid y) \,\&\, z = (x \mid y) \,\&\, (x \mid z)$;
(b) $x \,\&\, (y \mid z) = x \,\&\, y \mid x \,\&\, z$;
(c) $x \,\&\, y \mid x \,\&\, z \mid y \,\&\, z = x \,\&\, y \mid (x \,\hat{}\, y) \,\&\, z$.

The remaining results of this section are proved more naturally by induction.

Lemma 3.10 *For all $x \in \mathbb{Z}$, $y \in \mathbb{Z}$, and $n \in \mathbb{N}$,*

(a) $2^n x \,\&\, y = 2^n (x \,\&\, \lfloor y/2^n \rfloor)$;
(b) $2^n x \mid y = 2^n (x \mid \lfloor y/2^n \rfloor) + y \bmod 2^n$;
(c) $2^n x \,\hat{}\, y = 2^n (x \,\hat{}\, \lfloor y/2^n \rfloor) + y \bmod 2^n$.

Proof
(a) The claim is trivial if $x = 0$, $y = 0$, or $y = 2^n x$; otherwise, by Definition 3.1 (a), induction, and Lemma 1.2,

$$2^n x \,\&\, y = 2\left(\lfloor 2^n x/2 \rfloor \,\&\, \lfloor y/2 \rfloor\right) + (2^n x[0]) \,\&\, y[0]$$

$$= 2(2^{n-1} x \,\&\, \lfloor y/2 \rfloor) + 0$$

$$= 2\left(2^{n-1}\left(x \,\&\, \left\lfloor \lfloor y/2 \rfloor / 2^{n-1} \right\rfloor\right)\right)$$

$$= 2^n (x \,\&\, \lfloor y/2^n \rfloor).$$

(b) Similarly,

$$2^n x \mid y = 2\left(\lfloor 2^n x/2 \rfloor \mid \lfloor y/2 \rfloor\right) + (2^n x)[0] \mid y[0]$$

$$= 2(2^{n-1} x \mid \lfloor y/2 \rfloor) + y[0]$$

$$= 2\left(2^{n-1}\left(x \mid \left\lfloor \lfloor y/2 \rfloor / 2^{n-1} \right\rfloor\right) + \lfloor y/2 \rfloor \bmod 2^{n-1}\right) + y[0]$$

$$= 2^n (x \mid \lfloor y/2^n \rfloor) + 2(\lfloor y/2 \rfloor \bmod 2^{n-1}) + y[0],$$

where, by Lemmas 2.1 (c) and 2.11,

$$2(\lfloor y/2 \rfloor \bmod 2^{n-1}) + y[0] = 2\lfloor y/2 \rfloor[n-2:0] + y[0] = 2y[n-1:1] + y[0]$$
$$= y[n-1:0]$$
$$= y \bmod 2^n.$$

The proof of (c) is similar to that of (b).   □

Corollary 3.11 *Let $x \in \mathbb{Z}$ and let y be an n-bit vector, where $n \in \mathbb{N}$. Then*

$$2^n x \mid y = 2^n x + y.$$

Proof By Lemmas 3.10 and 1.11 and Definition 3.1 (b),

$$2^n x \mid y = 2^n(x \mid \lfloor y/2^n \rfloor) + y \bmod 2^n = 2^n(x \mid 0) + y = 2^n x + y. \qquad \Box$$

Lemma 3.12 *For all $x \in \mathbb{Z}$ and $n \in \mathbb{N}$, $2^n \mid x = \begin{cases} x & \text{if } x[n] = 1 \\ x + 2^n & \text{if } x[n] = 0. \end{cases}$*

Proof The claim holds trivially if $x = 0$ or $x = 2^n$. In the remaining case, by Definition 3.1,

$$2^n \mid x = 2 \cdot (\lfloor 2^{n-1} \rfloor \mid \lfloor x/2 \rfloor) + 2^n[0] \mid x[0].$$

If $n = 0$, this reduces to

$$2\lfloor x/2 \rfloor + 1 = \begin{cases} 2\lfloor x/2 \rfloor + x[0] = x & \text{if } x[0] = 1 \\ 2\lfloor x/2 \rfloor + 1 + x[0] = x + 1 & \text{if } x[0] = 0. \end{cases}$$

If $n > 0$, then by induction, the expression reduces to

$$2 \cdot (2^{n-1} \mid \lfloor x/2 \rfloor) + x[0] = \begin{cases} 2\lfloor x/2 \rfloor + x[0] = x & \text{if } x[n] = 1 \\ 2(\lfloor x/2 \rfloor + 2^{n-1}) + x[0] = x + 2^n & \text{if } x[n] = 1. \end{cases} \qquad \Box$$

The logical "and" operator may be used to extract a bit slice:

Lemma 3.13 *Let $x \in \mathbb{Z}$, $n \in \mathbb{N}$, and $k \in \mathbb{N}$. If $k < n$, then*

$$x \mathbin{\&} (2^n - 2^k) = 2^k x[n-1:k].$$

Proof The proof is by induction on n. If $k = 0$, then by induction and Lemmas 2.11 and 2.16 (k),

$$x \mathbin{\&} (2^n - 2^k) = x \mathbin{\&} (2^n - 1)$$
$$= 2(\lfloor x/2 \rfloor \mathbin{\&} \lfloor (2^n - 1)/2 \rfloor) + x[0] \mathbin{\&} (2^n - 1)[0]$$
$$= 2(\lfloor x/2 \rfloor \mathbin{\&} (2^{n-1} - 1)) + x[0] \mathbin{\&} 1$$

$$= 2\lfloor x/2\rfloor[n-2:0] + x[0]$$
$$= 2x[n-1:1] + x[0]$$
$$= x[n-1:0].$$

In the remaining case, $n > k > 0$ and

$$x \mathbin{\&} (2^n - 2^k) = 2(\lfloor x/2\rfloor \mathbin{\&} \lfloor (2^n - 2^k)/2\rfloor) + x[0] \mathbin{\&} (2^n - 2^k)[0]$$
$$= 2(\lfloor x/2\rfloor \mathbin{\&} (2^{n-1} - 2^{k-1})) + x[0] \mathbin{\&} 0$$
$$= 2 \cdot 2^{k-1}\lfloor x/2\rfloor[n-2:k-1]$$
$$= 2^k x[n-1:k]. \qquad \qquad \square$$

Corollary 3.14 *For all $x \in \mathbb{Z}$ and $n \in \mathbb{N}$, $x \mathbin{\&} 2^n = 2^n x[n]$.*

Proof By Lemma 3.13,

$$x \mathbin{\&} 2^n = x \mathbin{\&} (2^{n+1} - 2^n) = x[n:n] = x[n]. \qquad \qquad \square$$

3.2 Complement

We have a simple arithmetic definition of the logical complement, which we denote with the symbol $\sim$ as a prefix:

Definition 3.2 For all $x \in \mathbb{Z}$, $\sim x = -x - 1$.

Lemma 3.15 *For all $x \in \mathbb{Z}$, $\sim(\sim x) = x$.*

Proof By Definition 3.2, $\sim(\sim x) = -(-x - 1) - 1 = x$. $\qquad \qquad \square$

Lemma 3.16 *If $x \in \mathbb{Z}$ and $k \in \mathbb{N}$, then $\sim(2^k x) = 2^k(\sim x) + 2^k - 1$.*

Proof By Definition 3.2,

$$2^k(\sim x) + 2^k - 1 = 2^k(-x - 1) + 2^k - 1 = -2^k x - 1 = \sim(2^k x). \qquad \square$$

Lemma 3.17 *If $x \in \mathbb{Z}$, $n \in \mathbb{N}$, and $n > 0$, then $\sim\lfloor x/n\rfloor = \lfloor \sim x/n\rfloor$.*

Proof By Definition 3.2 and Lemma 1.5,

$$\lfloor (\sim x)/n\rfloor = \left\lfloor \frac{-x-1}{n}\right\rfloor = \left\lfloor -\frac{x+1}{n}\right\rfloor = -\left\lfloor \frac{x}{n}\right\rfloor - 1 = \sim\lfloor x/n\rfloor. \qquad \square$$

Notation For the purpose of resolving ambiguous expressions, the complement has higher precedence than any of the binary logical operations as well as the bit slice operator, e.g., $\sim x[i : j] = (\sim x)[i : j]$.

Lemma 3.18 *If $x \in \mathbb{Z}$, $i \in \mathbb{N}$, $j \in \mathbb{N}$, and $j \leq i$, then*

$$\sim x[i : j] = 2^{i+1-j} - x[i : j] - 1.$$

Proof By Definition 3.2 and Lemmas 2.8, 1.5, and 1.18,

$$\sim x[i : j] = \left\lfloor \frac{\sim x}{2^j} \right\rfloor \bmod 2^{i+1-j}$$

$$= \left\lfloor \frac{-(x+1)}{2^j} \right\rfloor \bmod 2^{i+1-j}$$

$$= \left(-\left\lfloor \frac{x}{2^j} \right\rfloor - 1 \right) \bmod 2^{i+1-j}$$

$$= \left(2^{i+1-j} - 1 - \left\lfloor \frac{x}{2^j} \right\rfloor \right) \bmod 2^{i+1-j}$$

$$= 2^{i+1-j} - 1 - x[i : j]. \qquad \square$$

The usual bit-wise characterization of the complement is a special case of Lemma 3.18:

Corollary 3.19 *If $x \in \mathbb{Z}$ and $n \in \mathbb{N}$, then $\sim x[n] \neq x[n]$.*

Proof By Lemma 3.18, $\sim x[n] = 2^{n+1-n} - x[n] - 1 = 1 - x[n]$. $\qquad \square$

Lemma 3.20 *For all $x \in \mathbb{Z}$, $\sim x = x \wedge (-1)$.*

Proof By Lemmas 3.19, 2.16 (f), and 3.4 (d), for all $k \in \mathbb{N}$,

$$\sim x[k] = 1 \Leftrightarrow x[k] = 0 \Leftrightarrow x[k] \neq (-1)[k] \Leftrightarrow x[k] \wedge (-1)[k] = (x \wedge (-1))[k] = 1,$$

i.e., $\sim x[k] = (x \wedge (-1))[k]$. The claim follows from Lemma 2.22. $\qquad \square$

The remaining results of this section are properties of complements of bit slices that have proved useful in manipulating expressions derived from RTL designs.

Lemma 3.21 *Let $x \in \mathbb{Z}$, $i \in \mathbb{N}$, $j \in \mathbb{N}$, $k \in \mathbb{N}$, and $\ell \in \mathbb{N}$. If $\ell \leq k \leq i - j$, then*

$$\sim (x[i : j])[k : \ell] = \sim x[k + j : \ell + j].$$

Proof By Lemmas 3.18 and 2.15,

$$\sim(x[i:j])[k:\ell] = 2^{k+1-\ell} - x[i:j][k:\ell] - 1$$
$$= 2^{(k+j)+1-(\ell+j)} - x[k+j:\ell+j] - 1$$
$$= \sim x[k+j:\ell+j]. \qquad \square$$

Lemma 3.22 *If $x \in \mathbb{Z}$ and y in an n-bit vector, where $n \in \mathbb{N}$, then*

$$\sim(x[n-1:0]) \ \& \ y = \sim x[n-1:0] \ \& \ y.$$

Proof By Lemma 3.21, $\sim(x[n-1:0])[n-1:0] = \sim x[n-1:0]$, and hence by Lemmas 3.1, 2.3, and 3.4 (c),

$$\sim(x[n-1:0]) \ \& \ y = (\sim(x[n-1:0]) \ \& \ y) \,[n-1:0]$$
$$= \sim(x[n-1:0])[n-1:0] \ \& \ y[n-1:0]$$
$$= \sim x[n-1:0] \ \& \ y. \qquad \square$$

Lemma 3.23 *Let $x \in \mathbb{Z}$, $i \in \mathbb{N}$, $j \in \mathbb{N}$, $k \in \mathbb{N}$, and $\ell \in \mathbb{N}$. If $\ell \leq k \leq i - j$, then*

$$\sim(\sim x[i:j])[k:\ell] = x[k+j:\ell+j].$$

Proof By Lemmas 3.18, 2.15, and 3.15,

$$\sim(\sim x[i:j])[k:\ell] = 2^{k+1-\ell} - \sim x[i:j][k:\ell] - 1$$
$$= 2^{(k+j)+1-(\ell+j)} - \sim x[k+j:\ell+j] - 1$$
$$= \sim(\sim x)[k+j:\ell+j]$$
$$= x[k+j:\ell+j]. \qquad \square$$

Part II
Floating-Point Arithmetic

According to the IEEE floating-point standard [9], each of the elementary arithmetic operations of addition, multiplication, division, square root extraction, and fused multiplication-addition

> ... shall be performed as if it first produced an intermediate result correct to infinite precision and then rounded that result according to one of the [supported] modes ...

Since the operands (or operand, in the case of square root) and final result are represented as bit vectors under some format, the relationship between inputs and outputs that is implicit in this specification, which is known as the *principle of correct rounding*, is as diagrammed in Fig. 3.1. That is, the prescribed output is the result of the following procedure:

(1) Derive the numerical value(s) represented by the bit vector operand(s) under the given format.
(2) Apply the pure mathematical operation to be implemented to these values.
(3) Compute a rounded approximation of the result of this operation that is representable in the format.
(4) Derive the bit vector representation of the approximation.

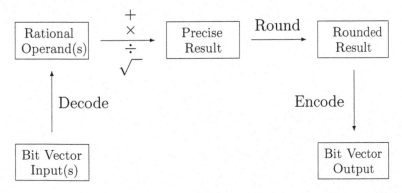

Fig. 3.1 Principle of correct rounding

Of course, an implementation is not actually required to generate the output by such a procedure. In fact, in general an explicit intermediate representation of the precise unrounded result is infeasible, as the number of bits required may be prohibitive or even, in the case of division or square root, infinite. In practice, an internal representation of the result is derived from the inputs with sufficient accuracy to allow its transformation to the encoding of the correctly rounded approximation. A variety of algorithms for computing the intermediate result for each of the operations of interest are presented in Part III; the other steps of the process are addressed in the present Part II. In Chaps. 4 and 5, we describe the common schemes for encoding real numbers as bit vectors. Chapter 6 addresses the problem of rounding, discussing the various *rounding modes* of interest and their bit-level implementations.

As discussed in the preface, the formalization of the principle of correct rounding in the ACL2 logic, which is limited to rational arithmetic, presents a challenge in the case of the square root operation. In Chap. 7, we discuss our solution to this technical problem, which is based on a function that computes a rational approximation $\sqrt[(66)]{x}$ of the real square root of a rational number x, with the property that it rounds to the same value as does the precise square root with respect to any IEEE-prescribed rounding mode and any format of interest. This chapter may be omitted by the reader who is not interested in the formalization problem.

Chapter 4
Floating-Point Numbers

The designation *floating-point number* is a contextual term used to refer to a rational number that is representable with respect to a particular format, as described in Chap. 5. In this chapter, we discuss the properties of real numbers that are relevant to these representations.

4.1 Decomposition

Floating-point arithmetic is based on the observation that every nonzero $x \in \mathbb{R}$ admits a unique representation of the form

$$x = \pm m \cdot 2^e,$$

where e is an integer, called the *exponent* of x, and m is a number in the range $1 \leq m < 2$, called the *significand* of x. These components are defined as follows.

Definition 4.1 Let $x \in \mathbb{R}$. If $x \neq 0$, then

(a) $sgn(x) = \frac{x}{|x|} = \begin{cases} 1 \text{ if } x > 0 \\ -1 \text{ if } x < 0; \end{cases}$

(b) $expo(x)$ is the unique integer that satisfies $2^{expo(x)} \leq |x| < 2^{expo(x)+1}$;

(c) $sig(x) = |x|2^{-expo(x)}$.

If $x = 0$, then $sgn(x) = expo(x) = sig(x) = 0$.

The decomposition property is immediate:

Lemma 4.1 *For all $x \in \mathbb{R}$, $x = sgn(x)sig(x)2^{expo(x)}$.*

© The Author(s), under exclusive license to Springer Nature Switzerland AG 2022
D. M. Russinoff, *Formal Verification of Floating-Point Hardware Design*,
https://doi.org/10.1007/978-3-030-87181-9_4

Lemma 4.2 *For all* $x \in \mathbb{R}$, $y \in \mathbb{R}$, *and* $n \in \mathbb{Z}$,

(a) *If* $2^n \leq x$, *then* $n \leq expo(x)$;
(b) *If* $0 < |x| < 2^{n+1}$, *then* $expo(x) \leq n$;
(c) *If* $0 < |x| \leq |y|$, *then* $expo(x) \leq expo(y)$;
(d) *If* $x \geq 1$, *then* $expo(\lfloor x \rfloor) = expo(x)$;
(e) $expo(2^n) = n$;
(f) *If* $x \in \mathbb{N}$ *and* $x > 0$, *then* $x[expo(x)] = 1$.

Proof

(a) If $n > expo(x)$, then $n \geq expo(x) + 1$, which implies $|x| < 2^{expo(x)+1} \leq 2^n$.
(b) If $n < expo(x)$, then $n + 1 \leq expo(x)$, which implies $2^{n+1} \leq 2^{expo(x)} \leq |x|$.
(c) Since $|x| \leq |y| < 2^{expo(y)+1}$, this follows from (b).
(d) Let $expo(x) = n$. Then $expo(\lfloor x \rfloor) \geq n$. Since $n \geq expo(1) = 0$, $2^n \in \mathbb{Z}$, and since $x \geq 2^n$, $\lfloor x \rfloor \geq 2^n$, which implies $expo(\lfloor x \rfloor) \geq n$.
(e) This is an immediate consequence of Definition 4.1.
(f) This follows from Lemma 2.17, substituting $expo(x)$ for k $expo(x)+1$ for n. $\square$

Lemma 4.3 *Let* $x \in \mathbb{R}$.

(a) *If* $x \neq 0$, *then* $1 \leq sig(x) < 2$;
(b) *If* $1 \leq x < 2$, *then* $sig(x) = x$;
(c) $sig(sig(x)) = sig(x)$.

Proof

(a) Definition 4.1 yields

$$1 = 2^{expo(x)}/2^{expo(x)} \leq |x|/2^{expo(x)} < 2^{expo(x)+1}/2^{expo(x)} = 2.$$

(b) Since $2^0 \leq |x| = x < 2^1$, $expo(x) = 0$, and hence, $sig(x) = x/2^0 = x$.
(c) follows from (a) and (b).

$\square$

Lemma 4.4 *Let* $x \in \mathbb{R}$. *If* $|x| = 2^e m$, *where* $e \in \mathbb{Z}$, $m \in \mathbb{R}$, *and* $1 \leq m < 2$, *then* $m = sig(x)$ *and* $e = expo(x)$.

Proof Since $1 \leq m < 2$, $2^e \leq 2^e m < 2^e 2 = 2^{e+1}$, where $2^e m = |x|$. It follows from Definition 4.1 that $e = expo(x)$, and therefore, $sig(x) = |x|/2^e = m$. $\square$

Changing the sign of a number does not affect its exponent or significand.

Lemma 4.5 *For all* $x \in \mathbb{R}$,

(a) $sgn(-x) = -sgn(x)$;
(b) $expo(-x) = expo(x)$;
(c) $sig(-x) = sig(x)$.

A shift does not affect the sign or significand.

Lemma 4.6 *If $x \in \mathbb{R}$, $x \neq 0$, and $n \in \mathbb{Z}$, then*

(a) $sgn(2^n x) = sgn(x)$;
(b) $expo(2^n x) = expo(x) + n$;
(c) $sig(2^n x) = sig(x)$.

Proof

(a) $sgn(2^n x) = 2^n x / |2^n x| = x / |x| = sgn(x)$.
(b) $2^{expo(x)} \leq |x| < 2^{expo(x)+1} \Rightarrow 2^{expo(x)+n} \leq |2^n x| < 2^{expo(x)+n+1}$.
(c) $sig(2^n x) = |2^n x| 2^{-(expo(x)+n)} = |x| 2^{-expo(x)} = sig(x)$.

$\square$

We have the following formulas for the components of a product.

Lemma 4.7 *Let $x \in \mathbb{R}$ and $y \in \mathbb{R}$. If $xy \neq 0$, then*

(a) $sgn(xy) = sgn(x)sgn(y)$;

(b) $expo(xy) = \begin{cases} expo(x) + expo(y) & \text{if } sig(x)sig(y) < 2 \\ expo(x) + expo(y) + 1 & \text{if } sig(x)sig(y) \geq 2 \end{cases}$;

(c) $sig(xy) = \begin{cases} sig(x)sig(y) & \text{if } sig(x)sig(y) < 2 \\ sig(x)sig(y)/2 & \text{if } sig(x)sig(y) \geq 2. \end{cases}$

Proof

(a) $sgn(xy) = xy/|xy| = (x/|x|)(y/|y|) = sgn(x)sgn(y)$.
(b) Since $2^{expo(x)} \leq |x| < 2^{expo(x)+1}$ and $2^{expo(y)} \leq |y| < 2^{expo(y)+1}$, we have

$$2^{expo(x)+expo(y)} = 2^{expo(x)} 2^{expo(y)}$$
$$\leq |xy|$$
$$< 2^{expo(x)+1} 2^{expo(y)+1}$$
$$= 2^{expo(x)+expo(y)+2}.$$

If $sig(x)sig(y) = |x|2^{-expo(x)}|y|2^{-expo(y)} < 2$, then

$$|xy| < 2 \cdot 2^{expo(x)} 2^{expo(y)} = 2^{expo(x)+expo(y)+1},$$

and by Definition 4.1, $expo(xy) = expo(x) + expo(y)$.
On the other hand, if $sig(x)sig(y) \geq 2$, then similarly,

$$|xy| \geq 2^{expo(x)+expo(y)+1},$$

and Definition 4.1 yields $expo(xy) = expo(x) + expo(y) + 1$.
(c) If $sig(x)sig(y) < 2$, then

$$sig(xy) = |xy|2^{-expo(xy)}$$
$$= |xy|2^{-(expo(x)+expo(y))}$$

$$= |x|2^{-expo(x)}|y|2^{-expo(y)}$$

$$= sig(x)sig(y).$$

Otherwise,

$$sig(xy) = |xy|2^{-expo(xy)}$$

$$= |xy|2^{-(expo(x)+expo(y)+1)}$$

$$= |x|2^{-expo(x)}|y|2^{-expo(y)}/2$$

$$= sig(x)sig(y)/2. \qquad \qquad \square$$

4.2 Exactness

In order for a significand, which has the form

$$(1.\beta_1\beta_2\cdots)_2,$$

to be represented by an n-bit field, we must have $\beta_k = 0$ for all $k \geq n$, or equivalently, a right shift of the radix point by $n - 1$ places must result in an integer. This motivates our definition of *exactness*.

Definition 4.2 If $x \in \mathbb{R}$ and $n \in \mathbb{N}$, then x is n-exact $\Leftrightarrow sig(x)2^{n-1} \in \mathbb{Z}$.

The definition may be restated in various ways:

Lemma 4.8 *Let $x \in \mathbb{R}$, $n \in \mathbb{N}$, and $k \in \mathbb{Z}$. The following are equivalent:*

(a) *x is n-exact;*
(b) *$-x$ is n-exact;*
(c) *$sig(x)$ is n-exact;*
(d) *$2^k x$ is n-exact;*
(e) *$x2^{n-1-expo(x)} \in \mathbb{Z}$.*

It is clear that if $sig(x)$ is representable in a given field of bits, then it is also representable in any wider field.

Lemma 4.9 *For all $x \in \mathbb{R}$, $n \in \mathbb{Z}$, and $m \in \mathbb{Z}$, if $m < n$ and x is m-exact, then x is n-exact.*

Proof Since $2^{n-m} \in \mathbb{Z}$ and $x \cdot 2^{m-1-expo(x)} \in \mathbb{Z}$,

$$x \cdot 2^{n-1-expo(x)} = 2^{n-m} \cdot x \cdot 2^{m-1-expo(x)} \in \mathbb{Z}. \qquad \qquad \square$$

A power of 2 has a 1-bit significand:

Lemma 4.10 *If $n \in \mathbb{Z}$ and $m \in \mathbb{Z}^+$, then 2^n is m-exact.*

Proof By Lemma 4.2 (d),

$$2^n \cdot 2^{m-1-expo(2^n)} = 2^n \cdot 2^{m-1-n} = 2^{m-1} \in \mathbb{Z}. \qquad \square$$

A bit vector is exact with respect to its width.

Lemma 4.11 *If x is a bit vector of width $n \in \mathbb{N}$, then x is n-exact.*

Proof Since 0 is n-exact for all $n \in \mathbb{Z}$, we may assume $x > 0$, and therefore, $n > 0$. Now since $0 < x < 2^n$, Lemma 4.2 implies $expo(x) \leq n - 1$, and hence, $x \cdot 2^{n-1-expo(x)} \in \mathbb{Z}$. $\qquad \square$

We have the following formula for the exactness of a product.

Lemma 4.12 *Let $x \in \mathbb{R}$, $y \in \mathbb{R}$, $m \in \mathbb{Z}$, and $n \in \mathbb{Z}$. If x is m-exact and y is n-exact, then xy is $(m + n)$-exact.*

Proof If $x2^{m-1-expo(x)}$ and $y2^{n-1-expo(y)}$ are integers, then by Lemma 4.7, so is

$$x2^{m-1-expo(x)}y2^{n-1-expo(y)}2^{expo(x)+expo(y)+1-expo(xy)} = xy2^{m+n-1-expo(xy)}. \quad \square$$

In particular, if x is n-exact, then x^2 is $2n$-exact. If x is rational, the converse also holds.

Lemma 4.13 *Let $x \in \mathbb{Q}$ and $n \in \mathbb{Z}$. If x^2 is 2n-exact, then x is n-exact.*

Proof We shall require an elementary fact of arithmetic:

$$\text{If } r \in \mathbb{Q} \text{ and } 2r^2 \in \mathbb{Z}, \text{ then } r \in \mathbb{Z}.$$

In order to establish this result, let $r = a/b$, where a and b are integers with no common prime factor, and assume that $2r^2 = 2a^2/b^2 \in \mathbb{Z}$. If b were even, say $b = 2c$, then $2a^2/b^2 = a^2/2c^2 \in \mathbb{Z}$, implying that a is even, which is impossible. Thus, b cannot have a prime factor that does not also divide a, and hence, $b = 1$.

Now to show that x is n-exact, i.e., $x2^{n-1-expo(x)} \in \mathbb{Z}$, it will suffice to show that

$$2\left(x2^{n-1-expo(x)}\right)^2 = x^22^{2n-1-2expo(x)} \in \mathbb{Z}.$$

But since x^2 is $2n$-exact, we have $x^22^{2n-1-expo(x^2)} \in \mathbb{Z}$, and since Lemma 4.7 implies $expo(x^2) \geq 2expo(x)$,

$$x^22^{2n-1-2expo(x)} = x^22^{2n-1-expo(x^2)}2^{expo(x^2)-2expo(x)} \in \mathbb{Z}. \qquad \square$$

Lemma 4.14 *Let $n \in \mathbb{N}$, $x \in \mathbb{R}$, and $y \in \mathbb{R}$. Assume that x and y are nonzero and k-exact for some $k \in \mathbb{N}$. If xy is n-exact, then so are x and y.*

Proof Let p and q be the smallest integers such that $x' = 2^p x$ and $y' = 2^q y$ are integers. Thus, x' and y' are odd. By Lemma 4.8, x, y, or xy, respectively, is n-exact iff x', y', or $x'y'$ is n-exact. Consequently, we may replace x and y with x' and y'. That is, we may assume without loss of generality that x and y are odd integers.

By Lemma 4.8, an odd integer z is n-exact iff $expo(z) \leq n - 1$, or, equivalently, according to Lemma 4.2, $|z| < 2^n$. Thus, since xy is n-exact, $xy < 2^n$, which implies $x < 2^n$ and $y < 2^n$, and hence, x and y are n-exact. $\qquad\square$

The next lemma gives a formula for exactness of a difference.

Lemma 4.15 *Let* $x \in \mathbb{R}$, $y \in \mathbb{R}$, $n \in \mathbb{Z}$, *and* $k \in \mathbb{Z}$. *Assume that* $n > 0$ *and* $n > k$. *If* x *and* y *are both* n-*exact and* $expo(x - y) + k \leq min(expo(x), expo(y))$, *then* $x - y$ *is* $(n - k)$-*exact.*

Proof Since x is n-exact and $expo(x - y) + k \leq expo(x)$,

$$x2^{n-1-(expo(x-y)+k)} = x2^{n-1-expo(x)}2^{expo(x)-(expo(x-y)+k)} \in \mathbb{Z}$$

by Lemma 4.8. Similarly, $y2^{n-1-(expo(x-y)+k)} \in \mathbb{Z}$. Thus,

$$(x - y)2^{(n-k)-1-expo(x-y)} = x2^{n-1-(expo(x-y)+k)} - y2^{n-1-(expo(x-y)+k)} \in \mathbb{Z}. \quad\square$$

Lemma 4.15 is often applied with $k = 0$, in the following weaker form.

Corollary 4.16 *Let* $x \in \mathbb{R}$, $y \in \mathbb{R}$, *and* $n \in \mathbb{Z}^+$. *If* x *and* y *are both* n-*exact and* $|x - y| \leq min(|x|, |y|)$, *then* $x - y$ *is* n-*exact.*

If x is positive and n-exact, then the least n-exact number exceeding x is computed as follows:

Definition 4.3 Let $x \in \mathbb{R}$, $x > 0$, and $n \in \mathbb{N}$, $n > 0$. Then

$$fp^+(x, n) = x + 2^{expo(x)+1-n}.$$

Lemma 4.17 *Let* $x \in \mathbb{R}$, $x > 0$, *and* $n \in \mathbb{N}$, $n > 0$. *If* x *is* n-*exact, then* $fp^+(x, n)$ *is* n-*exact.*

Proof Let $e = expo(x)$ and $x^+ = fp^+(x, n) = x + 2^{e+1-n}$. Since x is n-exact, $x \cdot 2^{n-1-e} \in \mathbb{Z}$, and by Lemma 4.2, $x < 2^{e+1}$. Thus,

$$x \cdot 2^{n-1-e} < 2^{e+1} \cdot 2^{n-1-e} = 2^n,$$

which implies $x \cdot 2^{n-1-e} \leq 2^n - 1$. Therefore, $x \leq 2^{e+1} - 2^{e+1-n}$ and $x^+ = x + 2^{e+1-n} \leq 2^{e+1}$. If $x^+ = 2^{e+1}$, then x^+ is n-exact by Lemma 4.10. Otherwise, $x^+ < 2^{e+1}$, $expo(x^+) = e$, and

$$x^+ \cdot 2^{n-1-expo(x^+)} = (x + 2^{e+1-n})2^{n-1-e} = x \cdot 2^{e+1-n} + 1 \in \mathbb{Z}. \quad\square$$

Lemma 4.18 *Let $x \in \mathbb{R}$, $y \in \mathbb{R}$, $y > x > 0$, and $n \in \mathbb{N}$, $n > 0$. If x and y are n-exact, then $y \geq fp^+(x, n)$.*

Proof Let $e = expo(x)$ and $x^+ = fp^+(x, n) = x + 2^{e+1-n}$. Since x is n-exact, $x \cdot 2^{n-1-e} \in \mathbb{Z}$. Similarly, since y is n-exact, $y \cdot 2^{n-1-expo(y)} \in \mathbb{Z}$. But $expo(y) \geq e$ by Lemma 4.2 (c), and hence,

$$y \cdot 2^{n-1-e} = y \cdot 2^{n-1-expo(y)} \cdot 2^{expo(y)-e} \in \mathbb{Z}$$

as well. Now since $y > x$, $y \cdot 2^{n-1-e} > x \cdot 2^{n-1-e}$, which implies

$$y \cdot 2^{n-1-e} \geq x \cdot 2^{n-1-e} + 1.$$

Thus,

$$y \geq (x \cdot 2^{n-1-e} + 1)/2^{n-1-e} = x + 2^{e+1-n} = x^+. \qquad \square$$

Corollary 4.19 *Let $x \in \mathbb{R}$, $x > 0$, and $n \in \mathbb{N}$, $n > 0$. If x is n-exact and $expo(fp^+(x, n)) \neq expo(x)$, then $fp^+(x, n) = 2^{expo(x)+1}$.*

Proof Since $fp^+(x, n) > x$, Lemma 4.2 (c) implies $expo(fp^+(x, n)) \geq expo(x)$. Therefore, we have $expo(fp^+(x, n)) > expo(x)$, which, according to Lemma 4.2, implies $fp^+(x, n) \geq 2^{expo(x)+1}$. On the other hand, since $2^{expo(x)+1}$ is n-exact by Lemma 4.10, Lemma 4.18 implies $2^{expo(x)+1} \geq fp^+(x, n)$. $\qquad \square$

Similarly, the following definition computes the greatest n-exact number that is less than a given n-exact number.

Definition 4.4 Let $x \in \mathbb{R}$, $x > 0$, and $n \in \mathbb{N}$, $n > 0$. Then

$$fp^-(x, n) = \begin{cases} x - 2^{expo(x)-n} & \text{if } x = 2^{expo(x)} \\ x \quad 2^{expo(x)+1-n} & \text{if } x \neq 2^{expo(x)}. \end{cases}$$

Lemma 4.20 *Let $x \in \mathbb{R}$, $x > 0$, and $n \in \mathbb{N}$, $n > 0$. If x is n-exact, then $fp^-(x, n)$ is n-exact.*

Proof Let $e = expo(x)$ and $x^- = fp^-(x, n)$.
Suppose first that $x = 2^e$. Then $x^- = x - 2^{e-n}$, and since

$$2^e > x^- \geq 2^e - 2^{e-1} = 2^{e-1},$$

Definition 4.1 implies $expo(x^-) = e - 1$. Therefore,

$$x^- 2^{n-1-expo(x^-)} = x^- 2^{n-e} = (x - 2^{e-n})2^{n-e} = (2^e - 2^{e-n})2^{n-e} = 2^n - 1 \in \mathbb{Z},$$

and x^- is n-exact by Lemma 4.8.

Now suppose $x \neq 2^e$. Then by Lemma 4.17,

$$x \geq fp^+(2^e, n) = 2^e + 2^{e+1-n}.$$

Now $x^- = x - 2^{e+1-n} \geq 2^e$, and hence, $expo(x^-) = e$. Thus,

$$x^- 2^{n-1-expo(x^-)} = x^- 2^{n-1-e} = (x - 2^{e+1-n}) 2^{n-1-e} = x 2^{n-1-e} - 1 \in \mathbb{Z},$$

which implies x^- is n-exact.   □

Lemma 4.21 *Let $x \in \mathbb{R}$, $x > 0$, and $n \in \mathbb{N}$, $n > 0$. If x is n-exact, then*

$$fp^+(fp^-(x, n), n) = fp^-(fp^+(x, n), n) = x.$$

Proof Let $e = expo(x)$, $x^- = fp^-(x, n)$, and $x^+ = fp^+(x, n)$. We shall prove first that $fp^+(x^-, n) = x$.

As noted in the proof of Lemma 4.17,

$$expo(x^-) = \begin{cases} e - 1 & \text{if } x = 2^e \\ e & \text{if } x > 2^e. \end{cases}$$

If $x = 2^e$, then

$$fp^+(x^-, n) = x^- + 2^{(e-1)+1-n} = x - 2^{e-n} + 2^{e-n} = x.$$

But if $x > 2^e$, then

$$fp^+(x^-, n) = x^- + 2^{e+1-n} = x - 2^{e+1-n} + 2^{e+1-n} = x.$$

Next, we prove that $fp^-(x^+, n) = x$. Note that $x^+ = x + 2^{e+1-n}$. If $x^+ < 2^{e+1}$, then $expo(x^+) = e$, and since $x^+ \neq 2^e$,

$$fp^-(x^+, n) = x^+ - 2^{e+1-n} = x + 2^{e+1-n} - 2^{e+1-n} = x.$$

We may assume, therefore, that $x^+ \geq 2^{e+1}$. By Lemma 4.17, since 2^{e+1} is n-exact, $2^{e+1} \geq x^+$, and hence, $x^+ = 2^{e+1}$. Thus,

$$fp^-(x^+, n) = x^+ - 2^{e+1-n} = x + 2^{e+1-n} - 2^{e+1-n} = x.$$ □

Lemma 4.22 *Let $x \in \mathbb{R}$, $y \in \mathbb{R}$, $x > y > 0$, and $n \in \mathbb{N}$, $n > 0$. If x and y are n-exact, then $y \leq fp^-(x, n)$.*

Proof Suppose $y > fp^-(x, n)$. Then since $fp^-(x, n)$ and y are both n-exact, Lemma 4.17 implies $y \geq fp^+(fp^-(x, n), n) = x$, a contradiction.   □

Chapter 5
Floating-Point Formats

A floating-point format is a scheme for representing a number as a bit vector consisting of three fields corresponding to its sign, exponent, and significand. In this chapter, we present a classification of such formats, including those prescribed by IEEE Standard 754 [9], and examine the characteristics of the numbers that they represent.

5.1 Classification of Formats

A floating-point format is characterized by the precision p with which representable numbers are differentiated, and the number q of bits allocated to the exponent, determining the range of representable numbers. Some formats represent all p bits of a number's significand explicitly, but a common optimization is to omit the integer bit. Thus, we define a format as a triple:

Definition 5.1 A floating-point format is a triple $F = \langle e, p, q \rangle$, where

(a) e is a boolean indication that the format is explicit ($e = 1$) or implicit ($e = 0$), i.e., whether or not the integer bit is explicitly represented;
(b) $p = prec(F)$ is an integer, $p \geq 2$, the precision of F;
(c) $q = expw(F)$ is an integer, $q \geq 2$, the exponent width of F.

The significand width of F is $sigw(F) = \begin{cases} p & \text{if } F \text{ is explicit} \\ p - 1 & \text{if } F \text{ is implicit.} \end{cases}$

In this chapter, every "format" will be understood to be a floating-point format.

Definition 5.2 An encoding for a format F is a bit vector of width $expw(F) + sigw(F) + 1$.

© The Author(s), under exclusive license to Springer Nature Switzerland AG 2022
D. M. Russinoff, *Formal Verification of Floating-Point Hardware Design*,
https://doi.org/10.1007/978-3-030-87181-9_5

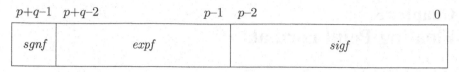

Fig. 5.1 A floating-point format with implicit integer bit

Definition 5.1 is sufficiently broad to include the formats that will be of interest to us. The most common implicit formats are the IEEE *basic single* ($p = 24$, $q = 8$) and *double precision* ($p = 53$, $q = 11$) formats, at least one of which must be implemented by any IEEE compliant floating-point unit. Explicit formats include most implementations of the *single extended* ($p = 32$, $q = 11$) and *double extended* ($p = 64$, $q = 15$) formats, as well as the higher-precision formats that are typically used for internal computations in floating-point units.

We establish the following notation for the formats that are used by the x86 and Arm elementary arithmetic operations discussed in Part IV:

Definition 5.3 The half, single, double, and (double) extended formats are as follows:

$$HP = \langle 0, 11, 5 \rangle;$$

$$SP = \langle 0, 24, 8 \rangle;$$

$$DP = \langle 0, 53, 11 \rangle;$$

$$EP = \langle 1, 64, 15 \rangle.$$

The *sign*, *exponent*, and *significand* fields of an encoding are defined as illustrated in Figs. 5.1 and 5.2. We also define the *mantissa* field as the significand field without the integer bit, if present:

Definition 5.4 If x is an encoding for a format F, then

(a) $sgnf(x, F) = x[expw(F) + sigw(F)]$;
(b) $expf(x, F) = x[expw(F) + sigw(F) - 1 : sigw(F)]$;
(c) $sigf(x, F) = x[sigw(F) - 1 : 0]$;
(d) $manf(x, F) = x[prec(F) - 2 : 0]$.

The encodings for a given format are partitioned into several classes determined primarily by the exponent field, as described in the next three sections.

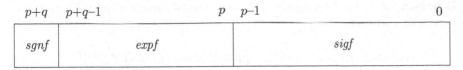

Fig. 5.2 A floating-point format with explicit integer bit

5.2 Normal Encodings

Encodings with exponent field 0 represent very small values according to a separate scheme (Sect. 5.3); those with maximal exponent field (all 1s) are reserved for representing non-numerical entities (Sect. 5.4). The remaining encodings are the subject of this section.

Definition 5.5 An encoding x for a format F is *normal* iff the following conditions hold:

(a) $0 < expf(x, F) < 2^{expw(F)} - 1$;
(b) If F is explicit, then $x[prec(F)-1] = 1$.

The implied integer bit of a normal encoding for an implicit format is 1, as reflected in Definition 5.8 below. For an explicit format, the case of a nonzero exponent field and a zero leading significand bit is an anomaly that should never be generated by hardware.

Definition 5.6 An encoding x for an explicit format F is *unsupported* iff the following conditions hold:

(a) $expf(x, F) > 0$;
(b) $x[prec(F)-1] = 0$.

Let x be a normal encoding for a format F with $prec(F) = p$ and $expw(F) = q$. The significand field $sigf(x, F)$ is interpreted as a p-exact value in the interval $[1, 2)$, i.e., with an implied binary point following the leading bit, which is 1, either explicitly or implicitly. The value encoded by x is the signed product of this value and a power of 2 determined by the exponent field. Since it is desirable for the range of exponents to be centered at 0, this field is interpreted with a bias of $2^{q-1} - 1$, i.e., the value of the exponent represented is

$$expf(x, F) - (2^{q-1} - 1),$$

which lies in the interval $[2 - 2^{q-1}, 2^{q-1}]$.

Definition 5.7 The exponent bias of a format F is $bias(F) = 2^{expw(F)-1} - 1$.

Thus, the decoding function is defined as follows. The definition applies to both implicit and explicit formats, although for the latter, it may be simplified by replacing $1 + 2^{1-p} manf(x, F)$ with $2^{1-p} sigf(x, F)$.

Definition 5.8 Let F be a format with $p = prec(F)$ and $b = bias(F)$. If x is a normal encoding for F, then

$$ndecode(x, F) = (-1)^{sgnf(x,F)} \left(1 + 2^{1-p} manf(x, F)\right) 2^{expf(x,F)-b}.$$

The following is a trivial consequence of Definition 5.8.

Lemma 5.1 *Let x be a normal encoding for a format F and let $\hat{x} = ndecode(x, F)$. Then*

(a) $sgn(\hat{x}) = (-1)^{sgnf(x,F)}$;
(b) $expo(\hat{x}) = expf(x, F) - bias(F)$;
(c) $sig(\hat{x}) = 1 + 2^{1-prec(F)} manf(x, F)$.

The numbers that admit normal encodings may be characterized as follows:

Definition 5.9 Let F be a format and let $r \in \mathbb{Q}$. Then r is a *normal value* of F iff the following conditions hold:

(a) $r \neq 0$;
(b) $0 < expo(r) + bias(F) < 2^{expw(F)} - 1$;
(c) r is $prec(F)$-exact.

The normal encoding of a normal value is derived as follows.

Definition 5.10 Let r be a normal value of F. Let

$$s = \begin{cases} 0 \text{ if } r > 0 \\ 1 \text{ if } r < 0, \end{cases}$$

$$e = expo(r) + bias(F),$$

and

$$m = 2^{prec(F)-1} sig(r).$$

Then

$$nencode(r, F) = \{1' s, expw(F)' e, sigw(F)' m\}.$$

The next two lemmas establish an inverse relation between the encoding and decoding functions, from which it follows that the numbers that admit normal encodings are precisely those that satisfy Definition 5.9.

Lemma 5.2 *If x is a normal encoding for a format F, then $ndecode(x, F)$ is a normal value of F, and*

$$nencode(ndecode(x, F), F) = x.$$

Proof Let $p = prec(F), q = expw(F), B = bias(F)$, and $\hat{x} = ndecode(x, F)$. It is clear from Definition 5.8 that $\hat{x} \neq 0$. By Lemma 5.1,

$$expo(\hat{x}) + B = expf(x, F)$$

is a q-bit vector, and

$$2^{p-1}sig(\hat{x}) = 2^{p-1}(1 + 2^{1-p}sigf(x, F)) = 2^{p-1} + sigf(x, F) \in \mathbb{Z},$$

i.e., $\hat{x}$ is p-exact. Thus, $\hat{x}$ is a normal value of F.

It is also clear from Definition 5.8 that $sgnf(x, F) = \begin{cases} 0 \text{ if } \hat{x} > 0 \\ 1 \text{ if } \hat{x} < 0. \end{cases}$

Suppose F is implicit. Then by Definitions 5.10 and 5.4 and Lemmas 2.26 and 2.3,

$$
\begin{aligned}
nencode(\hat{x}, F) &= \{sgnf(x, F), q'(expo(\hat{x}) + B)\}, (p-1)'(2^{p-1}(sig(\hat{x}) - 1))\} \\
&= \{sgnf(x, F), q'expf(x, F), (p-1)'sigf(x, p)\} \\
&= \{x[p+q-1], x[p+q-2:p-1], x[p-2:0]\} \\
&= x[p+q-1:0] \\
&= x.
\end{aligned}
$$

The explicit case is similar. □

Lemma 5.3 *If r is a normal value of a format F, then $nencode(r, F)$ is a normal encoding for F, and*

$$ndecode(nencode(r, F), F) = r.$$

Proof We give the proof for the implicit case; the explicit case is similar.

Let $p = prec(F), q = expw(F), B = bias(F)$, and $x = nencode(r, F)$. By Lemma 2.24, x is a $(p+q)$-bit vector, and by Lemma 2.28,

$$sgnf(x, F) = x[p+q-1] = \begin{cases} 0 \text{ if } r > 0 \\ 1 \text{ if } r < 0, \end{cases}$$

$$expf(x, F) = x[p+q-2:p-1] = (expo(r) + B)[q-1:0],$$

and

$$sigf(x, F) = x[p-2:0] = (2^{p-1}(sig(r) - 1))[p-2:0].$$

Since $expo(r) + B$ is a q-bit vector,

$$(expo(r) + B)[q-1 : 0] = expo(r) + B$$

by Lemma 2.3.

Since r is p-exact,

$$2^{p-1}(sig(r) - 1) = 2^{p-1}sig(r) - 2^{p-1} \in \mathbb{Z},$$

and by Lemma 4.3, $2^{p-1}(sig(r) - 1) < 2^{p-1}$, which implies

$$(2^{p-1}(sig(r) - 1))[p - 2 : 0] = 2^{p-1}(sig(r) - 1).$$

Finally, according to Definition 5.8,

$$\begin{aligned}
ndecode(x, F) &= (-1)^{sgnf(x,F)}(2^{p-1} + sigf(x, p))2^{expf(x,F)+1-p-bias(F)} \\
&= sgn(r)2^{p-1}sig(r)2^{expo(r)+B+1-p-B} \\
&= sgn(r)sig(r)2^{expo(r)} \\
&= r.
\end{aligned}$$

$\square$

We shall have occasion to refer to the smallest and largest positive numbers that admit normal representations.

Definition 5.11 The *smallest positive normal* of a format F is

$$spn(F) = 2^{1-bias(F)} = 2^{2-2^{expw(F)-1}}.$$

Lemma 5.4 *For any format F,*

(a) $spn(F) > 0$;
(b) $spn(F)$ is a normal value of F;
(c) If r is a normal value of F, then $|r| \geq spn(F)$.

Proof It is clear that $spn(F)$ is positive and satisfies Definition 5.9. Moreover, if $r > 0$ and r is a normal value of F, then since $expo(r) > -bias(F)$, $r \geq 2^{1-bias(F)}$ by Lemma 4.2.
$\square$

Definition 5.12 The *largest positive normal* of a format F is

$$lpn(F) = 2^{2^{expw(F)}-2-bias(F)}(2 - 2^{1-prec(F)}).$$

Lemma 5.5 *For any format F,*

(a) $lpn(F) > 0$;
(b) $lpn(F)$ is a normal value of F;
(c) If r is a normal value of F, then $r \leq lpn(F)$.

Proof It is clear that $lpn(F)$ is positive and satisfies Definition 5.9. Let $p = prec(F)$ and $q = expw(F)$. If r is a normal value of F, then by Definition 5.9, $expo(r) \leq 2^q - 2 - bias(F)$, and r is p-exact. Thus, $r < 2^{expo(r)+1} \leq 2^{2^q-1-bias(F)}$, and by Lemma 4.22,

$$r \leq fp^-(2^{2^q-1-bias(q)}, p) = lpn(F). \qquad \square$$

5.3 Denormals and Zeroes

An exponent field of 0 is used to encode numerical values that lie below the normal range. If the exponent and significand fields of an encoding are both 0, then the encoded value itself is 0 and the encoding is said to be a *zero*. If the exponent field is 0 and the significand field is not, then the encoding is either *denormal* or *pseudo-denormal*:

Definition 5.13 Let x be an encoding for a format F with $expf(x, F) = 0$.

(a) If $sigf(x, F) = 0$, then x is a *zero encoding for F*.
(b) If $sigf(x, F) \neq 0$ and either F is implicit or $x[prec(F)-1] = 0$, then x is a *denormal encoding for F*.
(c) If F is explicit and $x[prec(F)-1] = 1$, then x is a *pseudo-denormal* encoding for F.

Note that a zero can have either sign:

Definition 5.14 Let F be a format and let $s \in \{0, 1\}$. Then

$$zencode(s, F) = \{1's, (expw(F) + sigw(F))'0\}.$$

The numerical value of a denormal encoding is interpreted according to a separate scheme that provides a broader range (at the expense of lower precision) than would otherwise be possible. There are two differences between the decoding formulas for denormal and normal encodings:

(1) For a denormal encoding for an implicit format, the integer bit is taken to be 0 rather than 1.
(2) The power of 2 represented by the zero exponent field of a denormal or pseudo-denormal encoding is $2^{1-bias(F)}$ rather than $2^{0-bias(F)}$.

Definition 5.15 Let F be a format with $p = prec(F)$ and $B = bias(F)$. If x is an encoding for F with $expf(x, F) = 0$, then

$$ddecode(x, F) = (-1)^{sgnf(x,F)} \left(2^{1-p} sigf(x, F)\right) 2^{1-B}$$

$$= (-1)^{sgnf(x,F)} sigf(x, F) 2^{2-p-B}.$$

We also define a general decoding function:

Definition 5.16 Let x be an encoding for a format F. If $expf(x, F) \neq 2^{expw(F)} - 1$, then x is a *numerical* encoding, and

$$decode(x, F) = \begin{cases} ndecode(x, F) \text{ if } expf(x, F) \neq 0 \\ ddecode(x, F) \text{ if } expf(x, F) = 0. \end{cases}$$

Note that the function *ddecode* is applied to pseudo-denormal as well as denormal encodings. If x is a pseudo-denormal encoding for an explicit format F and x' is the normal encoding derived from x by replacing its 0 exponent field with 1, then expanding Definitions 5.8 and 5.15 and observing that $sigf(x, F) = 2^{p-1} + manf(x, F)$, we have

$$decode(x', F) = ndecode(x', F) = ddecode(x, F) = decode(x, F).$$

Thus, any value encoded as a pseudo-denormal admits an alternative encoding as a normal.

Lemma 5.6 *Let x be a denormal encoding for F and let $\hat{x} = ddecode(x, F)$.*

(a) $sgn(\hat{x}) = (-1)^{sgnf(x,F)}$.
(b) $expo(\hat{x}) = expo(sigf(x, F)) - bias(F) + 2 - prec(F)$.
(c) $sig(\hat{x}) = sig(sigf(x, F))$.

Proof (a) is trivial; (b) and (c) follow from Lemmas 4.5 and 4.6.   □

The class of numbers that are representable as denormal encodings is recognized by the following predicate.

Definition 5.17 Let F be a format and let $r \in \mathbb{Q}$. Then r is a *denormal value* of F iff the following conditions hold:

(a) $r \neq 0$;
(b) $2 - prec(F) \leq expo(r) + bias(F) \leq 0$;
(c) r is $(prec(F) + expo(r) - expo(spn(F)))$-exact.

Lemma 5.7 *If r is a denormal value of F, then*

(a) $|r| < spn(F)$;
(b) r is $prec(F)$-exact.

Proof

(a) Since $expo(r) \le bias(F)$,

$$|r| < 2^{expo(r)+1} \le 2^{1-bias(F)} = spn(F).$$

(b) This follows from Lemma 4.9, since

$$prec(F) + expo(r) - expo(spn(F)) \le prec(F) - bias(F) - (1 - bias(F))$$
$$< prec(F). \qquad \square$$

The encoding of a denormal value is constructed as follows.

Definition 5.18 If r is a denormal value of F with

$$s = \begin{cases} 0 \text{ if } r > 0 \\ 1 \text{ if } r < 0 \end{cases}$$

and

$$m = 2^{prec(F)-2+expo(r)+bias(F)} sig(r),$$

then

$$dencode(r, F) = \{1's, expw(F)'0, sigw(F)'m\}.$$

According to the next two lemmas, *ddecode* and *dencode* are inverse functions, and the denormal values are precisely those that are represented by denormal encodings.

Lemma 5.8 *If x is a denormal encoding for a format F, then $ddecode(x, F)$ is a denormal value of F, and*

$$dencode(ddecode(x, F), F) = x.$$

Proof Let $p = prec(F)$, $q = expw(F)$, $b = bias(F)$, $s = sgnf(x, F)$, $m = sigf(x, F)$, and $\hat{x} = ddecode(x, F)$. Since $1 \le m < 2^{p-1}$,

$$2^{2-p-b} \le |\hat{x}| = 2^{2-p-b}m < 2^{1-b},$$

and by Lemma 4.2,

$$2 - p - b \le expo(\hat{x}) < 1 - b,$$

which is equivalent to Definition 5.17 (b). In order to prove (c), we must show, according to Definition 4.2, that

$$2^{p+expo(r)-expo(spn(F)+expo(\hat{x})-1}sig(\hat{x}) = 2^{p-2+b+expo(\hat{x})}sig(\hat{x}) \in \mathbb{Z}.$$

But

$$2^{p-2+b+expo(\hat{x})}sig(\hat{x}) = 2^{p-2+b+expo(\hat{x})}|\hat{x}|2^{-expo(\hat{x})}$$

$$= 2^{p-2+b}|\hat{x}|$$

$$= 2^{p-2+b}(2^{2-p-b}m)$$

$$= m \in \mathbb{Z}.$$

This establishes that $\hat{x}$ is a denormal value.

Now by Definition 5.15, $s = \begin{cases} 0 \text{ if } \hat{x} > 0 \\ 1 \text{ if } \hat{x} < 0. \end{cases}$

Therefore, by Definitions 5.13, 5.18, and 5.4 and Lemmas 2.26 and 2.3,

$$dencode(\hat{x}, F) = \{1's, q'0, (2^{p-2+expo(\hat{x})+b}sig(\hat{x}))[p-2:0]\}$$

$$= \{1'sgnf(x, F), expw(F)'expf(x, F), sigw(F)'sigf(x, F)\}$$

$$= x. \qquad \qquad \qquad \qquad \qquad \qquad \qquad \qquad \Box$$

Lemma 5.9 *If r is a denormal value of F, then $dencode(r, F)$ is a denormal encoding for F, and*

$$ddecode(dencode(r, F), F) = r.$$

Proof Let $p = prec(F)$, $q = expw(F)$, $b = bias(F)$, and $x = dencode(r, F)$. By Lemma 2.24, x is a $(p + q)$-bit vector, and by Lemma 2.28,

$$sgnf(x, F) = x[p+q-1] = \begin{cases} 0 \text{ if } r > 0 \\ 1 \text{ if } r < 0, \end{cases}$$

$$expf(x, F) = x[p+q-2:p-1] = 0,$$

and

$$sigf(x, F) = x[p-2:0] = (2^{p-2+expo(r)+bias(F)}sig(r))[p-2:0].$$

Since r is $(p - 2 + 2^{q-1} + expo(r))$-exact,

$$2^{p-2+expo(r)+b}sig(r) = 2^{(p-2+2^{q-1}+expo(r))-1}sig(r) =\in \mathbb{Z}$$

and since $expo(r) + b \leq 0$,

$$2^{p-2+expo(r)+b} sig(r) < 2^{p-2} \cdot 2 = 2^{p-1}$$

by Lemma 4.3, which implies

$$(2^{p-2+expo(r)+b} sig(r))[p-2:0] = 2^{p-2+expo(r)+b} sig(r).$$

Finally, according to Definition 5.15,

$$ddecode(x, F) = (-1)^{sgnf(x,F)} sigf(x, F)2^{2-p-b}$$
$$= sgn(r)2^{p-2+expo(r)+b} sig(r)2^{2-p-b}$$
$$= sgn(r)sig(r)2^{expo(r)}$$
$$= r. \qquad \square$$

Definition 5.19 The smallest positive denormal of a format F is

$$spd(F) = 2^{2-bias(F)-prec(F)} = 2^{3-2^{expw(F)}-prec(F)}.$$

Lemma 5.10 *For any format F,*

(a) $spd(F) > 0$;
(b) $spd(F)$ is a denormal value of in F;
(c) If r is a denormal value of F, then $|r| \geq spd(F)$.

Proof Let $p = prec(F)$, $q = expw(F)$, and $b = bias(F)$. It is clear that $spd(F)$ is positive. To show that $spd(F)$ is $(p + expo(spd(F)) - expo(spn(F)))$-exact, we need only observe that

$$p + expo(spd(F)) - expo(spn(F)) = p + (2 - b - p) - (1 - b) = 1.$$

Finally, since

$$expo(spd(F)) + b = 2 - p < 0,$$

$spd(F)$ is a denormal value, and moreover, $spd(F)$ is the smallest positive r that satisfies $2 - p \leq expo(r) + b$. $\qquad \square$

Every number with a denormal representation is a multiple of the smallest positive denormal.

Lemma 5.11 *If $r \in \mathbb{Q}$ and let F be a format, then r is a denormal value of F iff $r = \pm m \cdot spd(F)$ for some $m \in \mathbb{N}$, $1 \leq m < 2^{prec(F)-1}$.*

Proof Let $p = prec(F)$ and $b = bias(F)$. For $1 \leq m \leq 2^{p-1}$, let $a_m = m \cdot spd(F)$. Then $a_1 = spd(F)$ and

$$a_{2^{p-1}} = 2^{p-1} spd(F) = 2^{p-1} 2^{2-b-p} = 2^{1-b} = spn(F).$$

We shall show, by induction on m, that a_m is a denormal value of F for $1 \leq m < 2^{p-1}$. First, note that for all such m,

$$fp^+(a_m, p + expo(a_m) - expo(spn(F)))$$

$$= a_m + 2^{expo(a_m)+1-(p+expo(a_m)-expo(spn(F)))}$$

$$= a_m + 2^{expo(spn(F))-(p-1)}$$

$$= a_m + spd(F)$$

$$= a_{m+1}.$$

Suppose that a_{m-1} is a denormal value for some m, $1 < m < 2^{p-1}$. Then a_{m-1} is $(p + expo(a_{m-1}) - expo(spn(F)))$-exact, and by Lemma 4.18, so is a_m. But since $expo(a_m) \geq expo(a_{m-1})$, it follows from Lemma 4.9 that a_m is also $(p + expo(a_m) - expo(spn(F)))$-exact. Since

$$a_m < a_{2^{p-1}} = spn(F) = 2^{1-b},$$

$expo(a_m) < 1 - b$, i.e., $expo(a_m) + b \leq 0$, and hence, a_m is a denormal value.

Now suppose that z is a denormal value. Let $m = \lfloor z/a_1 \rfloor$. Clearly, $1 \leq m < 2^{p-1}$, and $a_m \leq z < a_{m+1}$. It follows from Lemma 4.19 that $expo(z) = expo(a_m)$, and consequently, z is $(p+expo(a_m)-expo(spn(F)))$-exact. Thus, by Lemma 4.18, $z = a_m$.   □

5.4   Infinities and NaNs

The upper extreme value $2^{expw(F)} - 1$ of the exponent field is reserved for encoding non-numerical entities. According to Definition 5.6, an encoding for an explicit format with exponent field $2^{expw(F)} - 1$ is unsupported if its integer bit is 0. In all other cases, an encoding with this exponent field is an *infinity* if its mantissa field is 0 and a *NaN* ("Not a Number") otherwise. A NaN is further classified as an *SNaN* ("signaling NaN") or a *QNaN* ("quiet NaN") according to the most significant bit of its mantissa field:

Definition 5.20 Let x be an encoding for a format F with $expf(x, F) = 2^{expw(F)} - 1$, and assume that if F is explicit, then $x[prec(F)-1] = 1$.

(a) x is an infinity for F iff $manf(x, F) = 0$;
(b) x is a NaN for F iff $manf(x, F) \neq 0$;
(c) x is an SNaN for F iff x is a NaN and $x[prec(F) - 2] = 0$;
(d) x is a QNaN for F iff x is a NaN and $x[prec(F) - 2] = 1$.

An infinity is used to represent a computed value that lies above the normal range; when it occurs as an operand of an arithmetic operation, it is treated as $\pm\infty$. This function constructs an infinity with a given sign:

Definition 5.21 Let F be a format, let $s \in \{0, 1\}$, and let $e = 2^{expw(F)} - 1$. Then

$$iencode(s, F) = \begin{cases} \{1's, expw(F)'e, sigw(F)'0\} & \text{if } F \text{ is implicit} \\ \{1's, expw(F)'e, 1'1, (sigw(F) - 1)'0\} & \text{if } F \text{ is explicit.} \end{cases}$$

An SNaN operand of a CPU floating-point instruction always triggers an exception and is converted to a QNaN to be returned as the instruction value. SNaNs are not generated by hardware but may be written by software to indicate exceptional conditions. For example, a block of memory may be filled with SNaNs to guard against its access until it is initialized.

A QNaN may be generated by hardware to be returned by an instruction either by "quieting" an SNaN operand or as an indication of an invalid operation, such as an indeterminate form. A QNaN operand is generally propagated as the value of an instruction without signaling an exception.

The following function converts an SNaN to a QNaN:

Definition 5.22 If x is a NaN encoding for a format F, then

$$qnanize(x, F) = x \mid 2^{prec(F)-2}.$$

The following encoding, known as the *real indefinite QNaN*, is the default value used to signal an invalid operation:

Definition 5.23 Let F be a format and let $e = 2^{expw(F)} - 1$. Then

$$indef(F) = \begin{cases} \{1'0, expw(F)'e, 1'1, (sigw(F) - 1)'0\} & \text{if } F \text{ is implicit} \\ \{1'0, expw(F)'e, 2'3, (sigw(F) - 2)'0\} & \text{if } F \text{ is explicit.} \end{cases}$$

Infinities and NaNs will be discussed further in the context of elementary arithmetic operations in Part IV.

Chapter 6
Rounding

The objective of floating-point rounding is the approximation of a real number by one that is representable in a given floating-point format. Thus, a *rounding mode* is a function that computes an n-exact value $\mathcal{R}(x, n)$, given a real number x and precision n. In this chapter, we investigate the properties of a variety of rounding modes, including those that are prescribed by the IEEE standard and several others that are commonly used in the implementation of floating-point operations.

Considerations other than n-exactness are involved in the rounding of results that lie outside the normal range of the target format. In the case of *overflow*, which occurs when the result of a computation exceeds the representable range, the IEEE standard prescribes rounding either to the maximum representable number or to infinity. The rules that govern this choice, which are quite arbitrary from a mathematical perspective, are deferred to Part IV. The more interesting case of *underflow*, involving a denormal result, is addressed at the end of this chapter. We shall find it convenient in our discussion of the rounding of denormals (see Definition 6.9) to extend the notion of rounding to allow negative precisions. Thus, while the precision n is usually positive, we shall consider a rounding mode to be a mapping

$$\mathcal{R} : \mathbb{R} \times \mathbb{Z} \to \mathbb{R}.$$

Most of the results of this chapter pertaining to $\mathcal{R}(x, n)$ will be formulated with the hypothesis $n > 0$, with the exception of those that are required in the more general case in Sect. 6.6 (see Lemmas 6.82 and 6.90).

Every mode that we consider will be shown to satisfy the following axioms, for all $x \in \mathbb{R}$, $y \in \mathbb{R}$, $n \in \mathbb{Z}^+$, and $k \in \mathbb{Z}$:

(1) $\mathcal{R}(x, n)$ is n-exact.
(2) If x is n-exact, then $\mathcal{R}(x, n) = x$.
(3) If $x \leq y$, then $\mathcal{R}(x, n) \leq \mathcal{R}(y, n)$.
(4) $\mathcal{R}(2^k x, n) = 2^k \mathcal{R}(x, n)$.

© The Author(s), under exclusive license to Springer Nature Switzerland AG 2022
D. M. Russinoff, *Formal Verification of Floating-Point Hardware Design*,
https://doi.org/10.1007/978-3-030-87181-9_6

A critical consequence of these properties is that $\mathcal{R}$ is optimal in the sense that there can exist no n-exact number in the open interval between x and $\mathcal{R}(x, n)$. For example, if y is n-exact and $x < y$, then by (2) and (3), $\mathcal{R}(x, n) \leq \mathcal{R}(y, n) = y$. Since there exist arbitrarily small n-exact numbers, it follows that

$$sgn(\mathcal{R}(x, n)) = sgn(x).$$

Also note that (4) implies that $\mathcal{R}$ is determined by its behavior for $1 \leq |x| < 2$.

In the first two sections of this chapter, we examine the basic *directed* rounding modes *RTZ* ("round toward zero") and *RAZ* ("round away from zero"), character- ized by the inequalities

$$|RTZ(x, n)| \leq |x|$$

and

$$|RAZ(x, n)| \geq |x|.$$

It is clear that for any rounding mode $\mathcal{R}$ and arguments x and n, either $\mathcal{R}(x, n) = RTZ(x, n)$ or $\mathcal{R}(x, n) = RAZ(x, n)$. It is natural, therefore, to define other rounding modes in terms of these two. In Sect. 6.3, we discuss two versions of "rounding to nearest", *RNE* ("round to nearest even") and *RNA* ("round to nearest away from zero"), both of which select the more accurate of the two approximations but which handle the ambiguous case of a midpoint between consecutive representable numbers differently.

Two properties that are shared by some, but not all, of the rounding modes of interest are *symmetry*,

$$\mathcal{R}(-x, n) = -\mathcal{R}(x, n),$$

and *decomposability*, the property that for $m < n$,

$$\mathcal{R}(\mathcal{R}(x, n), m) = \mathcal{R}(x, m).$$

We shall see, for example, that both versions of rounding to nearest fail to satisfy the latter condition. This accounts for the phenomenon of "double rounding": when the result of a computation undergoes a preliminary rounding to be temporarily stored in a register that is wider than the target format of an instruction, care must be taken to ensure the accuracy of the final rounding. In Sect. 6.4, we discuss the mode *RTO* ("round to odd"), which is commonly used internally by floating-point units to address this problem.

Another desirable property of a rounding mode is that the expected (i.e., average) error that it incurs in rounding a representative set of values to a given precision is 0, so that errors generated over a long sequence of computations tend to cancel one another. Given a rounding precision n, one choice of "representative set" is the set of all m-exact numbers for some $m > n$. In order to limit the error computation to finite sets, we seek to identify a partition of the real numbers into intervals such that for all $m > n$, the expected error over the m-exact numbers within each interval may be shown to be 0. In practice, we find that a suitable partition is given by the set of intervals bounded by $(n-1)$-exact numbers. If we assume that $\mathcal{R}$ is symmetric, we need only consider positive values. Thus, we consider the set of m-exact numbers that lie between two consecutive positive $(n-1)$-exact values, x_0 and $fp^+(x_0, n-1) = x_0 + 2^{expo(x_0)+2-n}$, i.e.,

$$\{x_0 + 2^{expo(x_0)+1-m}k \mid 0 \le k < 2^{n+1-m}\}.$$

Clearly, the expected error over this set is

$$\frac{1}{2^{m+1-n}} \sum_{k=0}^{2^{m+1-n}-1} (\mathcal{R}(x_k, n) - x_k),$$

where $x_k = x_0 + 2^{expo(x_0)+1-m}k$. This leads to the following definition:

Definition 6.1 A symmetric rounding mode $\mathcal{R}$ is *unbiased* if for all $n \in \mathbb{N}$, $m \in \mathbb{N}$, and $x_0 \in \mathbb{R}$, if $m > n > 1$, $x_0 > 0$, and x_0 is $(n-1)$-exact, then

$$\sum_{k=0}^{2^{m+1-n}-1} (\mathcal{R}(x_k, n) - x_k) = 0,$$

where $x_k = x_0 + 2^{expo(x_0)+1-m}k$.

We shall show that the modes *RNE* and *RTO* both satisfy this definition.

In Sect. 6.5, we define the two other directed rounding modes that are prescribed by the IEEE standard and collect the properties that are shared by all IEEE modes. We present several techniques that are commonly employed in the implementation of rounding by commercial floating-point units.

The rounding of denormals is the subject of Sect. 6.6. In Sect. 6.7, we present implementation techniques for detecting underflow as defined by both the x86 and Arm architectures.

6.1 Rounding Toward Zero

The most basic rounding mode, which the IEEE standard calls "round toward 0", may be described as the truncation of the significand of x to $n-1$ fractional bits, i.e., replacing $sig(x)$ with

$$sig(x)^{(n-1)} = \lfloor 2^{n-1} sig(x) \rfloor 2^{1-n}.$$

Definition 6.2 For all $x \in \mathbb{R}$ and $n \in \mathbb{Z}$,

$$RTZ(x, n) = sgn(x) \lfloor 2^{n-1} sig(x) \rfloor 2^{expo(x)-n+1}.$$

Example Let $x = 45/8 = (101.101)_2$ and $n = 5$. Then $sgn(x) = 1$, $expo(x) = 2$, $sig(x) = (1.01101)_2$,

$$\lfloor 2^{n-1}sig(x)\rfloor 2^{1-n} = \lfloor(10110.1)_2\rfloor 2^{-4} = 10110 \cdot 2^{-4} = 1.011,$$

and

$$RTZ(x, n) = \lfloor 2^{n-1}sig(x)\rfloor 2^{1-n}2^{expo(x)} = 1.011 \cdot 2^2 = 101.1.$$

Note that this value is the largest 5-exact number that does not exceed x.

The second argument of any rounding mode is normally positive; for this mode, the negative-precision case is trivial:

Lemma 6.1 *For all $x \in \mathbb{R}$ and $n \in \mathbb{Z}$, if $n \leq 0$, then $RTZ(x, n) = 0$.*

Proof By Lemma 4.3, $0 < 2^{n-1}sig(x) \leq sig(x)/2 < 1$, which implies $\lfloor 2^{n-1}sig(x)\rfloor = 0$.   □

RTZ preserves exponent:

Lemma 6.2 *If $x \in \mathbb{R}$ and $n \in \mathbb{Z}^+$, then $expo(RTZ(x, n)) = expo(x)$.*

Proof Since $|RTZ(x, n)| \leq |x|$, Lemma 4.2(c) implies $expo(RTZ(x, n)) \leq expo(x)$. But by Lemmas 4.3 and 1.1,

$$|RTZ(x, n)| = \lfloor 2^{n-1}sig(x)\rfloor 2^{expo(x)-n+1} \geq 2^{n-1}2^{expo(x)-n+1} = 2^{expo(x)},$$

and hence, by Lemma 4.2, $expo(RTZ(x, n)) \geq expo(x)$.   □

The following inequality is the defining characteristic of *RTZ*:

Lemma 6.3 *For all $x \in \mathbb{R}$ and $n \in \mathbb{Z}^+$, $|RTZ(x, n)| \leq |x|$.*

Proof By Definition 1.1 and Lemma 4.1,

$$|RTZ(x, n)| \leq 2^{n-1}sig(x)2^{expo(x)-n+1} = sig(x)2^{expo(x)} = |x|.$$ □

The following complements Lemma 6.3, confining $RTZ(x, n)$ to an interval.

Lemma 6.4 *If $x \in \mathbb{R}$, $x \neq 0$, and $n \in \mathbb{Z}^+$, then*

$$|RTZ(x, n)| > |x| - 2^{expo(x)-n+1} \geq |x|(1 - 2^{1-n}).$$

Proof By Definitions 6.2 and 1.1 and Lemma 4.1,

$$|RTZ(x, n)| > (2^{n-1}sig(x) - 1)2^{expo(x)-n+1} = |x| - 2^{expo(x)-n+1}.$$

The second inequality follows from Definition 4.1.   □

Corollary 6.5 *For all $x \in \mathbb{R}$ and $n \in \mathbb{Z}^+$,*

$$|x - RTZ(x, n)| < 2^{expo(x)-n+1} \leq 2^{1-n}|x|.$$

Proof By Lemma 6.3,

$$|x - RTZ(x, n)| = ||x| - |RTZ(x, n)|| = |x| - |RTZ(x, n)|.$$

The corollary now follows from Lemma 6.4.   □

Lemmas 6.6–6.12 establish that *RTZ* has all the properties listed at the beginning of this chapter, including symmetry and decomposability.

Lemma 6.6 *For all $x \in \mathbb{R}$ and $n \in \mathbb{Z}$, $RTZ(-x, n) = -RTZ(x, n)$.*

Proof This is an immediate consequence of Definition 6.2.   □

Lemma 6.7 *For all $x \in \mathbb{R}$ and $n \in \mathbb{Z}^+$, $RTZ(x, n)$ is n-exact.*

Proof Since $expo(RTZ(x, n)) = expo(x)$, it suffices to observe that

$$RTZ(x, n)2^{n-1-expo(x)} = sgn(x)\lfloor 2^{n-1} sig(x) \rfloor \in \mathbb{Z}.$$ □

Lemma 6.8 *Let $x \in \mathbb{R}$ and $n \in \mathbb{Z}^+$. If x is n-exact, then $RTZ(x, n) = x$.*

Proof By Definition 4.2 and Lemma 1.1, $\lfloor 2^{n-1} sig(x) \rfloor = 2^{n-1} sig(x)$, and hence by Definition 6.2 and Lemma 4.1,

$$\begin{aligned}
RTZ(x, n) &= sgn(x)\lfloor 2^{n-1} sig(x) \rfloor 2^{expo(x)-n+1} \\
&= sgn(x)2^{n-1} sig(x)2^{expo(x)-n+1} \\
&= sgn(x)sig(x)2^{expo(x)} \\
&= x.
\end{aligned}$$ □

Lemma 6.9 *Let $x \in \mathbb{R}$, $a \in \mathbb{R}$, and $n \in \mathbb{Z}^+$. If a is n-exact and $a \leq x$, then $a \leq RTZ(x, n)$.*

Proof If $x < 0$, then $x \leq RTZ(x, n)$ by Lemma 6.3. Therefore, we may assume that $x \geq 0$. Suppose $a > RTZ(x, n)$. Then by Lemma 4.18,

$$RTZ(x, n) \leq a - 2^{expo(RTZ(x,n))-n+1} = a - 2^{expo(x)-n+1} \leq x - 2^{expo(x)-n+1},$$

contradicting Lemma 6.4.   □

Lemma 6.10 *Let $x \in \mathbb{R}$, $y \in \mathbb{R}$, and $n \in \mathbb{Z}^+$. If $x \leq y$, then*

$$RTZ(x, n) \leq RTZ(y, n).$$

Proof First suppose $x > 0$. By Lemma 6.3, $RTZ(x, n) \leq x \leq y$. Since $RTZ(x, n)$ is n-exact by Lemma 6.7, Lemma 6.9 implies $RTZ(x, n) \leq RTZ(y, n)$.

Now suppose $x \leq 0$. We may assume that $x \leq y < 0$. Thus, since $0 < -y \leq -x$, we have $RTZ(-y, n) \leq RTZ(-x, n)$, and by Lemma 6.6,

$$RTZ(x, n) = -RTZ(-x, n) \leq -RTZ(-y, n) = RTZ(y, n). \qquad \square$$

Lemma 6.11 *For all* $x \in \mathbb{R}$, $n \in \mathbb{Z}^+$, *and* $k \in \mathbb{Z}$, $RTZ(2^k x, n) = 2^k RTZ(x, n)$.

Proof By Lemma 4.6,

$$
\begin{aligned}
RTZ(2^k x, n) &= sgn(2^k x) \lfloor 2^{n-1} sig(2^k x) \rfloor 2^{expo(2^k x)-n+1} \\
&= sgn(x) \lfloor 2^{n-1} sig(x) \rfloor 2^{expo(x)+k-n+1} \\
&= 2^k RTZ(x, n). \qquad \square
\end{aligned}
$$

Lemma 6.12 *Let* $x \in \mathbb{R}$, $m \in \mathbb{Z}^+$, *and* $n \in \mathbb{Z}^+$. *If* $m \leq n$, *then*

$$RTZ(RTZ(x, n), m) = RTZ(x, m).$$

Proof We assume $x \geq 0$; the case $x < 0$ follows easily from Lemma 6.6. Let $e = expo(x)$. By Lemmas 6.2 and 1.2,

$$
\begin{aligned}
RTZ(RTZ(x, n), m) &= \lfloor 2^{m-1-e}(\lfloor 2^{n-1-e} x \rfloor 2^{e+1-n}) \rfloor 2^{e+1-m} \\
&= \lfloor \lfloor 2^{n-1-e} x \rfloor / 2^{n-m} \rfloor 2^{e+1-m} \\
&= \lfloor 2^{n-1-e} x / 2^{n-m} \rfloor 2^{e+1-m} \\
&= \lfloor 2^{m-1-e} x \rfloor 2^{e+1-m} \\
&= RTZ(x, m). \qquad \square
\end{aligned}
$$

If x is $(n + 1)$-exact but not n-exact, then x is equidistant from two successive n-exact numbers. In this case, we have an explicit formula for $RTZ(x, n)$.

Lemma 6.13 *Let* $x \in \mathbb{R}$ *and* $n \in \mathbb{N}$. *If* x *is* $(n + 1)$-*exact but not* n-*exact, then*

$$RTZ(x, n) = x - sgn(x) 2^{expo(x)-n}.$$

Proof By Lemma 6.6, we may assume $x > 0$. For the case $n = 0$, we have $sig(x) = 1$ by Definition 4.2, and by Lemmas 4.1 and 6.1,

$$x - 2^{expo(x)-n} = 2^{expo(x)} - 2^{expo(x)} = 0 = RTZ(x, n).$$

Thus, we may assume $n > 0$. Let $a = x - 2^{expo(x)-n}$ and $b = x + 2^{expo(x)-n}$. Since $x > 2^{expo(x)}$, $x \geq 2^{expo(x)} + 2^{expo(x)+1-n}$ by Lemma 4.18, and hence, $a \geq 2^{expo(x)}$ and $expo(a) = expo(x)$. It follows that $b = fp^+(a, n)$.

By hypothesis, $x2^{n-expo(x)} \in \mathbb{Z}$ but $x2^{n-expo(x)}/2 = x2^{n-1-expo(x)} \notin \mathbb{Z}$, and therefore, $x2^{n-expo(x)}$ is odd. Let $x2^{n-expo(x)} = 2k + 1$. Then

$$a2^{n-1-expo(a)} = (x - 2^{expo(x)-n})2^{n-1-expo(x)} = (2k+1)/2 - 1/2 = k \in \mathbb{Z}.$$

Thus, a is n-exact, and by Lemma 4.17, so is b. Now by Lemma 6.9, $a \leq RTZ(x, n)$, but if $a < RTZ(x, n)$, then since $RTZ(x, n)$ is n-exact, Lemma 4.18 would imply $b \leq RTZ(x, n)$, contradicting $x < b$. Therefore, $a = RTZ(x, n)$. $\square$

Figure 6.1 is provided as a visual aid in understanding the following lemma, which formulates the conditions under which a truncated sum may be computed by truncating one of the summands in advance of the addition.

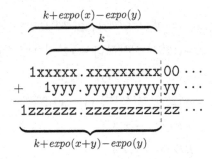

Fig. 6.1 Lemma 6.14

Lemma 6.14 *Let* $x \in \mathbb{R}$, $y \in \mathbb{R}$, *and* $k \in \mathbb{Z}$. *If* $x \geq 0$, $y \geq 0$, *and* x *is* $(k + expo(x) - expo(y))$-*exact, then*

$$x + RTZ(y, k) = RTZ(x + y, k \mid expo(x + y) - expo(y)).$$

Proof Let $n = k + expo(x) - expo(y)$. Since x is n-exact,

$$x2^{k-1-expo(y)} = x2^{n-1-expo(x)} \in \mathbb{Z}.$$

Let $k' = k + expo(x + y) - expo(y)$. Then by Lemma 1.1,

$$
\begin{aligned}
x + RTZ(y, k) &= x + \lfloor 2^{k-1-expo(y)}y \rfloor 2^{expo(y)+1-k} \\
&= (x2^{k-1-expo(y)} + \lfloor 2^{k-1-expo(y)}y \rfloor)2^{expo(y)+1-k} \\
&= \lfloor 2^{k-1-expo(y)}(x + y) \rfloor 2^{expo(y)+1-k} \\
&= \lfloor 2^{k'-1-expo(x+y)}(x + y) \rfloor 2^{expo(x+y)+1-k'} \\
&= RTZ(x + y, k').
\end{aligned}
$$

$\square$

Lemma 6.14 holds for subtraction as well but only if the summands are properly ordered.

Lemma 6.15 *Let $x \in \mathbb{R}$, $y \in \mathbb{R}$, and $k \in \mathbb{Z}$. If $y > x > 0$, $k + expo(x - y) - expo(y) > 0$, and x is $(k + expo(x) - expo(y))$-exact, then*

$$x - RTZ(y, k) = RTZ(x - y, k + expo(x - y) - expo(y)).$$

Proof Let $n = k + expo(x) - expo(y)$. Since x is n-exact,

$$x2^{k-1-expo(y)} = x2^{n-1-expo(x)} \in \mathbb{Z}.$$

Let $k' = k + expo(x - y) - expo(y)$. Then by Lemma 1.1,

$$\begin{aligned}
x - RTZ(y, k) &= x - \lfloor 2^{k-1-expo(y)} y \rfloor 2^{expo(y)+1-k} \\
&= -(\lfloor 2^{k-1-expo(y)} y \rfloor - x2^{k-1-expo(y)}) 2^{expo(y)+1-k} \\
&= -\lfloor 2^{k-1-expo(y)} (y - x) \rfloor 2^{expo(y)+1-k} \\
&= -\lfloor 2^{k'-1-expo(y-x)} (y - x) \rfloor 2^{expo(y-x)+1-k'} \\
&= -RTZ(y - x, k') \\
&= RTZ(x - y, k').
\end{aligned}$$
$\square$

Truncation of a bit vector may be described as a shifted bit slice.

Lemma 6.16 *Let $x \in \mathbb{N}$ and $k \in \mathbb{N}$, and let $n = expo(x) + 1$. If $0 < k \leq n$, then*

$$RTZ(x, k) = 2^{n-k} x[n-1 : n - k].$$

Proof By Lemmas 2.3 and 2.11,

$$\begin{aligned}
RTZ(x, k) &= \lfloor 2^{k-1-expo(x)} x \rfloor 2^{expo(x)+1-k} \\
&= \lfloor x/2^{n-k} \rfloor 2^{n-k} \\
&= 2^{n-k} \lfloor x/2^{n-k} \rfloor [k-1 : 0] \\
&= 2^{n-k} x[n-1 : n - k].
\end{aligned}$$
$\square$

Corollary 6.17 *Let $x \in \mathbb{N}$, $m \in \mathbb{N}$, and $k \in \mathbb{N}$, and let $n = expo(x) + 1$. If $x \geq 0$ and $0 < k < m \leq n$, then*

$$RTZ(x, m) = RTZ(x, k) + 2^{n-m} x[n - k - 1 : n - m].$$

Proof By Lemmas 6.16 and 2.10,

$$RTZ(x, m) = 2^{n-m}x[n-1 : n - m]$$
$$= 2^{n-m}(2^{m-k}x[n-1 : n - k] + x[n - k-1 : n - m])$$
$$= 2^{n-k}x[n-1 : n - k] + 2^{n-m}x[n - k-1 : n - m]$$
$$= RTZ(x, k) + 2^{n-m}x[n - k-1 : n - m]. \qquad \square$$

Corollary 6.18 *Let $x \in \mathbb{N}$, $m \in \mathbb{N}$, and $k \in \mathbb{N}$, and let $n = expo(x)$. If $0 < k < n \leq m$, then*

$$RTZ(x, k) = x \mathbin{\&} (2^m - 2^{n-k}).$$

Proof By Lemmas 6.16 and 3.13,

$$RTZ(x, k) = 2^{n-k}x[n-1 : n - k] = x \mathbin{\&} (2^n - 2^{n-k}),$$

and by Lemmas 3.1, 2.3, and 3.4(c),

$$x \mathbin{\&} (2^n - 2^{n-k}) = \left(x \mathbin{\&} (2^n - 2^{n-k})\right)[n-1 : 0]$$
$$= x[n-1 : 0] \mathbin{\&} (2^n - 2^{n-k})[n-1 : 0]$$
$$= x \mathbin{\&} (2^n - 2^{n-k})[n-1 : 0].$$

But since

$$2^m - 2^{n-k} = 2^n(2^{m-n} - 1) + 2^n - 2^{n-k},$$

$$(2^m - 2^{n-k})[n-1 : 0] = (2^m - 2^{n-k}) \bmod 2^n = 2^n - 2^{n-k}. \qquad \square$$

6.2 Rounding Away from Zero

The dual of truncation is defined similarly using the ceiling instead of the floor.

Definition 6.3 For all $x \in \mathbb{R}$ and $n \in \mathbb{Z}$,

$$RAZ(x, n) = sgn(x)\lceil 2^{n-1}sig(x)\rceil 2^{expo(x)-n+1}.$$

Example Let $x = 45/8 = (101.101)_2$ and $n = 5$. Then $sgn(x) = 1$, $expo(x) = 2$, $sig(x) = (1.01101)_2$,

$$\lceil 2^{n-1}sig(x)\rceil 2^{1-n} = \lceil(10110.1)_2\rceil 2^{-4} = (10111)_2 \cdot 2^{-4} = (1.0111)_2,$$

and

$$RAZ(x, n) = \lceil 2^{n-1} sig(x) \rceil 2^{1-n} 2^{expo(x)} = (1.0111)_2 \cdot 2^2 = (101.11)_2.$$

Note that this value is the smallest 5-exact number not exceeded by x.

_____

The negative-precision case is less than intuitive.

Lemma 6.19 *For all $x \in \mathbb{R}$ and $n \in \mathbb{Z}$, if $n \leq 0$, then*

$$RAZ(x, n) = sgn(x)2^{expo(x)+1-n}.$$

Proof By Lemma 4.3, $0 < 2^{n-1} sig(x) \leq sig(x)/2 < 1$, and hence, by Lemma 1.6,

$$RAZ(x, n) = sgn(x)\lceil 2^{n-1} sig(x) \rceil 2^{expo(x)+1-n} = sgn(x)2^{expo(x)+1-n}. \qquad \square$$

We have the following bounds on $RAZ(x, n)$.

Lemma 6.20 *For all $x \in \mathbb{R}$ and $n \in \mathbb{Z}^+$, $|RAZ(x, n)| \geq |x|$.*

Proof By Lemmas 1.6 and 4.1,

$$|RAZ(x, n)| \geq 2^{n-1} sig(x)2^{expo(x)-n+1} = sig(x)2^{expo(x)} = |x|. \qquad \square$$

Lemma 6.21 *If $x \in \mathbb{R}$, $x \neq 0$, and $n \in \mathbb{Z}^+$, then*

$$|RAZ(x, n)| < |x| + 2^{expo(x)-n+1} \leq |x|(1 + 2^{1-n}).$$

Proof By Definitions 6.3 and 1.1 and Lemma 4.1,

$$|RAZ(x, n)| < (2^{n-1} sig(x) + 1)2^{expo(x)-n+1} = |x| + 2^{expo(x)-n+1}.$$

The second inequality follows from Definition 4.1. $\square$

Corollary 6.22 *For all $x \in \mathbb{R}$ and $n \in \mathbb{Z}^+$,*

$$|RAZ(x, n) - x| < 2^{expo(x)-n+1} \leq 2^{1-n}|x|.$$

Proof By Lemmas 6.20,

$$|RAZ(x, n) - x| = ||RAZ(x, n)| - |x|| = |RAZ(x, n)| - |x|.$$

The corollary now follows from Lemma 6.21. $\square$

Unlike *RTZ*, *RAZ* is not guaranteed to preserve the exponent of its argument, but the only exception is the case in which a number is rounded up to a power of 2.

Lemma 6.23 *For all $x \in \mathbb{R}$ and $n \in \mathbb{Z}^+$, if $|RAZ(x, n)| \neq 2^{expo(x)+1}$, then*

$$expo(RAZ(x, n)) = expo(x).$$

Proof By Lemma 4.3,

$$\begin{aligned}
|RAZ(x, n)| &= \lceil 2^{n-1} sig(x) \rceil 2^{expo(x)-n+1} \\
&\leq \lceil 2^n \rceil 2^{expo(x)-n+1} \\
&= 2^n 2^{expo(x)-n+1} \\
&= 2^{expo(x)+1}.
\end{aligned}$$ □

The claim now follows from Lemma 6.20 and 4.2(c).

The standard rounding mode properties may now be derived.

Lemma 6.24 *For all $x \in \mathbb{R}$ and $n \in \mathbb{Z}$, $RAZ(-x, n) = -RAZ(x, n)$.*

Proof This is an immediate consequence of Definition 6.2. □

Lemma 6.25 *If $x \in \mathbb{R}$ and $n \in \mathbb{Z}^+$, then $RAZ(x, n)$ is n-exact.*

Proof By Lemmas 6.23 and 4.10, we may assume that

$$expo(RAZ(x, n)) = expo(x).$$

Consequently, it suffices to observe that

$$RAZ(x, n)2^{n-1-expo(x)} = sgn(x)\lceil 2^{n-1} sig(x) \rceil \in \mathbb{Z}.$$ □

Lemma 6.26 *Let $x \in \mathbb{R}$ and $n \in \mathbb{Z}^+$. If x is n-exact, then $RAZ(x, n) = x$.*

Proof The proof is essentially the same as that of Lemma 6.8. □

Lemma 6.27 *Let $x \in \mathbb{R}$, $a \in \mathbb{R}$, and $n \in \mathbb{Z}^+$. If a is n-exact and $a \geq x$, then $a \geq RAZ(x, n)$.*

Proof If $x < 0$, then $x = -|x| \geq -|RAZ(x, n)| = RAZ(x, n)$ by Lemma 6.20. Therefore, we may assume that $x \geq 0$. Suppose $a < RAZ(x, n)$. Then by Lemmas 4.18 and 4.2(c),

$$RAZ(x, n) \geq a + 2^{expo(a)-n+1} \geq x + 2^{expo(x)-n+1},$$

contradicting Lemma 6.21. □

Corollary 6.28 *Let $x \in \mathbb{R}$, $a \in \mathbb{R}$, and $n \in \mathbb{Z}^+$. If a is n-exact and $a > |RTZ(x, n)|$, then $a \geq |RAZ(x, n)|$.*

Proof We may assume that $x > 0$ and x is not n-exact. By Lemma 6.9, $a > x$, and by Lemma 6.27, $a \geq RAZ(x, n)$. □

Lemma 6.29 *Let $x \in \mathbb{R}$, $y \in \mathbb{R}$, and $n \in \mathbb{Z}^+$. If $x \leq y$, then*

$$RAZ(x, n) \leq RAZ(y, n).$$

Proof First suppose $x > 0$. By Lemma 6.20, $RAZ(y, n) \geq y \geq x$. Since $RAZ(x, n)$ is n-exact by Lemma 6.25, Lemma 6.27 implies

$$RAZ(y, n) \geq RAZ(x, n).$$

Now suppose $x \leq 0$. We may assume that $x \leq y < 0$. Thus, since $0 < -y \leq -x$, we have $RAZ(-y, n) \leq RAZ(-x, n)$, and by Lemma 6.24,

$$RAZ(x, n) = -RAZ(-x, n) \leq -RAZ(-y, n) = RAZ(y, n). \qquad \square$$

Lemma 6.30 *For all $x \in \mathbb{R}$, $n \in \mathbb{Z}^+$, and $k \in \mathbb{Z}^+$,*

$$RAZ(2^k x, n) = 2^k RAZ(x, n).$$

Proof By Lemma 4.6,

$$
\begin{aligned}
RAZ(2^k x, n) &= sgn(2^k x) \lceil 2^{n-1} sig(2^k x) \rceil 2^{expo(2^k x) - n + 1} \\
&= sgn(x) \lceil 2^{n-1} sig(x) \rceil 2^{expo(x) + k - n + 1} \\
&= 2^k RAZ(x, n). \qquad \square
\end{aligned}
$$

Lemma 6.31 *Let $x \in \mathbb{R}$, $m \in \mathbb{N}$, and $n \in \mathbb{N}$. If $0 < m \leq n$, then*

$$RAZ(RAZ(x, n), m) = RAZ(x, m).$$

Proof We may assume $x > 0$. Consider first the case

$$RAZ(x, n) = 2^{expo(x) + 1}.$$

In this case, $RAZ(x, n)$ is m-exact, so that

$$RAZ(RAZ(x, n), m) = RAZ(x, n) = 2^{expo(x) + 1}.$$

By Lemma 6.23, we need only show that $RAZ(x, m) \geq 2^{expo(x) + 1}$. But since $m \leq n$, $RAZ(x, m)$ is n-exact, and since $RAZ(x, m) \geq x$, $RAZ(x, m) \geq RAZ(x, n)$ by Lemma 6.27.

Thus, we may assume $RAZ(x, n) < 2^{expo(x)+1}$. Let $e = expo(x)$. By Corollary 6.23, $expo(RAZ(x, n)) = e$, and hence by Lemma 1.8,

$$
\begin{aligned}
RAZ(RAZ(x, n), m) &= \lceil 2^{m-1-e}(\lceil 2^{n-1-e}x \rceil 2^{e+1-n}) \rceil 2^{e+1-m} \\
&= \lceil \lceil 2^{n-1-e}x \rceil / 2^{n-m} \rceil 2^{e+1-m} \\
&= \lceil 2^{n-1-e}x / 2^{n-m} \rceil 2^{e+1-m} \\
&= \lceil 2^{m-1-e}x \rceil 2^{e+1-m} \\
&= RAZ(x, m). \qquad\qquad\qquad\qquad \square
\end{aligned}
$$

The next three results correspond to Lemmas 6.13, 6.14, and 6.15 of the preceding section.

Lemma 6.32 *Let $x \in \mathbb{R}$ and $n \in \mathbb{N}$. If x is $(n + 1)$-exact but not n-exact, then*

$$RAZ(x, n) = x + sgn(x)2^{expo(x)-n}.$$

Proof By Lemma 6.24, we may assume $x > 0$. For the case $n = 0$, we have $sig(x) = 1$ by Definition 4.2, and by Lemmas 4.1 and 6.19,

$$x + 2^{expo(x)-n} = 2^{expo(x)} + 2^{expo(x)} = 2^{expo(x)+1} = RAZ(x, n).$$

Thus, we may assume $n > 0$. Let $a = x - 2^{expo(x)-n}$ and $b = x + 2^{expo(x)-n}$. As noted in the proof of Lemma 6.13, a and b are both n-exact and $b = fp^+(a, n)$. Now by Lemma 6.27, $b \geq RAZ(x, n)$, but if $b = fp^+(a, n) > RAZ(x, n)$, then since $RAZ(x, n)$ is n-exact, Lemma 4.18 would imply $a \geq RAZ(x, n)$, contradicting $a < x$. Therefore, $a = RAZ(x, n)$. $\qquad\qquad \square$

Lemma 6.33 *Let $x \in \mathbb{R}$, $y \in \mathbb{R}$, and $k \in \mathbb{Z}$. If $x \geq 0$, $y \geq 0$, and x is $(k + expo(x) - expo(y))$-exact, then*

$$x + RAZ(y, k) = RAZ(x + y, k + expo(x + y) - expo(y)).$$

Proof Let $n = k + expo(x) - expo(y)$. Since x is n-exact,

$$x2^{k-1-expo(y)} = x2^{n-1-expo(x)} \in \mathbb{Z}.$$

Let $k' = k + expo(x + y) - expo(y)$. Then by Lemma 1.7(d),

$$
\begin{aligned}
x + RAZ(y, k) &= x + \lceil 2^{k-1-expo(y)}y \rceil 2^{expo(y)+1-k} \\
&= (x2^{k-1-expo(y)} + \lceil 2^{k-1-expo(y)}y \rceil)2^{expo(y)+1-k} \\
&= \lceil 2^{k-1-expo(y)}(x + y) \rceil 2^{expo(y)+1-k} \\
&= \lceil 2^{k'-1-expo(x+y)}(x + y) \rceil 2^{expo(x+y)+1-k'} \\
&= RAZ(x + y, k'). \qquad\qquad\qquad\qquad\qquad \square
\end{aligned}
$$

Lemma 6.34 *Let $x \in \mathbb{R}$, $y \in \mathbb{R}$, and $k \in \mathbb{Z}$. If $x > y > 0$, $k + expo(x - y) - expo(y) > 0$, and x is $(k + expo(x) - expo(y))$-exact, then*

$$x - RTZ(y, k) = RAZ(x - y, k + expo(x - y) - expo(y)).$$

Proof Let $n = k + expo(x) - expo(y)$. Since x is n-exact,

$$x 2^{k-1-expo(y)} = x 2^{n-1-expo(x)} \in \mathbb{Z}.$$

Let $k' = k + expo(x - y) - expo(y)$. Then by Lemma 1.1,

$$
\begin{aligned}
x - RTZ(y, k) &= x - \lfloor 2^{k-1-expo(y)} y \rfloor 2^{expo(y)+1-k} \\
&= -(\lfloor 2^{k-1-expo(y)} y \rfloor - x 2^{k-1-expo(y)}) 2^{expo(y)+1-k} \\
&= -\lfloor 2^{k-1-expo(y)}(y - x) \rfloor 2^{expo(y)+1-k} \\
&= -\lfloor 2^{k'-1-expo(y-x)}(y - x) \rfloor 2^{expo(y-x)+1-k'} \\
&= -RTZ(y - x, k') \\
&= RTZ(x - y, k').
\end{aligned}
$$
$\square$

The following result combines Lemmas 6.15 and 6.34.

Corollary 6.35 *Let $x \in \mathbb{R}$ and $y \in \mathbb{R}$ such that $x \neq 0$, $y \neq 0$, and $x + y \neq 0$. Let $k \in \mathbb{Z}$,*

$$k' = k + expo(x) - expo(y),$$

and

$$k'' = k + expo(x + y) - expo(y).$$

If $k'' > 0$ and x is k'-exact, then

$$
x + RTZ(y, k) = \begin{cases} RTZ(x + y, k'') & \text{if } sgn(x + y) = sgn(y) \\ RAZ(x + y, k'') & \text{if } sgn(x + y) \neq sgn(y). \end{cases}
$$

Proof By Lemmas 6.6 and 6.24, we may assume that $x > 0$. The case $y > 0$ is handled by Lemma 6.14. For the case $y < 0$, Lemmas 6.15 and 6.34 cover the subcases $-y > x$ and $-y < x$, respectively. $\square$

We turn now to the problem of bit-level implementation of rounding. *RTZ*, according to Lemma 6.16, is equivalent to a bit slice operation, which may be implemented as a logical operation using Corollary 6.18. Other rounding modes may be reduced to the case of truncation by a method known as *constant injection*. Let x be m-exact with $expo(x) = e$,

$$x = (1.\beta_1\beta_2 \cdots \beta_{m-1})_2 \cdot 2^e,$$

to be rounded to n bits, where $n \leq m$, according to a rounding mode $\mathcal{R}$. Our goal is to construct a *rounding constant* $\mathcal{C}$, depending on m, n, e, and $\mathcal{R}$, such that

$$\mathcal{R}(x, n) = RTZ(x + \mathcal{C}, n).$$

Since the result of an arithmetic operation is commonly computed as a sum of two or more terms, the expense of inserting an additional term is minimal.

The appropriate constant for the case $\mathcal{R} = RAZ$ is

$$\mathcal{C} = 2^e(2^{-(n-1)} - 2^{-(m-1)}) = 2^{e+1}(2^{-n} - 2^{-m}),$$

which consists of a string of 1s at the bit positions corresponding to the least significant $m - n$ bits of x, as illustrated below:

$$
\begin{array}{rcccccccc}
 & 1 \, . & \beta_1\beta_2 & \cdots & \beta_{n-1}\beta_n & \cdots & \beta_{m-1} & \times & 2^e \\
+ & 0 \, . & 0 \;\; 0 & \cdots & 0 \;\; 1 & \cdots & 1 & \times & 2^e \\
\hline
\end{array}
$$

As suggested by the diagram, the addition $x + \mathcal{C}$ generates a carry into the position of β_{n-1} unless $\beta_n = \cdots = \beta_{m-1} = 0$, i.e., unless x is n-exact, and x is rounded up accordingly. This observation is formalized by the following lemma.

Lemma 6.36 *Let $x \in \mathbb{R}$, $m \in \mathbb{Z}^+$, and $n \in \mathbb{Z}^+$. If x is m-exact, $x > 0$, and $m \geq n$, then*

$$RAZ(x, n) = RTZ(x + 2^{expo(x)+1}(2^{-n} - 2^{-m}), n).$$

Proof Let $a = RTZ(x + 2^{expo(x)+1}(2^{-n} - 2^{-m}), n)$. Since a and $RAZ(x, n)$ are both n-exact and

$$a < x + 2^{expo(x)+1-n} \leq RAZ(x, n) + 2^{expo(RAZ(x,n))+1-n},$$

$a \leq RAZ(x, n)$ by Lemma 4.18.

If x is n-exact, then $a \geq RTZ(x, n) = x = RAZ(x, n)$, and hence $a = RAZ(x, n)$. Thus, we may assume x is not n-exact. But then since $x > RTZ(x, n)$ and x is m-exact, $x \geq RTZ(x, n) + 2^{expo(x)+1-m}$, and hence

$$x + 2^{expo(x)+1}(2^{-n} - 2^{-m}) \geq RTZ(x, n) + 2^{expo(x)+1-n} = RAZ(x, n),$$

which implies $a \geq RAZ(x, n)$. $\qquad\square$

6.3   Rounding to Nearest

Next, we examine the mode *RNE*, which may round in either direction, selecting the representable number that is closest to its argument. This mode is known as "rounding to nearest even" because of the manner in which it resolves the ambiguous case of a number that is equidistant from two successive representable numbers.

Definition 6.4 Given $x \in \mathbb{R}$ and $n \in \mathbb{Z}$, let $z = \lfloor 2^{n-1}sig(x) \rfloor$, and $f = 2^{n-1}sig(x) - z$. Then

$$RNE(x, n) = \begin{cases} RTZ(x, n) & \text{if } f < 1/2 \text{ or } f = 1/2 \text{ and } z \text{ is even} \\ RAZ(x, n) & \text{if } f > 1/2 \text{ or } f = 1/2 \text{ and } z \text{ is odd.} \end{cases}$$

Example Let $x = (101.101)_2$ and $n = 5$. Then

$$z = \lfloor 2^{n-1}sig(x) \rfloor = \lfloor (10110.1)_2 \rfloor = (10110)_2$$

and

$$f = 2^{n-1}sig(x) - z = (10110.1)_2 - (10110)_2 = (0.1)_2 = 1/2,$$

indicating a "tie", i.e., that x is equidistant from two successive 5-exact numbers. Since z is even, the tie is broken in favor of the lesser of the two:

$$RNE(x, n) = RTZ(x, n) = (101.10)_2.$$

Like all rounding modes, the value of *RNE* is always that of either *RTZ* or *RAZ*. We list several properties of *RNE* that may be derived from this observation and the corresponding properties of *RTZ* and *RAZ*.

Lemma 6.37 *If $x \in \mathbb{R}$ and $n \in \mathbb{Z}^+$, then $RNE(x, n)$ is n-exact.*

Lemma 6.38 *Let $x \in \mathbb{R}$ and $n \in \mathbb{Z}^+$. If x is n-exact, then $RNE(x, n) = x$.*

Lemma 6.39 *Let $x \in \mathbb{R}$, $a \in \mathbb{R}$, and $n \in \mathbb{N}$. Suppose a is n-exact.*

(a) *If $a \geq x$, then $a \geq RNE(x, n)$;*
(b) *If $a \leq x$, then $a \leq RNE(x, n)$.*

Lemma 6.40 *For all $x \in \mathbb{R}$ and $n \in \mathbb{N}$, if $|RNE(x, n)| \neq 2^{expo(x)+1}$, then*

$$expo(RNE(x, n)) = expo(x).$$

Lemma 6.41 *For all* $x \in \mathbb{R}$, $n \in \mathbb{Z}^+$, *and* $k \in \mathbb{Z}$,

$$RNE(2^k x, n) = 2^k RNE(x, n).$$

Proof It is clear from Definition 6.4 that the choice between $RTZ(x, n)$ and $RAZ(x, n)$ depends only on $sig(x)$. Thus, for example, if $RNE(x, n) = RTZ(x, n)$, then since $sig(2^k x) = sig(x)$, $RNE(2^k x, n) = RTZ(2^k x, n)$ as well, and by Lemma 6.11,

$$RNE(2^k x, n) = RTZ(2^k x, n) = 2^k RTZ(x, n) = 2^k RNE(x, n). \qquad \square$$

Lemma 6.42 *For all* $x \in \mathbb{R}$ *and* $n \in \mathbb{Z}$, $RNE(-x, n) = -RNE(x, n)$.

Proof This may be derived from Lemmas 6.6 and 6.24 by following the same reasoning as used in the proof of Lemma 6.41. $\qquad \square$

In computing $RNE(x, n)$, the choice between $RTZ(x, n)$ and $RAZ(x, n)$ is governed by their relative distances from x.

Lemma 6.43 *Let* $x \in \mathbb{R}$ *and* $n \in \mathbb{Z}^+$.

(a) *If* $|x - RTZ(x, n)| < |x - RAZ(x, n)|$, *then* $RNE(x, n) = RTZ(x, n)$.
(b) *If* $|x - RTZ(x, n)| > |x - RAZ(x, n)|$, *then* $RNE(x, n) = RAZ(x, n)$.

Proof We may assume that $2^{n-1} sig(x) \notin \mathbb{Z}$, for otherwise

$$RTZ(x, n) = RAZ(x, n) = RNE(x, n) = x.$$

Let $f = 2^{n-1} sig(x) - \lfloor 2^{n-1} sig(x) \rfloor$. Then

$$
\begin{aligned}
|x - RTZ(x, n)| &= |x| - |RTZ(x, n)| \\
&= 2^{expo(x)+1-n}(2^{n-1} sig(x) - \lfloor 2^{n-1} sig(x) \rfloor) \\
&= 2^{expo(x)+1-n} f
\end{aligned}
$$

and

$$
\begin{aligned}
|x - RAZ(x, n)| &= |RAZ(x, n)| - |x| \\
&= 2^{expo(x)+1-n}(\lceil 2^{n-1} sig(x) \rceil - 2^{n-1} sig(x)) \\
&= 2^{expo(x)+1-n}(1 - f).
\end{aligned}
$$

Thus, (a) and (b) correspond to $f < 1/2$ and $f > 1/2$, respectively. $\qquad \square$

No n-exact number can be closer to x than is $RNE(x, n)$:

Lemma 6.44 *Let $x \in \mathbb{R}$, $y \in \mathbb{R}$, and $n \in \mathbb{Z}^+$. If y is n-exact, then*

$$|x - y| \geq |x - RNE(x, n)|.$$

Proof Assume $|x - y| < |x - RNE(x, n)|$. We shall only consider the case $x > 0$, as the case $x < 0$ is handled similarly.

First suppose $RNE(x, n) = RTZ(x, n)$. Since $RNE(x, n) \leq x$, we must have $y > RNE(x, n)$, and hence $y > x$ by Lemma 6.9. But since $RAZ(x, n) - x \geq x - RNE(x, n)$ by Lemma 6.43, we also have $y < RAZ(x, n)$, and hence $y < x$ by Lemma 6.27.

In the remaining case, $RNE(x, n) = RAZ(x, n) > x$. Now $y < RNE(x, n)$ and by Lemma 6.27, $y < x$. But in this case, Lemma 6.43 implies $y > RTZ(x, n)$, and hence $y > x$ by Lemma 6.9.   □

Consequently, the maximum *RNE* rounding error is half the distance between successive representable numbers.

Lemma 6.45 *If $x \in \mathbb{R}$ and $n \in \mathbb{Z}^+$, then*

$$|x - RNE(x, n)| \leq 2^{expo(x)-n} \leq 2^{-n}|x|.$$

Proof By Lemma 6.42, we may assume $x > 0$. Let

$$a = RTZ(x, n) + 2^{expo(x)+1-n} = fp^+(RTZ(x, n), n).$$

If the statement fails, then since $RTZ(x, n)$ and $RAZ(x, n)$ are both n-exact, Lemma 6.44 implies

$$RTZ(x, n) < x - 2^{expo(x)-n} < x + 2^{expo(x)-n} < RAZ(x, n),$$

and hence $a < RAZ(x, n)$. Then by Lemmas 4.18 and 6.27, we have $a < x$, contradicting Lemma 6.4.   □

Lemma 6.46 *Let $x \in \mathbb{R}$, $y \in \mathbb{R}$, and $n \in \mathbb{Z}^+$. If $x \neq 0$, y is n-exact, and $|x - y| < 2^{expo(x)-n}$, then $y = RNE(x, n)$.*

Proof We shall consider the case $x > 0$; the case $x < 0$ then follows from Lemmas 4.8(b) and 6.42.

Let $e = expo(x)$ and $z = RNE(x, n)$. By Lemma 6.40, $expo(z) \geq e$. We also have $expo(y) \geq e$, for otherwise $y < 2^e$, and by Lemma 4.22,

$$y \leq fp^-(2^e, n) = 2^e - 2^{e-n} \leq x - 2^{e-n},$$

contradicting $|x - y| < 2^{e-n}$. By Lemma 6.45,

$$|y - z| \leq |x - y| + |x - z| < 2^{e-n} + 2^{e-n} = 2^{e+1-n}.$$

Since

$$z < y + 2^{e+1-n} \leq y + 2^{expo(y)+1-n} = fp^{+}(y, n),$$

Lemma 4.18 implies $z \leq y$. Similarly, $y < fp^{+}(z, n)$, implying $y \leq z$.   □

The one remaining rounding mode axiom, monotonicity, may be derived as a consequence of Lemma 6.44:

Lemma 6.47 Let $x \in \mathbb{R}$, $y \in \mathbb{R}$, and $n \in \mathbb{Z}^{+}$. If $x \leq y$, then

$$RNE(x, n) \leq RNE(y, n).$$

Proof Let $x' = RNE(x, n)$ and $y' = RNE(y, n)$. Suppose $x < y$ and $y' > x'$. By Lemma 6.44, $|x - x'| \leq |x - y'|$ and $|y - y'| \leq |y - x'|$. It follows that $x > y'$, for otherwise $x \leq y' < x'$, which implies $|x - y'| < |x - x'|$. Similarly, $y < x'$, and consequently $y' < x < y < x'$. Thus,

$$y - y' \leq x' - y < x' - x \leq x - y' < y - y',$$

a contradiction.   □

A number that is $(n + 1)$-exact but not n-exact is a *rounding boundary* with respect to *RNE*:

Lemma 6.48 Let $x \in \mathbb{R}$, $y \in \mathbb{R}$, and $n \in \mathbb{N}$. If $0 < x < y$ and $RNE(x, n) \neq RNE(y, n)$, then for some $a \in \mathbb{R}$, $x \leq a \leq y$ and a is $(n + 1)$-exact but not n-exact.

Proof By Lemma 6.47, $RNE(x, n) < RNE(y, n)$. Let $e = expo(RNE(x, n))$,

$$a = fp^{+}(RNE(x, n), n + 1) = RNE(x, n) + 2^{e-n}$$

and

$$b = fp^{+}(RNE(x, n), n) = RNE(x, n) + 2^{e+1-n} = a + 2^{e-n}.$$

Since $RNE(x, n)$ and $RNE(y, n)$ are both n-exact by Lemmas 6.37, 4.17 implies that a is $(n + 1)$-exact and b is n-exact, and consequently, by Lemma 4.18, a is not n-exact and $RNE(y, n) \geq b$. Thus,

$$RNE(x, n) < a < b \leq RNE(y, n).$$

Moreover, $x \leq a \leq y$, for if $x > a$, then $|x - b| < |x - RNE(x, n|$, contradicting Lemma 6.44, and similarly if $y < a$, then $|y - RNE(x, n)| < |y - RNE(y, n)|$.   □

Lemma 6.49 *Let $x \in \mathbb{R}$, $y \in \mathbb{R}$, $a \in \mathbb{R}$, $n \in \mathbb{Z}^+$, and $k \in \mathbb{Z}^+$ with $n \geq k$. If a is $(n+1)$-exact, $0 < a < x$, and $0 < y < a + 2^{expo(a)-n}$, then*

$$RNE(x, k) \geq RNE(y, k).$$

Proof By Lemma 6.47, we may assume $x < y$, so that $a < x < y < a + 2^{expo(a)-n}$. By Lemmas 4.17 and 4.18, a and $a + 2^{expo(a)-n}$ are successive $(n+1)$-exact numbers, and hence $RNE(x, k) = RNE(y, k)$ by Lemma 6.48.   □

The meaning of "round to nearest even" is that in the case of a midpoint x between two n-exact numbers, $RNE(x, n)$ is defined to be the one that is also $(n-1)$-exact:

Lemma 6.50 *Let $n \in \mathbb{N}$, $n > 1$, and $x \in \mathbb{R}$. If x is $(n+1)$-exact but not n-exact, then $RNE(x, n)$ is $(n-1)$-exact.*

Proof Again we may assume $x > 0$. Let $z = \lfloor 2^{n-1} sig(x) \rfloor$ and $f = 2^{n-1} sig(x) - z$. Since $2^{n-1} sig(x) \notin \mathbb{Z}$, $0 < f < 1$. But $2^n sig(x) = 2z + 2f \in \mathbb{Z}$; hence $2f \in \mathbb{Z}$ and $f = \frac{1}{2}$.

If z is even, then $RNE(x, n) = RTZ(x, n) = z2^{expo(x)+1-n}$, and by Lemma 6.2,

$$2^{n-2-expo(RNE(x,n))} RNE(x, n) = 2^{n-2-expo(x)} z2^{expo(x)+1-n} = z/2 \in \mathbb{Z}.$$

If z is odd, then $RNE(x, n) = RAZ(x, n) = (z+1)2^{expo(x)+1-n}$. We may assume $RAZ(x, n) \neq 2^{expo(x)+1}$, and hence by Corollary 6.23,

$$2^{n-2-expo(RNE(x,n))} RNE(x, n) = 2^{n-2-expo(x)}(z+1)2^{expo(x)+1-n}$$

$$= (z+1)/2$$

$$\in \mathbb{Z}.$$ □

As we have noted, RNE does not satisfy the property of decomposability:

Example Let $x = 43 = (101011)_2$.

$$RNE(RNE(x, 5), 3) = RNE((101100)_2, 3) = (110000)_2 = 48,$$

whereas

$$RNE(x, 3) = (101000)_2 = 40.$$

A consequence of Lemma 6.50 is that a midpoint is sometimes rounded up and sometimes down, and therefore, over the course of a long series of computations and approximations, rounding error is less likely to accumulate to a significant degree in one direction than it would be if the choice were made more consistently. In fact, *RNE* is strictly *unbiased* in the sense of Definition 6.1 and for this reason is identified by the standard as the "default" rounding mode.

Lemma 6.51 *RNE is an unbiased rounding mode.*

Proof Let $e = expo(x_0)$, where $x_0 > 0$ and x_0 is $(n-1)$-exact, and let $N = 2^{m+1-n}$. For $0 \leq k \leq N$, let $x_k = x_0 + 2^{e+1-m}k$ and $\epsilon_k = \mathcal{R}(x_k, n) - x_k$. Since x_0 and $x_{N/2}$ are both n-exact, $\epsilon_0 = \epsilon_{N/2} = 0$ and the sum in Definition 6.1 may be expressed as

$$\mathcal{E}(\mathcal{R}, n, m, x_0) = \frac{1}{2^{n+1-m}}(\Sigma_1 + \epsilon_{N/4} + \Sigma_2 + \Sigma_3 + \epsilon_{3N/4} + \Sigma_4),$$

where

$$\Sigma_1 = \sum_{k=1}^{\frac{N}{4}-1} \epsilon_k, \quad \Sigma_2 = \sum_{k=\frac{N}{4}+1}^{\frac{N}{2}-1} \epsilon_k, \quad \Sigma_3 = \sum_{k=\frac{N}{2}}^{\frac{3N}{4}-1} \epsilon_k, \quad \text{and } \Sigma_4 = \sum_{k=\frac{3N}{4}+1}^{N-1} \epsilon_k.$$

It follows easily from Lemmas 4.17 and 4.18 that x_N is $(n-1)$-exact, $x_{N/2}$ is n-exact but not $(n-1)$-exact, $x_{N/4}$ and $x_{3N/4}$ are $(n+1)$-exact but not n-exact, and no other n-exact numbers lie between x_0 and x_N. By Lemmas 6.37, 6.45, and 6.50, $RNE(x_{N/4}, n) = x_0$ and $RNE(x_{3N/4}, n) = x_N$, and hence

$$\epsilon_{N/4} + \epsilon_{3N/4} = (x_0 - x_{N/4}) + (x_N - x_{3N/4}) = -2^{e-n} + 2^{e-n} = 0.$$

By Lemma 6.46, $RNE(x_k, n) = x_0$ for $0 < k < N/4$, and $RNE(x_k, n) = x_{N/2}$ for $N/4 < k < N/2$. Therefore,

$$\Sigma_1 = \sum_{k=1}^{\frac{N}{4}-1} (x_0 - x_k) = \sum_{k=1}^{\frac{N}{4}-1} (-2^{e+1-m}k) = -2^{e+1-m} \sum_{k=1}^{\frac{N}{4}-1} k$$

and

$$\Sigma_2 = \sum_{k=\frac{N}{4}+1}^{\frac{N}{2}} (x_{N/2} - x_k)$$

$$= \sum_{k=\frac{N}{4}+1}^{\frac{N}{2}} ((x_{N/2} - x_0) + (x_0 - x_k))$$

$$= \frac{N}{4}(2^{e+1-n}) - 2^{e+1-m} \sum_{k=\frac{N}{4}+1}^{\frac{N}{2}} k$$

$$= 2^{e+1-m} \left(\frac{N}{4}\frac{N}{2} - \sum_{k=\frac{N}{4}+1}^{\frac{N}{2}} k \right).$$

Thus,

$$\Sigma_1 + \Sigma_2 = 2^{e+1-m} \left(-\sum_{k=1}^{\frac{N}{4}-1} k + \frac{N}{4}\frac{N}{2} - \sum_{k=\frac{N}{4}+1}^{\frac{N}{2}} k \right)$$

$$= 2^{e+1-m} \left(\frac{N^2}{8} + \frac{N}{4} - \sum_{k=1}^{\frac{N}{2}} k \right)$$

$$= 2^{e+1-m} \left(\frac{N^2}{8} + \frac{N}{4} - \frac{1}{2}\frac{N}{2}\left(\frac{N}{2} + 1 \right) \right)$$

$$= 0.$$

Similarly, $\Sigma_3 + \Sigma_4 = 0$ and the lemma follows. $\square$

The cost of this feature is a complicated definition requiring an expensive implementation. When the goal of a computation is provable accuracy rather than IEEE compliance, a simpler version of "round to nearest" may be appropriate. The critical feature of this mode then becomes the relative error bound guaranteed by Lemma 6.45, since this is likely to be the basis for any formal error analysis. The following definition presents an alternative to *RNE* known as "round to nearest away from zero". This variant, which is mentioned but not prescribed by the standard, respects the same error bound as *RNE* (see Lemma 6.62) but admits a simpler implementation and is therefore commonly used for internal floating-point calculations.

Definition 6.5 Given $x \in \mathbb{R}$ and $n \in \mathbb{Z}$, let $z = \lfloor 2^{n-1}sig(x) \rfloor$, and $f = 2^{n-1}sig(x) - z$. Then

$$RNA(x,n) = \begin{cases} RTZ(x,n) & \text{if } f < 1/2 \\ RAZ(x,n) & \text{if } f \geq 1/2. \end{cases}$$

Example: Let $x = (101.101)_2$ and $n = 5$. Since

$$f = 2^{n-1}sig(x) - \lfloor 2^{n-1}sig(x) \rfloor = (10110.1)_2 - (10110)_2 = (0.1 = 1/2)_2,$$

$$RNA(x,n) = RAZ(x,n) = (101.11)_2.$$

———

Naturally, many of the properties of *RNE* are held by *RNA* as well. We list some of them here, omitting the proofs, which are essentially the same as those given above for *RNE*.

Lemma 6.52 *For all $x \in \mathbb{R}$ and $n \in \mathbb{N}$, $RNA(x, n)$ is n-exact.*

Lemma 6.53 *Let $x \in \mathbb{R}$ and $n \in \mathbb{N}$. If x is n-exact, then $RNA(x, n) = x$.*

Lemma 6.54 *Let $x \in \mathbb{R}$, $a \in \mathbb{R}$, and $n \in \mathbb{N}$. Suppose a is n-exact.*

(a) If $a \geq x$, then $a \geq RNA(x, n)$;
(b) If $a \leq x$, then $a \leq RNA(x, n)$.

Lemma 6.55 *Let $x \in \mathbb{R}$, $y \in \mathbb{R}$, and $n \in \mathbb{Z}^+$. If $x \leq y$, then*

$$RNA(x, n) \leq RNA(y, n).$$

Lemma 6.56 *For all $x \in \mathbb{R}$, $n \in \mathbb{Z}^+$, and $k \in \mathbb{Z}$, $RNA(2^k x, n) = 2^k RNA(x, n)$.*

Lemma 6.57 *For all $x \in \mathbb{R}$ and $n \in \mathbb{N}$, if $|RNA(x, n)| \neq 2^{expo(x)+1}$, then*

$$expo(RNA(x, n)) = expo(x).$$

Lemma 6.58 *For all $x \in \mathbb{R}$ and $n \in \mathbb{Z}$, $RNA(-x, n) = -RNA(x, n)$.*

Lemma 6.59 *Let $x \in \mathbb{R}$ and $n \in \mathbb{N}$.*

(a) If $|x - RTZ(x, n)| < |x - RAZ(x, n)|$, then $RNA(x, n) = RTZ(x, n)$.
(b) If $|x - RTZ(x, n)| > |x - RAZ(x, n)|$, then $RNA(x, n) = RAZ(x, n)$.

Lemma 6.60 *Let $x \in \mathbb{R}$, $y \in \mathbb{R}$, and $n \in \mathbb{Z}^+$. If y is n-exact, then*

$$|x - y| \geq |x - RNA(x, n)|.$$

Lemma 6.61 *Let $x \in \mathbb{R}$, $y \in \mathbb{R}$, and $n \in \mathbb{Z}^+$. If $x \neq 0$, y is n-exact, and $|x - y| < 2^{expo(x)-n}$, then $y = RNA(x, n)$.*

Lemma 6.62 *If $x \in \mathbb{R}$, $n \in \mathbb{Z}^+$, then $|x - RNA(x, n)| \leq 2^{expo(x)-n}$.*

Lemma 6.63 *Let $x \in \mathbb{R}$, $y \in \mathbb{R}$, and $n \in \mathbb{N}$. If $0 < x < y$ and $RNA(x, n) \neq RNA(y, n)$, then for some $a \in \mathbb{R}$, $x \leq a \leq y$ and a is $(n + 1)$-exact but not n-exact.*

Lemma 6.64 *Let $x \in \mathbb{R}$, $y \in \mathbb{R}$, $a \in \mathbb{R}$, $n \in \mathbb{Z}^+$, and $k \in \mathbb{Z}^+$ with $n \geq k$. If a is $(n+1)$-exact, $0 < a < x$, and $0 < y < a+2^{expo(a)-n}$, then $RNA(x, k) \geq RNA(y, k)$.*

The difference between *RNE* and *RNA* is that the latter always rounds a midpoint away from 0.

Lemma 6.65 *Let $x \in \mathbb{R}$ and $n \in \mathbb{N}$. If x is $(n + 1)$-exact but not n-exact, then*

$$RNA(x, n) = RAZ(x, n) = x + sgn(x)2^{expo(x)-n}.$$

Proof By Lemmas 6.42 and 6.24, we may assume $x > 0$. Let $z = \lfloor 2^{n-1}sig(x) \rfloor$ and $f = 2^{n-1}sig(x) - z$. Since $2^{n-1}sig(x) \notin \mathbb{Z}$, $0 < f < 1$. But $2^n sig(x) = 2z + 2f \in \mathbb{Z}$; hence $2f \in \mathbb{Z}$ and $f = \frac{1}{2}$. Therefore, according to Definition 6.5, $RNA(x, n) = RAZ(x, n)$. The second inequality is a restatement of Lemma 6.32.

$\square$

RNA is not decomposable, as illustrated by the same example that we used for *RNE*:

Example Let $x = 43 = (101011)_2$.

$$RNA(RNA(x, 5), 3) = RNA((101100)_2, 3) = (110000)_2 = 48,$$

whereas

$$RNA(x, 3) = (101000)_2 = 40.$$

There is one case of a midpoint for which *RNE* and *RNA* are guaranteed to produce the same result: if the greater of the two representable numbers that are equidistant from x is a power of 2, i.e., $x = 2^{expo(x)+1} - 2^{expo(x)-n}$, then both modes round to this number.

Lemma 6.66 *Let $n \in \mathbb{Z}^+$ and $x \in \mathbb{R}$, $x > 0$. If $x + 2^{expo(x)-n} \geq 2^{expo(x)+1}$, then*

$$RNE(x, n) = RNA(x, n) = 2^{expo(x)+1} = RTZ(x + 2^{expo(x)-n}, n).$$

Proof Suppose $RNE(x, n) \neq 2^{expo(x)+1}$. By Lemma 6.46, $x = 2^{expo(x)+1} - 2^{expo(x)-n}$, which is easily shown to be $(n+1)$-exact but not n-exact. Suppose $n > 1$. Let

$$z = x - 2^{expo(x)-n} = 2^{expo(x)+1} - 2^{expo(x)-n+1}.$$

Then $expo(z) = expo(x)$, z is n-exact but not $(n + 1)$-exact, and $fp^+(z, n) = 2^{expo(x)+1}$. Since $RNE(x, n)$ is n-exact, Lemma 4.18 implies $RNE(x, n) = z$, contradicting Lemma 6.50. On the other hand, if $n = 1$, then

$$x = 2^{expo(x)+1} - 2^{expo(x)-1} = 2^{expo(x)-1} \cdot 3$$

and

$$RNE(x, 1) = 2^{expo(x)-1}RNE(3, 1) = 2^{expo(x)-1} \cdot 4 = 2^{expo(x)+1},$$

a contradiction.

Now suppose $RNA(x, n) \neq 2^{expo(x)+1}$. By Lemma 6.61, $x = 2^{expo(x)+1} - 2^{expo(x)-n}$, and the contradiction follows from Lemma 6.50.

Finally, suppose $2^{expo(x)+1} \neq RTZ(x + 2^{expo(x)-n}, n)$. Since $2^{expo(x)+1}$ is n-exact, $2^{expo(x)+1} < RTZ(x + 2^{expo(x)-n}, n)$ by Lemma 6.9. But then by Lemma 4.18,

$$RTZ(x + 2^{expo(x)-n}, n) \geq 2^{expo(x)+1} + 2^{expo(x)+2-n} > x + 2^{expo(x)-n},$$

contradicting Lemma 6.3.   □

The additive property shared by *RTZ* and *RAZ* that is described in Lemmas 6.14 and 6.33, respectively, does not hold for *RNE* in precisely the same form.

Example Let $x = 2 = (10)_2$, $y = 5 = (101)_2$, and $k = 2$. Then

$$k + expo(x) - expo(y) = 2 + 1 - 2 = 1$$

and

$$k + expo(x + y) - expo(y) = 2 + 2 - 2 = 2.$$

Although x is clearly $(k + expo(x) - expo(y))$-exact,

$$x + RNE(y, k) = (10)_2 + RNE((101)_2, 2) = (10)_2 + (100)_2 = (110)_2,$$

while

$$RNE(x + y, k + expo(x + y) - expo(y)) = RNE((111)_2, 2) = (1000)_2.$$

However, this property is shared by *RNA*, and a slightly weaker version holds for *RNE*.

Lemma 6.67 *Let $x \in \mathbb{R}$, $y \in \mathbb{R}$, and $k \in \mathbb{Z}$ with $x \geq 0$ and $y \geq 0$. Let*

$$k' = k + expo(x) - expo(y)$$

and

$$k'' = k + expo(x + y) - expo(y).$$

(a) If x is $(k' - 1)$-exact, then

$$x + RNE(y, k) = RNE(x + y, k'').$$

(b) If x is k'-exact, then

$$x + RNA(y, k) = RNA(x + y, k'').$$

Proof

(a) Applying Lemma 6.14 and 6.33, we need only show that either

$$RNE(y, k) = RTZ(y, k) \text{ and } RNE(x + y, k'') = RTZ(x + y, k'')$$

or

$$RNE(y, k) = RAZ(y, k) \text{ and } RNE(x + y, k'') = RAZ(x + y, k'').$$

Let $z_1 = \lfloor 2^{k-1} sig(y) \rfloor$, $f_1 = 2^{k-1} sig(y) - z_1$, $z_2 = \lfloor 2^{k''-1} sig(x + y) \rfloor$, and $f_2 = 2^{k''-1} sig(x + y) - z_2$. According to Definition 6.4, it will suffice to show that

$$2^{k''-1} sig(x + y) - 2^{k-1} sig(y) = 2\ell,$$

for some $\ell \in \mathbb{Z}$, for then Lemma 1.1 will imply that $z_2 = z_1 + 2\ell$ and $f_2 = f_1$. But

$$\begin{aligned}
2^{k''-1} sig(x + y) - 2^{k-1} sig(y) &= 2^{k''-expo(x+y)-1}(x + y) - 2^{k-expo(y)-1} y \\
&= 2^{k-expo(y)-1}(x + y) - 2^{k-expo(y)-1} y \\
&= 2^{k-expo(y)-1} x \\
&= 2^{k'-expo(x)-1} x \\
&= 2\ell,
\end{aligned}$$

where $\ell = 2^{(k'-1)-expo(x)-1} \in \mathbb{Z}$ by Lemma 4.8.

(b) Here we must show that either

$$RNA(y, k) = RTZ(y, k) \text{ and } RNA(x + y, k'') = RTZ(x + y, k'')$$

or

$$RNA(y, k) = RAZ(y, k) \text{ and } RNA(x + y, k'') = RAZ(x + y, k'').$$

According to Definition 6.5, this is true whenever $f_1 = f_2$. Thus, we need only show that

$$2^{k''-1} sig(x + y) - 2^{k-1} sig(y) = 2^{k'-expo(x)-1} x \in \mathbb{Z},$$

which is equivalent to the hypothesis that x is k'-exact.   □

The rounding constant $\mathcal{C}$ (see the discussion preceding Lemma 6.36) for both of the modes RNE and RNA is a simple power of 2, equal to half the value of the least significant bit of the rounded result. That is, if the rounding precision is n and the unrounded result is

$$x = (1.\beta_1\beta_2\cdots)_2 \times 2^e,$$

then $C = 2^{e-n}$, as illustrated below:

$$
\begin{array}{llllllll}
 & 1 \,.\, \beta_1\beta_2 & \cdots & \beta_{n-1}\beta_n & \cdots & \times & 2^e \\
+ & 0 \,.\, 0 \;\; 0 & \cdots & 0 \;\; 1 & & \times & 2^e \\
\hline
\end{array}
$$

The following lemma exposes the extra expense of implementing *RNE* as compared to *RNA*. While the correctly rounded result is given by $RTZ(x + C, n)$ in most cases, special attention is required for the computation of *RNE* in the case where it differs from *RNA*, i.e., when x is $(n+1)$-exact and $\beta_n = 1$. In this case, the least significant bit must be forced to 0. This is accomplished by truncating $x + C$ to $n - 1$ bits rather than n.

Lemma 6.68 *If $n \in \mathbb{Z}^+$, $x \in \mathbb{R}$, $x > 0$, and $\mathcal{R}$ is either RNE or RNA, then*

$$\mathcal{R}(x, n) = RTZ(x + 2^{expo(x)-n}, \nu),$$

where

$$
\nu = \begin{cases} n - 1 & \text{if } \mathcal{R} = RNE,\ n > 1,\ \text{and } x \text{ is } (n+1)\text{-exact but not } n\text{-exact} \\ n & \text{otherwise.} \end{cases}
$$

Proof If $x + 2^{expo(x)-n} \geq 2^{expo(x)+1}$, then by Lemma 6.66,

$$RNE(x, n) = RNA(x, n) = 2^{expo(x)+1} = RTZ(x + 2^{expo(x)-n}, n).$$

But if $n > 1$, then by Lemmas 6.12, 4.10, and 6.8,

$$
\begin{aligned}
RTZ(x + 2^{expo(x)-n}, n - 1) &= RTZ(RTZ(x + 2^{expo(x)-n}, n), n - 1) \\
&= RTZ(2^{expo(x)+1}, n - 1) \\
&= 2^{expo(x)+1} \\
&= RTZ(x + 2^{expo(x)-n}, n).
\end{aligned}
$$

Thus, we may assume $x + 2^{expo(x)-n} < 2^{expo(x)+1}$, and it follows from Lemmas 6.40, 6.45, 6.57, and 6.62 that

$$expo(RNE(x, n)) = expo(RNA(x, n)) = expo(x + 2^{expo(x)-n}) = expo(x).$$

Case 1: x is *n*-exact.

By Lemma 6.9, $RTZ(x + 2^{expo(x)-n}, n) \geq x$. But since

$$RTZ(x + 2^{expo(x)-n}, n) \leq x + 2^{expo(x)-n} < x + 2^{expo(x)+1-n},$$

Lemma 4.18 yields $RTZ(x + 2^{expo(x)-n}, n) \leq x$, and hence

$$RTZ(x + 2^{expo(x)-n}, n) = x = RNE(x, n) = RNA(x, n).$$

Case 2: x is not $(n + 1)$-exact.

We have $RNE(x, n) > x - 2^{expo(x)-n}$, for otherwise Lemma 6.45 would imply

$$RNE(x, n) = x - 2^{expo(x)-n},$$

and since $RNE(x, n)$ is $(n + 1)$-exact, so would be

$$RNE(x, n) + 2^{expo(RNE(x,n))-n} = x - 2^{expo(x)-n} + 2^{expo(x)-n} = x.$$

Since $RNE(x, n) \leq x + 2^{expo(x)-n}$, $RNE(x, n) \leq RTZ(x + 2^{expo(x)-n}, n)$ by Lemma 6.9. On the other hand,

$$RTZ(x + 2^{expo(x)-n}, n) \leq x + 2^{expo(x)-n} < RNE(x, n) + 2^{expo(x)+1-n}.$$

If $RNE(x, n) \neq 2^{expo(x)+1}$, then $RNE(x, n) = expo(x)$ by Lemma 6.40 and

$$RTZ(x + 2^{expo(x)-n}, n) \leq RNE(x, n)$$

by Lemma 4.18. But if $RNE(x, n) = 2^{expo(x)+1}$, the same conclusion holds by hypothesis.

The same argument applies to $RNA(x, n)$ but with Lemma 6.62 invoked in place of Lemma 6.45.

Case 3: x is $(n + 1)$-exact but not *n*-exact.

The identity for $RNA(x, n)$ is given by Lemma 6.65. To prove the claim for RNE, we first note that if $n = 1$, then $x = 2^{expo(x)} \cdot \frac{3}{2}$, and

$$\begin{aligned}
RNE(x, 1) &= 2^{expo(x)} RNE\left(\frac{3}{2}, 1\right) \\
&= 2^{expo(x)} \cdot 2 \\
&= 2^{expo(x)+1} \\
&= RTZ(2^{expo(x)+1}, 1) \\
&= RTZ(x + 2^{expo(x)-1}, 1).
\end{aligned}$$

Thus, we may assume $n > 1$. Suppose $RNE(x, n) > x$. Since $RNE(x, n)$ is $(n + 1)$-exact, $RNE(x, n) \geq x + 2^{expo(x)-n}$, hence $RNE(x, n) = x + 2^{expo(x)-n}$, and by Lemma 6.50,

$$RTZ(x + 2^{expo(x)-n}, n - 1) = RTZ(RNE(x, n), n - 1) = RNE(x, n).$$

On the other hand, suppose $RNE(x, n) < x$. Then $RNE(x, n) < x + 2^{expo(x)-n}$ implies $RNE(x, n) \leq RTZ(x + 2^{expo(x)-n}, n - 1)$. But since

$$RTZ(x + 2^{expo(x)-n}, n - 1) \leq x + 2^{expo(x)-n}$$

$$= x - 2^{expo(x)-n} + 2^{expo(x)+1-n}$$

$$< RNE(x, n) + 2^{expo(x)+2-n},$$

we have $RTZ(x + 2^{expo(x)-n}, n - 1) \leq RNE(x, n)$. $\square$

6.4 Odd Rounding

A landmark paper[1] of 1946 by von Neumann et al. contains an early discussion of rounding [3]:

> ... the round-off is intended to produce satisfactory n-digit approximations for the product xy and the quotient x/y of two n-digit numbers. Two things are wanted of the round-off: (1) The approximation should be good, i.e., its variance from the "true" xy or x/y should be as small as practical; (2) The approximation should be unbiased, i.e., its mean should be equal to the "true" xy or x/y.

The authors are willing to relax the exclusion of bias, noting that it generally incurs a cost in efficiency and concluding that "we shall not complicate the machine by introducing such corrections". Two methods of rounding are recommended:

> The first class is characterized by its ignoring all digits beyond the nth, and even the nth digit itself, which it replaces by a 1. The second class is characterized by the procedure of adding one unit in the $n + 1^{st}$ digit, performing the carries which this may induce, and then keeping only the first n digits.

The second of these is the mode that we have designated RNA in the preceding section. The first, which has come to be known as *von Neumann rounding*, may be formulated as

$$\mathcal{R}(x, n) = RTZ(x, n - 1) + sgn(x)2^{expo(x)+1-n}.$$

---

[1] A notable aspect of this paper, which has been cited as the "birth certificate of computer science", is its position, expounded in Sect. 5.3, that floating-point arithmetic is generally a bad idea.

This mode has twice the error range of rounding to nearest but is much easier to implement, involving neither carry propagation nor operand analysis. On the other hand, not only is it slightly biased, but it is in violation of our second axiom, since $\mathcal{R}(x, n) \neq x$ when x is $(n-1)$-exact. Just as modern computing has replaced *RNA* with the less efficient but strictly unbiased *RNE*, we have the following variant of von Neumann rounding, known as *sticky* [23] or *odd* [1] rounding:

Definition 6.6 If $x \in \mathbb{R}$, $n \in \mathbb{N}$, and $n > 1$, then

$$RTO(x, n) = \begin{cases} x & \text{if } x \text{ is } (n-1)\text{-exact} \\ RTZ(x, n - 1) + sgn(x)2^{expo(x)+1-n} & \text{otherwise.} \end{cases}$$

Lemma 6.69 *RTO is an unbiased rounding mode.*

Proof Let x_0, e, n, m, N, and x_k be defined as in the proof of Lemma 6.51. Then for $0 < k < N$, $x_0 < x_k < x_N$, where x_0 and x_N are successive $(n-1)$-exact numbers, and therefore

$$RTO(x_k, n) = RTZ(x_k, n - 1) + 2^{expo(x_k)+1-n} = x_0 + 2^{e+1-n} = x_{N/2}.$$

Thus,

$$\sum_{k=0}^{N-1} (RTO(x_k, n) - x_k) = \sum_{k=1}^{N-1} (x_{N/2} - x_k)$$

$$= \sum_{k=1}^{N-1} (x_{N/2} - x_0 - 2^{e+1-m}k)$$

$$= (N - 1)(x_{N/2} - x_0) - 2^{e+1-m} \sum_{k=1}^{N-1} k$$

$$= (N - 1)2^{e+1-n} - 2^{e+1-m} \cdot \frac{N}{2}(N - 1)$$

$$= (N - 1)(2^{e+1-n} - 2^{e+1-m} \cdot 2^{m-n})$$

$$= 0. \qquad \square$$

We shall see that odd rounding holds all of the properties discussed at the beginning of this chapter, as well as others that are relevant to the double rounding problem (Lemma 6.79) and the implementation of floating-point addition (Lemma 6.80).

Lemma 6.70 *Let $x \in \mathbb{R}$ and $n \in \mathbb{Z}^+$. If x is n-exact, then $RTO(x, n) = x$.*

Proof We may assume that x is not $(n-1)$-exact, and hence, by Lemma 6.13,

$$
\begin{aligned}
RTO(x, n) &= RTZ(x, n-1) + sgn(x)2^{expo(x)+1-n} \\
&= (x - sgn(x)2^{expo(x)+1-n}) + sgn(x)2^{expo(x)+1-n} \\
&= x. \qquad \square
\end{aligned}
$$

Lemma 6.71 *For all $x \in \mathbb{R}$ and $n \in \mathbb{N}$, $RTO(-x, n) = -RTO(x, n)$.*

Proof If x is $(n-1)$-exact, then so is $-x$, and

$$
RTO(-x, n) = -x = -RTO(x, n).
$$

Otherwise, by Lemmas 6.6 and 4.5,

$$
\begin{aligned}
RTO(-x, n) &= RTZ(-x, n-1) + sgn(-x)2^{expo(-x)+1-n} \\
&= -RTZ(x, n-1) - sgn(x)2^{expo(x)+1-n} \\
&= -RTO(x, n). \qquad \square
\end{aligned}
$$

Lemma 6.72 *If $x \in \mathbb{R}$ and $n \in \mathbb{Z}^+$, then $expo(RTO(x, n)) = expo(x)$.*

Proof By Lemma 6.71, we may assume that $x > 0$. We may also assume that x is not $(n-1)$-exact. By Lemma 6.2, $2^{expo(x)} \leq RTZ(x, n-1) < 2^{expo(x)+1}$. Since $RTZ(x, n-1)$ and $2^{expo(x)+1}$ are both $(n-1)$-exact, Lemma 4.18 implies

$$
\begin{aligned}
2^{expo(x)+1} &\geq RTZ(x, n-1) + 2^{expo(x)+1-(n-1)} \\
&> RTZ(x, n-1) + 2^{expo(x)+1-n} \\
&= RTO(x, n),
\end{aligned}
$$

and the claim follows. $\qquad \square$

Lemma 6.73 *For all $x \in \mathbb{R}$ and $n \in \mathbb{N}$, $RTO(x, n)$ is n-exact.*

Proof We may assume that $x > 0$ and x is not $(n-1)$-exact. Let $e = expo(x) = expo(RTZ(x, n-1)) = expo(RTO(x, n))$. Since $RTZ(x, n-1)$ is $(n-1)$-exact,

$$
2^{(n-1)-1-e}RTZ(x, n-1) = 2^{n-2-e}RTZ(x, n-1) \in \mathbb{Z}.
$$

Lemma 6.72 implies

$$
\begin{aligned}
2^{n-1-e}RTO(x, n) &= 2^{n-1-e}(RTZ(x, n-1) + 2^{e+1-n}) \\
&= 2^{n-1-e}RTZ(x, n-1) + 1 \\
&= 2 \cdot 2^{n-2-e}RTZ(x, n-1) + 1 \\
&\in \mathbb{Z}. \qquad \square
\end{aligned}
$$

Lemma 6.74 *Let $x \in \mathbb{R}$, $y \in \mathbb{R}$, and $n \in \mathbb{N}$. If $x \leq y$, then*

$$RTO(x, n) \leq RTO(y, n).$$

Proof Using Lemma 6.71, we may assume that $0 < x < y$. If x is $(n-1)$-exact, then by Lemmas 6.8 and 6.10,

$$RTO(x, n) = x = RTZ(x, n-1) = \leq RTZ(y, n-1) \leq RTO(y, n).$$

Similarly, if neither x nor y is $(n-1)$-exact, then Lemmas 6.10 and 4.2(c) imply

$$\begin{aligned}
RTO(x, n) &= RTZ(x, n-1) + 2^{expo(x)+1-n} \\
&\leq RTZ(y, n-1) + 2^{expo(y)+1-n} \\
&= RTO(y, n).
\end{aligned}$$

In the remaining case, y is $(n-1)$-exact and x is not. By Lemmas 6.7 and 4.9, $RTZ(x, n-1)$ and y are both n-exact, and hence, by Lemmas 4.18 and 6.2,

$$RTO(y, n) = y \geq RTZ(x, n-1) + 2^{expo(x)+1-n} = RTO(x, n). \qquad \square$$

Lemma 6.75 *For all $x \in \mathbb{R}$, $n \in \mathbb{Z}^+$, and $k \in \mathbb{Z}$, $RTO(2^k x, n) = 2^k RTO(x, n)$.*

Proof If x is $(n-1)$-exact, then so is $-x$, and

$$RTO(2^k x, n) = 2^k x = 2^k RTO(x, n).$$

Otherwise, by Lemmas 6.11 and 4.6,

$$\begin{aligned}
RTO(2^k x, n) &= RTZ(2^k x, n-1) + sgn(2^k x) 2^{expo(2^k x)+1-n} \\
&= 2^k RTZ(x, n-1) + 2^k sgn(x) 2^{expo(x)+1-n} \\
&= 2^k RTO(x, n).
\end{aligned}$$

$\qquad \square$

Lemma 6.76 *Let $m \in \mathbb{N}$, $n \in \mathbb{N}$, and $x \in \mathbb{R}$. If $n \geq m > 1$, then*

$$RTO(RTO(x, n), m) = RTO(x, m).$$

Proof We may assume that $x > 0$ and x is not $(n-1)$-exact. Let $a = RTZ(x, n-1)$, $e = expo(x) = expo(a)$, and

$$z = RTO(x, n) = a + 2^{e+1-n}.$$

Then $expo(z) = e$. Since a is $(n-1)$-exact and

$$a < z < a + 2^{e+2-n} = fp^+(a, n-1),$$

$RTZ(z, n-1) = a = RTZ(x, n-1)$. It follows that z is not $(n-1)$-exact, which implies z is not $(m-1)$-exact. By Lemma 6.12,

$$RTZ(z, m-1) = RTZ(RTZ(z, n-1), m-1)$$
$$= RTZ(RTZ(x, n-1), m-1)$$
$$= RTZ(x, m-1).$$

Thus,

$$RTO(RTO(x, n), m) = RTO(z, m)$$
$$= RTZ(z, m-1) + 2^{e+1-m}$$
$$= RTZ(x, m-1) + 2^{e+1-m}$$
$$= RTO(x, m). \qquad \square$$

The following property, which is not shared by any of the other modes that we have considered, is critical to this mode's utility.

Lemma 6.77 *Let $x \in \mathbb{R}$, $m \in \mathbb{Z}^+$, and $n \in \mathbb{Z}^+$. If $n > m$, then $RTO(x, n)$ is m-exact if and only if x is m-exact.*

Proof According to Lemma 6.71, we may assume that $x > 0$. But clearly we may also assume that x is not $(n-1)$-exact, and hence, by Lemma 4.9, x is not m-exact. On the other hand, $RTZ(x, n-1)$ is $(n-1)$-exact, and since

$$RTO(x, n) = RTZ(x, n-1) + 2^{expo(x)+1-n}$$
$$< RTZ(x, n-1) + 2^{expo(x)+1-(n-1)}$$
$$= fp^+(RTZ(x, n-1), n-1),$$

Lemma 4.18 implies that $RTO(x, n)$ is not $(n-1)$-exact. Applying Lemma 4.9 again, we conclude that $RTO(x, n)$ is not m-exact. $\qquad \square$

An important consequence of Lemma 6.77 is a generalization of Lemma 6.76: *a rounded result with respect to any of the modes considered thus far may be derived from an intermediate odd-rounded value.* In particular, for directed rounding, an m-bit rounded result may be derived from an $(m+1)$-bit odd rounding:

Lemma 6.78 *Let $m \in \mathbb{Z}^+$, $n \in \mathbb{Z}^+$, and $x \in \mathbb{R}$. If $n > m$, then*

(a) $RTZ(RTO(x, n), m) = RTZ(x, m)$;
(b) $RAZ(RTO(x, n), m) = RAZ(x, m)$.

Proof We present the proof of (a); (b) is similar. We may assume that $x > 0$ and x is not $(n-1)$-exact; the other cases follow trivially. First, note that by Lemmas 6.73 and 6.77, $RTO(x, n)$ is n-exact but not $(n-1)$-exact, and therefore, according to Lemmas 6.13 and 6.72,

$$RTZ(RTO(x, n), n - 1) = RTO(x, n) - 2^{expo(RTO(x,n))-(n-1)}$$

$$= RTO(x, n) - 2^{expo(x)+1-n}$$

$$= RTZ(x, n - 1).$$

Thus, by Lemma 6.12, for any $m < n$,

$$RTZ(RTO(x, n), m) = RTZ(RTZ(x, n - 1), m) = RTZ(x, m). \qquad \square$$

For rounding to nearest, one extra bit is required:

Lemma 6.79 *Let* $m \in \mathbb{Z}^+$, $n \in \mathbb{Z}^+$, *and* $x \in \mathbb{R}$. *If* $n > m + 1$, *then*

(a) $RNE(RTO(x, n), m) = RNE(x, m);$
(b) $RNA(RTO(x, n), m) = RNA(x, m).$

Proof In view of Lemma 6.78, it will suffice to show that if

$$RTZ(x, m + 1) = RTZ(y, m + 1) \text{ and } RAZ(x, m + 1) = RAZ(y, m + 1),$$

then

$$RNE(x, m) = RNE(y, m) \text{ and } RNA(x, m) = RNA(y, m).$$

Without loss of generality, we may assume $x \leq y$. According to Lemmas 6.48 and 6.63, if the desired conclusion fails to hold, then for some $(m + 1)$-exact a, $x \leq a \leq y$. But this implies $x = a$, for otherwise, by Lemmas 6.3 and 6.9,

$$RTZ(x, m + 1) \leq x < a \leq RTZ(y, m + 1).$$

Similarly, $y = a$, for otherwise $RAZ(x, m + 1) \leq a < y \leq RAZ(y, m + 1)$. Thus, $x = y$, a contradiction. $\qquad \square$

The following analog of Lemma 6.14 first appeared in [23]. This property is essential for the implementation of floating-point addition, as it allows a rounded sum or difference of unaligned numbers to be derived without computing the full sum explicitly. Figure 6.2 is provided as a visual aid to the proof. Note the minor departure from Fig. 6.1.

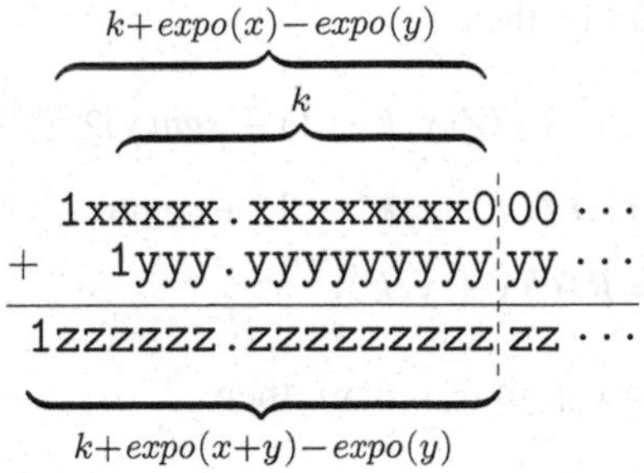

$$\overbrace{}^{k+expo(x)-expo(y)}$$

Fig. 6.2 Lemma 6.80

Lemma 6.80 *Let $x \in \mathbb{R}$ and $y \in \mathbb{R}$ such that $x \neq 0$, $y \neq 0$, and $x + y \neq 0$. Let $k \in \mathbb{N}$,*

$$k' = k + expo(x) - expo(y),$$

and

$$k'' = k + expo(x + y) - expo(y).$$

If $k > 1$, $k' > 1$, $k'' > 1$, and x is $(k' - 1)$-exact, then

$$x + RTO(y, k) = RTO(x + y, k'').$$

Proof Since x is $(k' - 1)$-exact,

$$2^{k-2-expo(y)} x = 2^{(k'-1)-1-expo(x)} x \in \mathbb{Z}.$$

Thus,

$$y \text{ is } (k - 1)\text{-exact} \Leftrightarrow 2^{k-2-expo(y)} y \in \mathbb{Z}$$
$$\Leftrightarrow 2^{k-2-expo(y)} y + 2^{k-2-expo(y)} x \in \mathbb{Z}$$
$$\Leftrightarrow 2^{k''-2-expo(x+y)} (x + y) \in \mathbb{Z}$$
$$\Leftrightarrow x + y \text{ is } (k'' - 1)\text{-exact}.$$

If y is $(k - 1)$-exact, then

$$x + RTO(y, k) = x + y = RTO(x + y, k'').$$

Thus, we may assume that y is not $(k - 1)$-exact. We invoke Corollary 6.35:

$$x + RTZ(y, k) = \begin{cases} RTZ(x + y, k'') & \text{if } sgn(x + y) = sgn(y) \\ RAZ(x + y, k'') & \text{if } sgn(x + y) \neq sgn(y). \end{cases}$$

Now if $sgn(x + y) = sgn(y)$, then

$$x + RTO(y, k) = x + RTZ(y, k - 1) + sgn(y)2^{expo(y)+1-k}$$
$$= RTZ(x + y, k'' - 1) + sgn(x + y)2^{expo(x+y)+1-k''}$$
$$= RTO(x + y, k'').$$

On the other hand, if $sgn(x + y) \neq sgn(y)$, then

$$x + RTO(y, k) = x + RTZ(y, k - 1) + sgn(y)2^{expo(y)+1-k}$$
$$= RAZ(x + y, k'' - 1) - sgn(x + y)2^{expo(x+y)+1-k''}$$
$$= RTZ(x + y, k'' - 1) + sgn(x + y)2^{expo(x+y)+1-k''}$$
$$= RTO(x + y, k'').$$

$\square$

6.5 IEEE Rounding

The IEEE standard prescribes four rounding modes: "round to nearest even" (*RNE*), "round toward 0" (*RTZ*), "round toward $+\infty$", and "round toward $-\infty$". The last two are formalized here by the following functions:

Definition 6.7 For all $x \in \mathbb{R}$ and $n \in \mathbb{N}$,

(a) $RUP(x, n) = \begin{cases} RAZ(x, n) & \text{if } x \geq 0 \\ RTZ(x, n) & \text{if } x < 0; \end{cases}$

(b) $RDN(x, n) = \begin{cases} RTZ(x, n) & \text{if } x \geq 0 \\ RAZ(x, n) & \text{if } x < 0. \end{cases}$

The essential properties of these modes are given by the following:

Lemma 6.81 *For all $x \in \mathbb{R}$ and $n \in \mathbb{N}$, $RUP(x, n) \geq x$ and $RDN(x, n) \leq x$.*

Proof This follows from Lemmas 6.3 and 6.20 and the observation that $sgn(RUP(x, n)) = sgn(RDN(x, n)) = sgn(x)$. For example, if $x < 0$, then

$$RUP(x, n) = RTZ(x, n) = -|RTZ(x, n)| \geq -|x| = x.$$

$\square$

In this section, we collect a set of general results that pertain to all of the IEEE rounding modes, many of which are essentially restatements of lemmas that are proved in earlier sections. Since these results also hold for two of the other modes that we have discussed, *RAZ* and *RNA*, we shall state them as generally as possible. For this purpose, we make the following definition:

Definition 6.8 The *common rounding modes* are *RTZ, RAZ, RNE, RNA, RUP*, and *RDN*.

Note that *RUP* and *RDN* do not share the symmetry property held by the other modes. A generalization is given by the following lemma.

Lemma 6.82 *Let $\mathcal{R}$ be a common rounding mode, and let*

$$\hat{\mathcal{R}} = \begin{cases} RDN & \text{if } \mathcal{R} = RUP \\ RUP & \text{if } \mathcal{R} = RDN \\ \mathcal{R} & \text{otherwise.} \end{cases}$$

For all $x \in \mathbb{R}$ and $n \in \mathbb{Z}$, $\mathcal{R}(-x, n) = -\hat{\mathcal{R}}(x, n)$.

Proof Suppose, for example, that $\mathcal{R} = RUP$ and $x > 0$. Then since $-x < 0$,

$$\mathcal{R}(-x, n) = RUP(-x, n) = RTZ(-x, n) = -RTZ(x, n) = -\hat{\mathcal{R}}(x, n).$$

The other cases are handled similarly. $\qquad\square$

Lemmas 6.83–6.92 are merely a collection of results of the preceding sections of this chapter.

Lemma 6.83 *Let $\mathcal{R}$ be a common rounding mode. For all $x \in \mathbb{R}$ and $n \in \mathbb{N}$, either $\mathcal{R}(x, n) = RTZ(x, n)$ or $\mathcal{R}(x, n) = RAZ(x, n)$.*

Lemma 6.84 *Let $\mathcal{R}$ be a common rounding mode. Let $n \in \mathbb{Z}$, $x \in \mathbb{R}$, and $a \in \mathbb{R}$.*

(a) $\mathcal{R}(x, n)$ is n-exact;
(b) If x is n-exact, then $\mathcal{R}(x, n) = x$;
(c) If $a \geq x$, then $a \geq \mathcal{R}(x, n)$;
(d) If $a \leq x$, then $a \leq \mathcal{R}(x, n)$.

Lemma 6.85 *If $x \in \mathbb{R}$, $n \in \mathbb{Z}^+$, and $\mathcal{R}$ is a common rounding mode, then*

$$sgn(\mathcal{R}(x, n)) = sgn(x).$$

Lemma 6.86 *Let $x \in \mathbb{R}$ and $n \in \mathbb{Z}^+$ and let $\mathcal{R}$ be a common rounding mode. Then*

$$|x - \mathcal{R}(x, n)| < 2^{expo(x)-n+1}.$$

Lemma 6.87 *Let $\mathcal{R}$ be a common rounding mode. For all $x \in \mathbb{R}$ and $n \in \mathbb{N}$, if $|\mathcal{R}(x, n)| \neq 2^{expo(x)+1}$, then $expo(\mathcal{R}(x, n)) = expo(x)$.*

Lemma 6.88 *Let $x \in \mathbb{R}$, $y \in \mathbb{R}$, and $n \in \mathbb{Z}^+$. Let $\mathcal{R}$ be a common rounding mode. If $x \leq y$, then $\mathcal{R}(x, n) \leq \mathcal{R}(y, n)$.*

Lemma 6.89 *Let $\mathcal{R}$ be a common rounding mode. For all $x \in \mathbb{R}$, $n \in \mathbb{N}$, and $k \in \mathbb{Z}$, $\mathcal{R}(2^k x, n) = 2^k \mathcal{R}(x, n)$.*

Lemma 6.90 *Let $\mathcal{R}$ be a common rounding mode, $x \in \mathbb{R}$, $y \in \mathbb{R}$, and $k \in \mathbb{Z}$ with $x \geq 0$ and $y \geq 0$. Let $k' = k + expo(x) - expo(y)$ and $k'' = k + expo(x+y) - expo(y)$. If x is $(k' - 1)$-exact, then*

$$x + \mathcal{R}(y, k) = \mathcal{R}(x + y, k + expo(x + y) - expo(y)).$$

Lemma 6.91 *Let $x \in \mathbb{R}$ and $n \in \mathbb{N}$ and let $\mathcal{R}$ be a common rounding mode. If $x \geq 0$, then*

$$RTZ(x, n) \leq \mathcal{R}(x, n) \leq RAZ(x, n).$$

Lemma 6.92 *Let $\mathcal{R}$ be a common rounding mode, $m \in \mathbb{Z}^+$, $n \in \mathbb{Z}^+$, and $x \in \mathbb{R}$. If $n \geq m + 2$, then*

$$\mathcal{R}(x, m) = \mathcal{R}(RTO(x, n), m).$$

Our analysis of the square root in Chap. 7 depends on the observation that for a given precision n, if x is not $(n + 1)$-exact and y is sufficiently close to x, then x and y round to the same value under any common rounding mode. We shall derive this result as a consequence of the following:

Lemma 6.93 *Let $x \in \mathbb{R}$, $y \in \mathbb{R}$, and $n \in \mathbb{Z}^+$ with $x > 0$. Let $\mathcal{R}$ be a common rounding mode. Assume that x is not $(n + 1)$-exact. If*

$$RTZ(x, n + 1) < y < RAZ(x, n + 1),$$

then $\mathcal{R}(x, n) = \mathcal{R}(y, n)$.

Proof We show the proof for the case $y < x$; the other case is similar.

By Lemma 6.88, we need only show that $\mathcal{R}(x, n) \leq \mathcal{R}(y, n)$. We also note that by Lemmas 6.12 and 6.3,

$$RTZ(x, n) = RTZ(RTZ(x, n + 1), n) \leq RTZ(x, n + 1) < y.$$

Case 1: $\mathcal{R} = RTZ$ or $\mathcal{R} = RDN$

By Lemma 6.7, $RTZ(x, n)$ is n-exact. By Lemma 6.9 (with $RTZ(x, n)$ and y substituted for a and x) and Definition 6.7,

$$\mathcal{R}(x, n) = RTZ(x, n) \leq RTZ(y, n) = \mathcal{R}(y, n).$$

Case 2: $\mathcal{R} = RAZ$ or $\mathcal{R} = RUP$

By Lemma 6.31,

$$RTZ(x, n) < y < x \leq RAZ(x, n + 1) \leq RAZ(RAZ(x, n + 1), n) = RAZ(x, n).$$

By Definition 6.7 and Lemmas 6.25, 6.27, and 6.28,

$$\mathcal{R}(y, n) = RAZ(y, n) = RAZ(x, n) = \mathcal{R}(x, n).$$

Case 3: $\mathcal{R} = RNE$ or $\mathcal{R} = RNA$
By Lemma 6.20,

$$RTZ(x, n + 1) < y < x < RAZ(x, n + 1) = fp^+(RTZ(x, n + 1)).$$

The claim follows from Lemmas 6.28, 6.49, and 6.64. □

Corollary 6.94 *Let $x \in \mathbb{R}$ and $n \in \mathbb{Z}^+$ with $x > 0$. Assume that x is not $(n + 1)$-exact. There exists $\epsilon \in \mathbb{R}$, $\epsilon > 0$, such that for all $y \in \mathbb{R}$ and every common rounding mode $\mathcal{R}$, if $|x - y| < \epsilon$, then $\mathcal{R}(x, n) = \mathcal{R}(y, n)$.*

Proof According to Lemma 6.93, this holds for

$$\epsilon = min(x - RTZ(x, n + 1), RAZ(x, n + 1) - x).$$

□

The remaining results of this section pertain to the implementation of rounding. For the sake of simplicity, our characterization of the method of constant injection is formulated in the context of bit vector rounding.

Lemma 6.95 *Let $n \in \mathbb{Z}^+$ and $M \in \mathbb{Z}^+$ with $e = expo(M)$, and let $\mathcal{R}$ be a common rounding mode. Assume that if $\mathcal{R} = RUP$ or $\mathcal{R} = RAZ$, then $e \geq n - 1$. Then*

$$\mathcal{R}(M, n) = RTZ(M + C, v),$$

where

$$C = \begin{cases} 2^{e-n} & \text{if } \mathcal{R} = RNE \text{ or } \mathcal{R} = RNA \\ 2^{e-n+1} - 1 & \text{if } \mathcal{R} = RUP \text{ or } \mathcal{R} = RAZ \\ 0 & \text{if } \mathcal{R} = RTZ \text{ or } \mathcal{R} = RDN \end{cases}$$

and

$$v = \begin{cases} n - 1 & \text{if } \mathcal{R} = RNE, \ n > 1, \text{ and } M \text{ is } (n + 1)\text{-exact but not } n\text{-exact} \\ n & \text{otherwise.} \end{cases}$$

Proof For the modes *RAZ* and *RUP*, the identity follows from Lemma 6.36, with $m = e+1$, and Lemma 4.11. For *RNE* and *RNA*, it reduces to Lemmas 6.68 and 6.68. For *RTZ* and *RDN*, the lemma is trivial. □

When rounding is based on constant injection, the constant is commonly inserted as an additional term of a sum, which means that the unrounded result M is not explicitly available and Lemma 6.95 is not applicable. The following result instead bases the rounding, as well as the detection of inexactness, on the injected sum $M + C$.

Corollary 6.96 *Let* $n \in \mathbb{Z}^+$ *and* $M \in \mathbb{Z}^+$ *with* $expo(M) = e > n$. *Let* $\mathcal{R}$ *be a common rounding mode, and let*

$$C = \begin{cases} 2^{e-n} & \text{if } \mathcal{R} = RNE \text{ or } \mathcal{R} = RNA \\ 2^{e-n+1} - 1 & \text{if } \mathcal{R} = RUP \text{ or } \mathcal{R} = RAZ \\ 0 & \text{if } \mathcal{R} = RTZ \text{ or } \mathcal{R} = RDN. \end{cases}$$

Let $\Sigma = M + C$, $\overline{\Sigma} = \Sigma[e + 1 : e + 1 - n]$, *and* $\underline{\Sigma} = \Sigma[e - n : 0]$. *Then*

(a) $\mathcal{R}(M, n) = \begin{cases} 2^{e+2-n}\overline{\Sigma}[n : 1] & \text{if } \mathcal{R} = RNE \text{ and } \underline{\Sigma} = 0 \\ 2^{e+1-n}\overline{\Sigma} & \text{otherwise;} \end{cases}$

(b) $\mathcal{R}(M, n) = M \Leftrightarrow \underline{\Sigma} = C$.

Proof Suppose first that $expo(\Sigma) > e$. Since $C < 2^{e+1-n}$, $2^{e+1} \leq \Sigma < 2^{e+1} + 2^{e+1-n}$, which implies

$$\overline{\Sigma} = \left\lfloor \frac{\Sigma}{2^{e+1-n}} \right\rfloor = \left\lfloor 2^n \left\lfloor \frac{\Sigma}{2^{e+1}} \right\rfloor \right\rfloor = 2^n$$

and

$$2^{e+2-n}\overline{\Sigma}[n : 1] = 2^{e+1-n}\overline{\Sigma} = 2^{e+1}.$$

It follows from Lemmas 6.95 and 6.2 that $expo(\mathcal{R}(M, n)) = e + 1$, and by Lemma 6.87, $\mathcal{R}(M, n) = 2^{e+1}$.

Thus, we may assume $expo(\Sigma) = e$. Note that

$$M \text{ is } n\text{-exact} \Leftrightarrow 2^{n-1-e}M \in \mathbb{Z} \Leftrightarrow M[e - n : 0] = 0$$

and similarly,

$$M \text{ is } (n + 1)\text{-exact} \Leftrightarrow M[e - n - 1 : 0] = 0.$$

Therefore, M is n-exact but not $(n+1)$-exact iff $M[e-n-1 : 0] = 0$ and $M[e-n] = 1$. If $\mathcal{R} = RNE$, this condition is equivalent to $\underline{\Sigma} = 0$. By Lemma 6.95,

$$\mathcal{R}(M, n) = RTZ(\Sigma, v), \tag{6.1}$$

where

$$v = \begin{cases} n - 1 & \text{if } \mathcal{R} = RNE, \ n \neq 1, \text{ and } \underline{\Sigma} = 0 \\ n & \text{otherwise.} \end{cases}$$

By Lemma 6.16,

$$RTZ(\Sigma, n - 1) = 2^{e+1-n}\Sigma[e : e - n + 1] = 2^{e+2-n}\overline{\Sigma}[n : 1]$$

and

$$RTZ(\Sigma, n) = 2^{e+1-n}\Sigma[e : e - n] = 2^{e+1-n+1}\overline{\Sigma}.$$

But if $\mathcal{R} = RNE$, $n = 1$, and $\underline{\Sigma} = 0$, then $M[e-n+1] = M[e] = 1$, which implies $\overline{\Sigma}[0] = \Sigma[e - n + 1] = 0$, and these two expressions are equal. Thus, (6.1) reduces to the desired result.

(b) Since $\underline{\Sigma} = (M + C) \bmod 2^{e-n+1}$,

$$\underline{\Sigma} = C \Leftrightarrow M \bmod 2^{e-n+1} = M[e - n : 0] = 0$$

$$\Leftrightarrow 2^{n-1-e}M \in \mathbb{Z}$$

$$\Leftrightarrow \mathcal{R}(M, n) = M. \qquad \square$$

Another common implementation of rounding is provided by the following result. Suppose our objective is a correct n-bit rounding (with respect to some common rounding mode) of a precise result $z > 0$ and that we have a bit vector representation of $x = \lfloor z \rfloor$. Assume that (a) $2^n \leq z$, so that x includes at least n bits following its leading 1, and (b) we have access to a "sticky" bit that indicates whether $z > x$. Then the desired result may be derived by rounding x in the direction specified by the following lemma. (This result is applied repeatedly in the correctness proofs of Part V.)

Lemma 6.97 *Let $\mathcal{R}$ be a common rounding mode, $z \in \mathbb{R}$, and $n \in \mathbb{N}$. Assume that $0 < n$ and $2^n \leq z$. Let $x = \lfloor z \rfloor$ and $e = expo(x)$.*

(a) z is n-exact iff $x[e - n : 0] = 0$ and $z \in \mathbb{Z}$.

(b) If any of the following conditions holds, then $\mathcal{R}(z, n) = fp^+(RTZ(x, n), n)$, and otherwise $\mathcal{R}(z, n) = RTZ(x, n)$:

– $\mathcal{R} = RUP$ *or* $\mathcal{R} = RAZ$ *and at least one of the following is true:*

* $x[e - n : 0] \neq 0$;
* $x \neq z$.

– $\mathcal{R} = RNA$ *and* $x[e - n] = 1$
– $\mathcal{R} = RNE$, $x[e - n] = 1$, *and at least one of the following is true:*

* $x[e - n-1 : 0] \neq 0$;
* $x \neq z$;
* $x[e + 1 - n] = 1$;

Proof It is clear that $expo(z) = e$. By Definition 6.2,

$$RTZ(z, e+1) = \lfloor 2^{(e+1)-1} sig(z) \rfloor 2^{e-(e+1)+1} = \lfloor 2^e sig(z) \rfloor = \lfloor z \rfloor = x,$$

and hence, by Lemma 6.12, $RTZ(z, n) = RTZ(x, n)$.

(a) According to Lemmas 6.7 and 6.8, it suffices to show that $z \neq RTZ(z, n)$ iff either $x[e - n : 0] \neq 0$ or $x \neq z$. By Lemma 6.3,

$$RTZ(x, n) \leq x = RTZ(z, e+1) \leq z,$$

and it is clear that $z \neq RTZ(z, n)$ iff either $x - RTZ(x, n) > 0$ or $x \neq z$. Now since x is $(e+1)$-exact, Lemma 6.7 and Corollary 6.17 imply

$$x - RTZ(x, n) = RTZ(x, e+1) - RTZ(x, n) = x[e - n : 0].$$

(b) Since $RDN(z, n) = RTZ(z, n)$ and $RUP(z, n) = RAZ(z, n)$, we need only consider RAZ, RNE, and RNA.

Suppose $\mathcal{R} = RAZ$. By Lemmas 6.28, 6.7, and 6.25,

$$RAZ(z, n) = \begin{cases} RTZ(x, n) & \text{if } z = RTZ(z, n) \\ fp^+(RTZ(x, n), n) & \text{if } z \neq RTZ(z, n). \end{cases}$$

Thus, $RAZ(z, n) = RTZ(x, n)$ iff $z = RTZ(z, n)$. But we have shown that this holds iff $x[e - n : 0] = 0$ and $z \in \mathbb{Z}$, as stated by the lemma.

For the remaining cases, RNE, and RNA, we refer directly to Definitions 6.4 and 6.5. Let

$$y = \lfloor 2^{n-1} sig(z) \rfloor = \lfloor 2^{n-1-e} z \rfloor = \lfloor x/2^{e+1-n} \rfloor$$

and

$$f = 2^{n-1} sig(z) - y = 2^{n-1-e} z - y.$$

By Definition 2.2, $y = x[e : 1 + e - n]$. By Definition 6.2, $RTZ(z, n) = 2^{1+e-n} y$, and consequently $2^{1+e-n} f = z - RTZ(z, n)$ and $f = 2^{n-e-1}(z - RTZ(z, n))$. By Corollary 6.17,

$$z - RTZ(z, n) = (RTZ(x, n+1) - RTZ(x, n)) + (z - RTZ(z, n+1))$$
$$= 2^{e-n} x[e - n] + (z - RTZ(z, n+1)).$$

Thus,

$$f = 2^{n-e-1}(z - RTZ(z, n)) = \frac{1}{2} x[e - n] + 2^{n-e-1}(z - RTZ(z, n+1)).$$

By Lemma 6.5,

$$2^{n-e-1}(z - RTZ(z, n+1)) < 2^{n-e-1}2^{e-n} = \frac{1}{2}.$$

This leads to the following observations:

(1) $f \geq \frac{1}{2}$ iff $x[e - n] = 1$.
(2) $f > \frac{1}{2}$ iff $x[e - n] = 1$ and $z - RTZ(z, n+1) > 0$, but by Corollary 6.17,

$$\begin{aligned}
z - RTZ(z, n+1) &= (z - x) + (x - RTZ(z, n+1)) \\
&= (z - x) + (RTZ(x, e+1) - RTZ(x, n+1)) \\
&= (z - x) + x[e - 1 - n : 0],
\end{aligned}$$

and hence $f > \frac{1}{2}$ iff $x[e - n] = 1$ and either $x[e - 1 - n : 0] \neq 0$ or $z \neq x$. Referring to Definition 6.5, we see that (1) is sufficient to complete the proof for the case $\mathcal{R} = RNA$.

For the case $\mathcal{R} = RNE$, it is clear from Definition 6.4 and (1) and (2) above that we need only show that y is even iff $x[1 + e - n] = 0$. But this is a consequence of the above equation $z = x[e : 1 + e - n]$.   □

The final result of this section is a variation of Lemma 6.97 that allows us to compute the absolute value of a rounded result when the unrounded value z is negative, given a signed integer encoding x of $\lfloor z \rfloor$, i.e., $x = \lfloor z \rfloor + 2^k$, where $z \geq -2^k$.

Lemma 6.98 *Let $\mathcal{R}$ be a common rounding mode, $z \in \mathbb{R}$, $n \in \mathbb{N}$, and $k \in \mathbb{N}$. Assume that $0 < n < k$ and $-2^k \leq z < -2^n$. Let $x = \lfloor z \rfloor + 2^k$, $\tilde{x} = 2^k - x - 1$, and $e = expo(\tilde{x})$.*

(a) *If $expo(z) \neq e$, then $z = -2^{e+1}$.*
(b) *z is n-exact iff $x[e - n : 0] = 0$ and $z \in \mathbb{Z}$.*
(c) *If any of the following conditions holds, then $|\mathcal{R}(z, n)| = fp^+(RTZ(\tilde{x}, n), n)$, and otherwise $|\mathcal{R}(z, n)| = RTZ(\tilde{x}, n)$:*

  - *$\mathcal{R} = RDN$ or $\mathcal{R} = RAZ$;*
  - *$\mathcal{R} = RUP$ or $\mathcal{R} = RTZ$ and both of the following are true:*

    * *$x[e - n : 0] = 0$;*
    * *$z \in \mathbb{Z}$;*

  - *$\mathcal{R} = RNA$ and at least one of the following is true:*

    * *$x[e - n] = 0$;*
    * *$x[e - n-1 : 0] = 0$ and $z \in \mathbb{Z}$;*

– $\mathcal{R} = RNE$ and at least one of the following is true:

* $x[e - n] = 0$
* $x[e + 1 - n] = x[e - n - 1 : 0] = 0$ and $z \in \mathbb{Z}$.

Proof Let $f = z - \lfloor z \rfloor$. Then $x = (z - f) + 2^k$,

$$\tilde{x} = 2^k - x - 1 = 2^k - (z - f + 2^k) - 1 = -z - (1 - f) = |z| - (1 - f),$$

and $|z| = \tilde{x} + (1 - f)$.

To prove (a), note that if $expo(z) \neq e$, then since $\tilde{x} \leq 2^{e+1} - 1$, we must have $\tilde{x} = 2^{e+1} - 1$, $f = 0$, and $|z| = 2^{e+1}$.

For the proof of (b) and (c), by Lemmas 2.10 and 6.16, we have

$$|z| = \tilde{x} + (1 - f)$$
$$= 2^{e+1-n}\tilde{x}[e : e + 1 - n] + \tilde{x}[e - n : 0] + (1 - f)$$
$$= RTZ(\tilde{x}, n) + \tilde{x}[e - n : 0] + (1 - f).$$

Suppose first that $f = x[e - n : 0] = 0$. Then

$$|z| = RTZ(\tilde{x}, n) + (2^{e+1-n} - 1) + 1 = fp^+(RTZ(\tilde{x}, n), n) \in \mathbb{Z}$$

and the lemma claims that (b) z is n-exact and (c) $|\mathcal{R}(z, n)| = fp^+(RTZ(\tilde{x}, n), n)$ for all $\mathcal{R}$. Both claims follow trivially from Lemmas 4.17, 6.7, and 6.84(b).

In the remaining case, $RTZ(\tilde{x}, n) < |z| < fp^+(RTZ(\tilde{x}, n), n)$. By Lemma 4.18, z is not n-exact, and (b) follows. The claim (c) will be derived from Lemma 6.97. Note that if $\hat{\mathcal{R}}$ is defined as in Lemma 6.82, then

$$|\mathcal{R}(z, n)| = -\mathcal{R}(z, n) = \hat{\mathcal{R}}(-z, n) = \hat{\mathcal{R}}(|z|, n),$$

and (c) may be restated as follows:

(c′) *If any of the following conditions holds, then* $\hat{\mathcal{R}}(|z|, n) = fp^+(RTZ(\tilde{x}, n), n)$, *and otherwise* $\hat{\mathcal{R}}(|z|, n) = RTZ(\tilde{x}, n)$:

– $\hat{\mathcal{R}} = RUP$ or $\hat{\mathcal{R}} = RAZ$;
– $\hat{\mathcal{R}} = RDN$ or $\hat{\mathcal{R}} = RTZ$ and both of the following are true:

* $x[e - n : 0] = 0$;
* $z \in \mathbb{Z}$;

– $\hat{\mathcal{R}} = RNA$ and at least one of the following is true:

* $x[e - n] = 0$;
* $x[e - n - 1 : 0] = 0$ and $z \in \mathbb{Z}$;

– $\hat{\mathcal{R}} = RNE$ and at least one of the following is true:

* $x[e - n] = 0$
* $x[e + 1 - n] = x[e - n - 1 : 0] = 0$ and $z \notin \mathbb{Z}$.

We invoke Lemma 6.97 with $|z|$ and $\hat{\mathcal{R}}$ substituted for z and $\mathcal{R}$, respectively. This yields the following:

(c″) *If any of the following holds, then* $\hat{\mathcal{R}}(|z|, n) = fp^+(RTZ(\lfloor |z| \rfloor, n), n)$, *and otherwise* $\hat{\mathcal{R}}(|z|, n) = RTZ(\lfloor |z| \rfloor, n)$:

– $\hat{\mathcal{R}} = RUP$ or $\hat{\mathcal{R}} = RAZ$ and at least one of the following is true:

* $\lfloor |z| \rfloor [e - n : 0] \neq 0$;
* $z \notin \mathbb{Z}$.

– $\hat{\mathcal{R}} = RNA$ and $\lfloor |z| \rfloor [e - n] = 1$
– $\hat{\mathcal{R}} = RNE$, $\lfloor |z| \rfloor [e - n] = 1$, and at least one of the following is true:

* $\lfloor |z| \rfloor [e - n - 1 : 0] \neq 0$;
* $z \notin \mathbb{Z}$;
* $\lfloor |z| \rfloor [e + 1 - n] = 1$;

We must show that (c″) implies (c′).

Suppose $f > 0$. Then $z \notin \mathbb{Z}$, and by Lemma 1.4,

$$\tilde{x} = 2^k - x - 1 = -\lfloor z \rfloor - 1 = \lfloor |z| \rfloor$$

and we may replace $\tilde{x}$ with $\lfloor |z| \rfloor$ in (c′). Furthermore, $|z|[e - n] = \tilde{x}[e - n]$, which implies $|z|[e - n] \neq x[e - n]$, and the claim (c′) follows trivially from (c″).

In the final case, $f = 0$ and $x[e - n : 0] \neq 0$. Thus, $|z| = \tilde{x} + 1$ and

$$\tilde{x}[e - n : 0] < 2^{e+1-n} - 1.$$

Since

$$|z| = \tilde{x} + 1$$
$$= 2^{e+1-n}\tilde{x}[e : e + 1 - n] + \tilde{x}[e - n : 0] + 1$$
$$\leq 2^{e+1-n}(2^n - 1) + (2^{e+1-n} - 2) + 1$$
$$= 2^{e+1} - 1,$$

$expo(z) = e$. Since

$$|z|[e : e + 1 - n] = \left\lfloor \frac{|z|}{2^{e+1-n}} \right\rfloor = \tilde{x}[e : e + 1 - n],$$

$|z|[e + 1 - n] = \tilde{x}[e + 1 - n]$, which implies $|z|[e + 1 - n] \neq x[e + 1 - n]$. Lemma 6.16 implies $RTZ(|z|, n) = RTZ(\tilde{x}, n)$, and again we may replace $\tilde{x}$ with $\lfloor |z| \rfloor$ in (c′). Furthermore,

$$|z|[e - n : 0] = |z| \bmod 2^{e+1-n} = \tilde{x}[e - n : 0] + 1 = 2^{e+1-n} - x[e - n : 0].$$

If $x[e - n - 1 : 0] = 0$, then $x[e - n] = 1$ and

$$|z|[e - n : 0] = 2^{e+1-n} - x[e - n : 0] = 2^{e+1-n} - 2^{e-n} = 2^{e-n},$$

which implies $|z|[e - n] = 1$ and $|z|[e - n - 1 : 0] = 0$. On the other hand, if $x[e - n - 1 : 0] \neq 0$, then $\tilde{x}[e - n - 1 : 0] < 2^{e-n} - 1$, which implies

$$|z|[e - n] = \tilde{x}[e - n] \neq x[e - n]$$

and

$$|z|[e - n - 1 : 0] = \tilde{x}[e - n - 1] + 1 \neq 0.$$

In either case, (c') again follows easily from (c'').   □

6.6 Denormal Rounding

As we saw in Sect. 5.3, in order for a number x to be representable as a denormal in a format F, it must be $(prec(F) + expo(x) - expo(spn(F)))$-exact. This suggests the following definition of denormal rounding. Note that its arguments include the format itself, both parameters of which are required to determine the precision of the result.

Definition 6.9 If $\mathcal{R}$ is a rounding mode, F is a format, and $x \in \mathbb{R}$, then

$$drnd(x, \mathcal{R}, F) = \mathcal{R}(x, prec(F) + expo(x) - expo(spn(F)).$$

While the conciseness of this formula is appealing, its computation for small x may involve negative-precision rounding, which has been observed to produce unintuitive results. In particular, according to Definitions 5.11 and 5.19,

$$prec(F) + expo(x) - expo(spn(F)) \leq 0 \Leftrightarrow |x| < |spd(F)|.$$

We shall find, however (Lemma 6.101 below), that the results of rounding such tiny values are not unexpected.

Naturally, denormal rounding inherits many of the properties of normal rounding. It is not true in general that $sgn(drnd(x, \mathcal{R}, F)) = sgn(x)$ because a nonzero denormal may be rounded to 0, but we do have the following analog of Lemma 6.82.

Lemma 6.99 *Let F be a format and let $x \in \mathbb{R}$, $|x| < spn(F)$. Let $\mathcal{R}$ be a common rounding mode, and let*

$$\hat{\mathcal{R}} = \begin{cases} RDN & \text{if } \mathcal{R} = RUP \\ RUP & \text{if } \mathcal{R} = RDN \\ \mathcal{R} & \text{otherwise.} \end{cases}$$

Then

$$drnd(-x, \mathcal{R}, F) = -drnd(x, \hat{\mathcal{R}}, F).$$

Proof This follows from Lemmas 6.82 and 4.5.   □

Definition 6.9 admits the following alternative formulation:

Lemma 6.100 *Let F be a format and let $x \in \mathbb{R}$, $|x| < spn(F)$. Let $p = prec(F)$ and $a = sgn(x)spn(F)$. Let $\mathcal{R}$ be a common rounding mode. Then*

$$drnd(x, \mathcal{R}, F) = \mathcal{R}(x + a, p) - a.$$

Proof We first consider the case $x \geq 0$ and apply Lemma 6.90, substituting $a = spn(F)$ for x, x for y, and $p + expo(x) - expo(spn(F))$ for k. Thus,

$$k' = k + expo(spn(F)) - expo(x) = p > 1,$$

and $spn(F)$ is $(k' - 1)$-exact by Lemma 4.10. Since

$$2^{expo(spn(F))} = spn(F) \leq spn(F) + x < 2 \cdot spn(F) = 2^{expo(spn(F))+1},$$

$expo(spn(F) + x) = expo(spn(F))$, and therefore

$$k'' = k + expo(spn(F) + x) - expo(x) = p$$

as well. Thus, we have

$$spn(F) + \mathcal{R}(x, k) = \mathcal{R}(spn(F) + x, k'')$$

and

$$drnd(x, \mathcal{R}, F) = \mathcal{R}(x, p + expo(x) - expo(spn(F)))$$
$$= \mathcal{R}(x, k)$$
$$= \mathcal{R}(spn(F) + x, k'') - spn(F)$$
$$= \mathcal{R}(spn(F) + x, n) - spn(F).$$

The result may be extended to $x < 0$ by invoking Lemmas 6.82 and 6.99: if $\hat{\mathcal{R}}$ is defined as in these lemmas, then

$$drnd(x, \mathcal{R}, p, q) = -drnd(-x, \hat{\mathcal{R}}, p, q)$$
$$= -(\hat{\mathcal{R}}(-x + sgn(-x)spn(F), p) - sgn(-x)spn(F))$$
$$= -\hat{\mathcal{R}}(-x + sgn(-x)spn(F), p) + sgn(-x)spn(F)$$
$$= \mathcal{R}(x + sgn(x)spn(F), p) - sgn(x)spn(F).$$ □

Lemma 6.100 is useful in characterizing the rounding of a denormal smaller than $spd(F)$:

Lemma 6.101 *Let F be a format, $x \in \mathbb{R}$, and $\mathcal{R}$ a common rounding mode.*

(a) If $0 < x < \frac{1}{2}spd(F)$, then

$$drnd(x, \mathcal{R}, F) = \begin{cases} spd(F) & \text{if } \mathcal{R} \in \{RAZ, RUP\} \\ 0 & \text{otherwise;} \end{cases}$$

(b) If $x = \frac{1}{2}spd(F)$, then

$$drnd(x, \mathcal{R}, F) = \begin{cases} spd(F) & \text{if } \mathcal{R} \in \{RAZ, RNA, RUP\} \\ 0 & \text{otherwise;} \end{cases}$$

(c) If $\frac{1}{2}spd(F) < x < spd(F)$, then

$$drnd(x, \mathcal{R}, F) = \begin{cases} 0 & \text{if } \mathcal{R} \in \{RTZ, RDN\} \\ spd(F) & \text{otherwise.} \end{cases}$$

Proof Let $p = prec(F)$, $q = expw(F)$, $a = spn(F) = 2^{2-2^{q-1}}$, and

$$b = fp^+(a, p) = a + 2^{expo(spn(F))+1-p} = a + 2^{3-2^{q-1}-p} = a + spd(F).$$

By Lemma 6.100,

$$drnd(x, \mathcal{R}, F) = \mathcal{R}(a + x, p) - a.$$

Case 1: $\mathcal{R} = RAZ$ or $\mathcal{R} = RUP$
    By Lemma 6.20,

$$\mathcal{R}(a + x, p) = RAZ(a + x, p) \geq a + x > a,$$

and hence, by Lemmas 6.25 and 4.18,

$$RAZ(a + x, p) \geq b.$$

On the other hand, since

$$b = a + spd(F) > a + x,$$

Lemma 6.27 implies $b \geq RAZ(a + x, p)$, and therefore $RAZ(a + x, p) = b$, and

$$drnd(x, \mathcal{R}, F) = \mathcal{R}(a + x, p) - a = b - a = spd(F).$$

Case 2: $\mathcal{R} = RTZ$ or $\mathcal{R} = RDN$

First note that by Lemma 4.21,

$$fp^-(b, p) = fp^-(fp^+(a, p), p) = a.$$

Now by Lemma 6.3,

$$\mathcal{R}(a + x, p) = RTZ(a + x, p) \leq a + x < b,$$

and hence, by Lemmas 6.7 and 4.22,

$$RAZ(a + x, p) \leq a.$$

Furthermore, Lemma 6.9 implies $a \leq RTZ(a + x, p)$, and therefore $RTZ(a + x, p) = a$ and

$$drnd(x, \mathcal{R}, F) = \mathcal{R}(a + x, p) - a = a - a = 0.$$

Case 3: $\mathcal{R} = RNE$ or $\mathcal{R} = RNA$

$\mathcal{R}(a + x, p)$ is either a or b, and hence $drnd(x, \mathcal{R}, F)$ is either 0 or $spn(F)$, respectively.

Since

$$|(a + x) - RAZ(a + x, p)| = b - (a + x) = spd(F) - x$$

and

$$|(a + x) - RTZ(a + x, p)| = |(a + x) - a| = x,$$

the claims (a) and (c) follow from Lemmas 6.43 and 6.59. For the proof of (c), suppose $x = \frac{1}{2} spd(F)$. Then

$$a + x = 2^{2-2^{q-1}} + 2^{2-2^{q-1}-p} = 2^{2-2^{q-1}}(1 + 2^{-p})$$

and $sig(a + x) = 1 + 2^{-p}$. Thus, $a + x$ is $(p + 1)$-exact but not p-exact. The case $\mathcal{R} = RNA$ now follows from Lemma 6.50, and since b is not $(p - 1)$-exact, the case $\mathcal{R} = RNE$ follows from Lemma 6.65. $\qquad\square$

As a consequence of Lemma 6.101(a), for any given rounding mode, two sufficiently small numbers produce the same rounded result.

Corollary 6.102 *Let $x \in \mathbb{R}$ and $y \in \mathbb{R}$. Let F be a format and $\mathcal{R}$ a common rounding mode. If $0 < x < \frac{1}{2} spd(F)$ and $0 < y < \frac{1}{2} spd(F)$, then*

$$drnd(x, \mathcal{R}, F) = drnd(y, \mathcal{R}, F).$$

A denormal is always rounded to a representable number, which may be denormal, 0, or the smallest representable normal.

Lemma 6.103 *Let F be a format and let $x \in \mathbb{R}$, $|x| < spn(F)$. Let $\mathcal{R}$ be a common rounding mode. Then one of the following is true:*

(a) $drnd(x, \mathcal{R}, F) = 0$;
(b) $drnd(x, \mathcal{R}, F) = sgn(x)spn(F)$;
(c) $drnd(x, \mathcal{R}, F)$ is representable as a denormal in F.

Proof By Lemmas 6.99 and 6.101, we may assume $x \geq spd(F)$. Let $p = prec(F)$. Then

$$expo(x) \geq expo(spn(F)) = 2 - bias(F) - p.$$

Since $x < spn(F) = 2^{1-bias(F)}$, $expo(x) \leq -bias(F)$, and

$$2 - p \leq expo(x) + bias(F) \leq 0.$$

Let $d = drnd(x, \mathcal{R}, F)$. By Lemma 6.84(a), d is $(p + expo(x) - expo(spn(F)))$-exact. If $expo(d) = expo(x)$, then d is representable as a denormal. If not, then Lemma 6.87 implies $d = 2^{expo(x)+1}$. In this case, either $expo(x) = -bias(F)$ and

$$d = 2^{1-bias(F)} = spn(F),$$

or $expo(x) + bias(F) < 0$ and

$$expo(d) + bias(F) = 1 + expo(x) + bias(F) \leq 0,$$

which implies that d is representable as a denormal.   □

Lemma 6.104 *If x is representable as a denormal in F and $\mathcal{R}$ is a common rounding mode, then $drnd(x, \mathcal{R}, F) = x$.*

Proof Let $p = prec(F)$. By Definition 5.17, x is $(p + expo(x) - expo(spn(F)))$-exact. Therefore, by Lemma 6.84(b),

$$drnd(x, \mathcal{R}, F) = \mathcal{R}(x, p + expo(x) - expo(spn(F))) = x.$$ □

Lemma 6.105 *Let F be a format and let $x \in \mathbb{R}$, $|x| < spn(F)$. Let a be representable as a denormal in F, and let $\mathcal{R}$ be a common rounding mode.*

(a) If $a \geq x$, then $a \geq drnd(x, \mathcal{R}, F)$.
(b) If $a \leq x$, then $a \leq drnd(x, \mathcal{R}, F)$.

Proof By Lemma 6.101, we may assume $|x| \geq spd(F)$. By Definition 5.17, a is $(prec(F) + expo(x) - expo(spn(F)))$-exact. The claim follows from Lemma 6.84.   □

Naturally, denormal rounding to nearest returns the representable number that is closest to its argument.

Lemma 6.106 *Let F be a format and let $x \in \mathbb{R}$, $|x| < spn(F)$. Let a be representable as a denormal in F. Then*

$$|x - drnd(x, RNE, F)| \leq |x - a|$$

and

$$|x - drnd(x, RNA, F)| \leq |x - a|.$$

Proof By Definition 5.17, a is $(prec(F) + expo(x) - expo(spn(F)))$-exact. The result follows from Lemmas 6.45 and 6.62. □

6.7 Underflow Detection

As discussed in detail in Part IV, an underflow exception may be reported if the result of a floating-point operation is denormal. Let x be the precise mathematical result of an arithmetic operation, to be rounded according to a mode $\mathcal{R}$ and encoded in a format F with precision p, and suppose x lies within the denormal range of F. Let $d = drnd(x, \mathcal{R}, F)$ and $r = \mathcal{R}(x, p)$. The result may be architecturally forced to 0, in which case the underflow exception flag is set. Otherwise, the returned value is d. If $d = x$, then the flag is not set. Note that in this case, $d = r$:

Lemma 6.107 *Let F be a format and let $x \in \mathbb{R}$, $|x| < spn(F)$. Let $\mathcal{R}$ be a common rounding mode. If $drnd(x, \mathcal{R}, F) = x$, then $\mathcal{R}(x, prec(F)) = x$.*

Proof Since $expo(x) < expo(spn(F))$, this follows from Definition 6.9 and Lemma 4.9. □

In the case $d \neq x$, the underflow flag is always set by an Arm instruction (Sect. 14.3), but for x86 instructions (Sect. 12.5), it is set only if r is also denormal. Detecting this condition requires extra logic, since r may be denormal even when d is normal. The following result obviates the need for an x86 implementation to compute r explicitly by providing a simple criterion, based on the number of leading ones of the mantissa of x, for predicting the denormal condition for both d and r. As a corollary, we shall see that whenever d is denormal, so is r.

Lemma 6.108 *Let F be a format with precision $p > 2$, let $S = spn(F)$, and let $x = \pm S \cdot m$, where $0 < m < 1$. Let $r = \mathcal{R}(x, p)$ and $d = drnd(x, \mathcal{R}, F)$, where $\mathcal{R}$ is a common rounding mode.*

(a) If $\mathcal{R} = RUP$ and $x < 0$, $\mathcal{R} = RDN$ and $x > 0$, or $\mathcal{R} = RTZ$, then

$$|r| < S \text{ and } |d| < S.$$

(b) If $\mathcal{R} = RAZ$, $\mathcal{R} = RUP$ and $x > 0$, or $\mathcal{R} = RDN$ and $x < 0$, then

$$|r| < S \Leftrightarrow m \leq 1 - 2^{-p}$$

and

$$|d| < S \Leftrightarrow m \leq 1 - 2^{1-p}.$$

(c) If $\mathcal{R} = RNE$ or $\mathcal{R} = RNA$, then

$$|r| < S \Leftrightarrow m < 1 - 2^{-p-1}$$

and

$$|d| < S \Leftrightarrow m < 1 - 2^{-p}.$$

Proof As a minor simplification, we assume that $x > 0$; the proof is easily extended to the negative case. We may also assume that $m \geq \frac{1}{2}$; otherwise, all the bounds on m hold and $expo(x) < expo(S) - 1$, which implies

$$expo(r) \leq expo(x) + 1 < expo(S),$$

and the same holds for $expo(d)$.

Thus, $expo(m) = -1$, $expo(x) = expo(S) - 1$, and by Definition 6.9,

$$d = \mathcal{R}(x, p + expo(x) - expo(S) = \mathcal{R}(x, p - 1).$$

Now

$$r = \mathcal{R}(x, p) = S \cdot \mathcal{R}(m, p) < S \Leftrightarrow \mathcal{R}(m, p) < 1$$

and

$$d = \mathcal{R}(x, p - 1) = S \cdot \mathcal{R}(m, p - 1) < S \Leftrightarrow \mathcal{R}(m, p - 1) < 1.$$

(a) This follows from Lemma 6.3.

(b) Suppose $\mathcal{R} \in \{RAZ, RUP\}$. If $m \leq 1 - 2^{-p}$, then by Lemma 6.21,

$$\mathcal{R}(m, p) < m + 2^{expo(m)-p+1} = m + 2^{-p} \leq 1.$$

Conversely, suppose $m > 1 - 2^{-p}$. Note that $expo(1 - 2^{-p}) = -1$ and since

$$2^{p-1-(-1)}(1 - 2^{-p}) = 2^p - 1 \in \mathbb{Z},$$

$1 - 2^{-p}$ is p-exact. But $\mathcal{R}(m, p)$ is also p-exact and

$$\mathcal{R}(m, p) \geq m > 1 - 2^{-p}.$$

Thus, by Lemma 4.18,

$$\mathcal{R}(m, p) \geq 1 - 2^{-p} + 2^{-1+1-p} = 1.$$

The proof of the statement about d is the same, with p replaced by $p - 1$.

(c) Suppose $\mathcal{R} \in \{RNE, RNA\}$. If $m < 1 - 2^{-p-1}$, then by Lemmas 6.45 and 6.62,

$$\mathcal{R}(m, p) \leq m + 2^{expo(m)-p} = m + 2^{-p-1} < 1.$$

Conversely, if $m \geq 1 - 2^{-p-1}$, then

$$m + 2^{expo(m)-p} = m + 2^{-p-1} \geq 1 = 2^{expo(m)+1}$$

and by Lemma 6.66, $\mathcal{R}(m, p) = 1$.

The proof of the statement about d is the same, with p replaced by $p - 1$. Note that the hypothesis $p > 2$ is required for the application of Lemma 6.66 to conclude that $\mathcal{R}(m, p - 1) = 1$. □

Corollary 6.109 *Let F be a format and let $x \in \mathbb{R}$, $|x| < S = spn(F)$. Let $\mathcal{R}$ be a common rounding mode. If*

$$|\mathcal{R}(x, prec(F))| = S,$$

then

$$|drnd(x, \mathcal{R}, F)| = S.$$

When rounding is based on constant injection (Lemma 6.95), the unrounded result is generally unavailable, and another method is needed to detect underflow. Let all parameters be as specified in Lemma 6.108, except that we now assume the range of m to be $0 < m < 2$, i.e., $0 < |x| < 2S$. Note that if $S \leq |x| < 2S$, then $expo(x) = expo(S)$, and

$$r = \mathcal{R}(x, p) = \mathcal{R}(x, expo(x) - expo(S) + p) = d.$$

Thus, d is the returned value in all cases. We shall also assume that for some $k \geq p + 2$, $M = 2^{k-1}m \in \mathbb{N}$. Thus, M is a k-bit vector and $x = 2^{1-k}S \cdot M$. (We may assume that M has been derived by appending a sticky bit to a truncated result.) Note that

$$|x| < S \Leftrightarrow M < 2^{k-1},$$

and by Lemmas 6.82 and 6.89,

$$|r| < S \Leftrightarrow \mathcal{R}(M, p) < 2^{k-1}.$$

We do not assume direct access to M but rather to the injected sum $\Sigma = M + \mathcal{C}$, where $\mathcal{C}$ is the rounding constant of Lemma 6.95. In order to address both Arm and x86 rounding conventions, we must detect inexactness ($d \neq x$) and determine the normality of both x and r by examining the bit vector Σ. The following lemma provides efficient solutions to these problems as well as a formula relating d and Σ.

Lemma 6.110 *Let F be a format with precision p, let $S = \mathrm{spn}(F)$, and let $x = \pm S \cdot m$, where $0 < m < 2$. Let $r = \mathcal{R}(x, p)$ and $d = \mathrm{drnd}(x, \mathcal{R}, F)$, where $\mathcal{R}$ is a common rounding mode. Let $k \in \mathbb{N}$, $k \geq p + 2$, and assume that $M = 2^{k-1}m \in \mathbb{N}$. Let $\Sigma = M + \mathcal{C}$, where*

$$\mathcal{C} = \begin{cases} 2^{k-p-1} & \text{if } \mathcal{R} = RNE \text{ or } \mathcal{R} = RNA \\ 2^{k-p} - 1 & \text{if } \mathcal{R} = RAZ, \ \mathcal{R} = RUP \text{ and } x > 0, \text{ or } \mathcal{R} = RDN \text{ and } x < 0 \\ 0 & \text{if } \mathcal{R} = RTZ, \ \mathcal{R} = RUP \text{ and } x < 0, \text{ or } \mathcal{R} = RDN \text{ and } x > 0. \end{cases}$$

Let $\overline{\Sigma} = \Sigma[k : k - p]$, $\underline{\Sigma} = \Sigma[k - p - 1 : 0]$, $G = \Sigma[k - p - 1]$, and $R = \Sigma[k - p - 2]$.

(a) $|d| = \begin{cases} 2^{2-p}S \cdot \overline{\Sigma}[p : 1] & \text{if } \mathcal{R} = RNE \text{ and } \underline{\Sigma} = 0 \\ 2^{1-p}S \cdot \overline{\Sigma} & \text{otherwise;} \end{cases}$

(b) $x = d \Leftrightarrow \underline{\Sigma} = \mathcal{C}$;

(c) *If $\overline{\Sigma} > 2^{p-1}$, then $|x| \geq S$ and $|r| \geq S$;*

(d) *If $\overline{\Sigma} < 2^{p-1}$, then $|x| < S$ and $|r| < S$;*

(e) *If $\overline{\Sigma} = 2^{p-1}$ and $\mathcal{R} = RTZ$, $\mathcal{R} = RUP$ and $x < 0$, or $\mathcal{R} = RDN$ and $x > 0$, then $|x| \geq S$ and $|r| \geq S$.*

(f) *If $\overline{\Sigma} = 2^{p-1}$ and $\mathcal{R} = RAZ$, $\mathcal{R} = RUP$ and $x > 0$, or $\mathcal{R} = RDN$ and $x < 0$, then*

  (i) $|x| < S \Leftrightarrow \underline{\Sigma} \neq \mathcal{C}$;

  (ii) $|r| < S \Leftrightarrow G = 0$;

(g) *If $\overline{\Sigma} = 2^{p-1}$ and $\mathcal{R} = RNE$ or $\mathcal{R} = RNA$, then*

  (i) $|x| < S \Leftrightarrow G = 0$;

  (ii) $|r| < S \Leftrightarrow G = R = 0$;

Proof As in the proof of Lemma 6.108, we shall assume $x > 0$. Note that since $x = Sm = 2^{1-k}SM$,

$$r = \mathcal{R}(2^{1-k}SM, p) = 2^{1-k}S\mathcal{R}(M, p),$$

$$r < S \Leftrightarrow \mathcal{R}(M, p) < 2^{k-1},$$

$$\mathrm{expo}(x) = \mathrm{expo}(S) + 1 - k + \mathrm{expo}(M),$$

and

$$d = \mathcal{R}(x, expo(x) - expo(S) + p) = 2^{1-k} S \cdot \mathcal{R}(M, expo(M) + p + 1 - k).$$

(a) The case $expo(M) \geq k - p$ is a direct consequence of Lemma 6.96(a) under the substitution $n = expo(M) + p + 1 - k$.

Thus, we may assume that $expo(M) < k - p$. This implies $M < 2^{k-p}$, $\overline{\Sigma} \leq 1$,

$$x = 2^{1-k} S M < 2^{1-k} S 2^{k-p} = 2^{1-p} S = spd(F),$$

and therefore, d is either 0 or $spd(F)$.

Suppose $\mathcal{R} = RNE$ and $\underline{\Sigma} = 0$. Then $M = 2^{k-p-1}$, $x = \frac{1}{2} spd(F)$, $\Sigma = 2^{k-p}$, and by Lemma 6.101,

$$d = 0 = 2^{2-p} S \cdot \overline{\Sigma}[p : 1].$$

In all other cases, we must show that $d = 2^{1-p} S \cdot \overline{\Sigma}$, where

$$2^{1-p} S \cdot \overline{\Sigma} = spd(F) \cdot \overline{\Sigma} = \begin{cases} spd(F) & \text{if } \overline{\Sigma} = 1 \\ 0 & \text{if } \overline{\Sigma} = 0, \end{cases}$$

i.e., we must show that $d = spd(F)$ iff $\overline{\Sigma} = 1$. This is easily verified in all cases by similarly referring to Lemma 6.101:

- If $\mathcal{R} = RNE$ and $\underline{\Sigma} > 0$, then

$$\overline{\Sigma} = 1 \Leftrightarrow M > 2^{k-p-1} \Leftrightarrow x > \frac{1}{2} spd(F) \Leftrightarrow d = spd(F);$$

- If $\mathcal{R} = RNA$, then

$$\overline{\Sigma} = 1 \Leftrightarrow M \geq 2^{k-p-1} \Leftrightarrow x \geq \frac{1}{2} spd(F) \Leftrightarrow d = spd(F);$$

- If $\mathcal{R} = RAZ$ or $\mathcal{R} = RUP$, then since $M > 0$, $\overline{\Sigma} = 1$ and $d = spd(F)$;
- If $\mathcal{R} = RTZ$ or $\mathcal{R} = RDN$, then $\overline{\Sigma} = 0$ and $d = 0$.

(b) The case $expo(M) \geq k - p$ is a direct consequence of Lemma 6.96(b) under the substitution $n = expo(M) + p + 1 - k$. Suppose $expo(M) < k - p$. Then $0 < x < spd(F)$ and $d \neq x$. Since

$$M \bmod 2^{k-p} = M \neq 0,$$

$$\underline{\Sigma} = (M + C) \bmod 2^{k-p} \neq C.$$

(c) Since $\overline{\Sigma} \in \mathbb{Z}$,

$$
\begin{aligned}
\overline{\Sigma} > 2^{p-1} &\Rightarrow \overline{\Sigma} \geq 2^{p-1} + 1 \\
&\Rightarrow \Sigma = M + C \geq (2^{p-1} + 1)2^{k-p} = 2^{k-1} + 2^{k-p} \\
&\Rightarrow M \geq 2^{k-1} + 2k - p - C > 2^{k-1} \\
&\Rightarrow x > S \\
&\Rightarrow r \geq S.
\end{aligned}
$$

(d) Since

$$
M \leq \Sigma \leq 2^{k-p}(2^{p-1} - 1)2^{k-p} + \underline{\Sigma} < (2^{p-1} - 1)2^{k-p} + 2^{k-p} = 2^{k-1},
$$

$x < S$. It follows from (a) that $d \leq 2^{1-k} S \Sigma < S$, and by Corollary 6.109, $r < S$ as well.

(e) Since $C = 0$, $M = \Sigma \geq 2^{k-1}$, which implies $x \geq S$ and $r \geq S$.

(f) Since $\Sigma = M + C = 2^{k-p} M[k-1 : k-p] + M[k-p-1 : 0] + C$,

$$
2^{p-1} = \overline{\Sigma} = \lfloor 2^{p-k} \Sigma \rfloor = \begin{cases} M[k-1 : k-p] & \text{if } M[k-p-1 : 0] + C < 2^{k-p} \\ M[k-1 : k-p] + 1 & \text{if } M[k-p-1 : 0] + C \geq 2^{k-p}. \end{cases}
$$

In this case, $C = 2^{k-p} - 1$, and hence,

$$
\begin{aligned}
x < S &\Leftrightarrow M < 2^{k-1} \\
&\Leftrightarrow M[k-1 : k-p] < 2^{p-1} \\
&\Leftrightarrow M[k-p-1 : 0] + C \geq 2^{k-p} \\
&\Leftrightarrow M[k-p-1 : 0] \neq 0 \\
&\Leftrightarrow \underline{\Sigma} \neq C.
\end{aligned}
$$

To prove (ii), first suppose $x \geq S$. Then $r \geq S$, $\underline{\Sigma} = C$, and

$$
G = \underline{\Sigma}[k-p-1] = C[k-p-1] = 1.
$$

Thus, we may assume $x < S$, which implies $M[k-1 : k-p] = 2^{p-1} - 1$, $M[k-p-1 : 0] > 0$, and

$$
M = 2^{k-p}(2^{p-1} - 1) + M[k-p-1 : 0] = 2^{k-1} - 2^{k-p} + M[k-p-1 : 0].
$$

Let $E = 2^{k-1} - 2^{k-p-1}$. Since $expo(E) = k - 2$,

$$
2^{p-1-expo(E)} E = 2^{p-k+1} E = 2^p - 1 \in \mathbb{Z},
$$

i.e., E is p-exact. Furthermore,

$$fp^+(E, p) = E + 2^{expo(E)+1-p} = E + 2^{k-p-1} = 2^{k-1}.$$

Thus, if $M \leq E$, then $RUP(M, p) \leq E < 2^{k-1}$, and if $M > E$, then $RUP(M, p) \geq 2^{k-1}$. We must show, therefore, that $M \leq E$ iff $G = 0$.

Suppose $M[k-p-1] = 0$. Then $M < 2^{k-1} - 2^{k-p} + 2^{k-p-1} = E$ and

$$2^{k-p} \leq M[k-p-1:0] + C = M[k-p-1:0] + 2^{k-p} - 1 < 2^{k-p-1} + 2^{k-p},$$

which implies

$$\Sigma = (M[k-p-1:0] + C) \bmod 2^{k-p} < 2^{k-p-1}$$

and $G = \Sigma[k - p - 1] = 0$.

On the other hand, suppose $M[k-p-1] = 1$. Then

$$M = 2^{k-1} - 2^{k-p} + 2^{k-p-1} + M[k-p-2:0] = E + M[k-p-2:0].$$

Furthermore,

$$2^{k-p} \leq M[k-p-1:0] + C < 2^{k-p} + 2^{k-p},$$

which implies

$$\begin{aligned}
\Sigma &= (M[k-p-1:0] + C) \bmod 2^{k-p} \\
&= M[k-p-1:0] + C - 2^{k-p} \\
&= (2^{k-p-1} + M[k-p-2:0]) + (2^{k-p} - 1) - 2^{k-p} \\
&= 2^{k-p-1} + M[k-p-2:0] - 1.
\end{aligned}$$

Thus,

$$\begin{aligned}
G = 1 &\Leftrightarrow \Sigma \geq 2^{k-p-1} \\
&\Leftrightarrow M[k - p - 2:0] > 0 \\
&\Leftrightarrow M > E.
\end{aligned}$$

(g) Arguing as in the preceding case, we have

$$x < S \Leftrightarrow M[k-p-1:0] + C \geq 2^{k-p}.$$

In this case, $C = 2^{k-p-1}$, and this condition is equivalent to $M[k-p-1] = 1$, or $G = 0$.

If $x \geq S$, then $r \geq S$ and $G = 1$. Therefore, for the proof of (ii), we may assume that $x < S$ and $G = 0$ and need only show that $r < S$ iff $R = 0$. Thus, $M[k-p-1] = 1$, and once again, $M[k-1 : k-p] = 2^{p-1} - 1$. Let E be defined as in the preceding case. Then

$$M = 2^{k-p}(2^{p-1} - 1) + 2^{k-p-1} + 2^{k-p-2}M[k-p-2] + M[k-p-3 : 0]$$
$$= E + 2^{k-p-2}R + M[k-p-3 : 0].$$

Thus,

$$R = 0 \Leftrightarrow M < E + 2^{k-p-2} = E + 2^{expo(M)-p},$$

and the claim follows from Lemmas 6.46, 6.61, and 6.66. $\qquad\square$

Chapter 7
IEEE-Compliant Square Root

This chapter addresses the problem of formally specifying and verifying the correctness of a square root operation in a logic based on rational arithmetic, such as ACL2, in which the square root function cannot be explicitly defined. This material should be useful to ACL2 users and may be of general academic interest, but it is not relevant to the sequel with the exception of Chap. 20 and, in particular, is not required for a reading of Chap. 10.

Many of the preceding results are propositions pertaining to real variables, which are formalized by ACL2 events in which these variables are restricted to the rational domain. Many of the lemmas of this chapter similarly apply to arbitrary real numbers, but in light of our present focus, these results are formulated to correspond more closely with their formal versions. Apart from the attendant informal discussion, the lemmas themselves contain no references to the real numbers or the square root function.

Establishing IEEE compliance of a floating-point square root module entails proving that the final value r computed for a given radicand x, rounding mode $\mathcal{R}$, and precision n satisfies

$$r = \mathcal{R}(\sqrt{x}, n). \tag{7.1}$$

We would like to formulate a proposition of rational arithmetic that is transparently equivalent to (7.1). This requirement is satisfied by the following criterion:

For all positive rational numbers ℓ and h, if $\ell^2 \leq x \leq h^2$, then

$$\mathcal{R}(\ell, n) \leq r \leq \mathcal{R}(h, n). \tag{7.2}$$

Obviously, the monotonicity of rounding (Lemma 6.88) and of the square root ensure that (7.1) implies (7.2). On the other hand, suppose that (7.2) holds. According to Lemma 6.94, either $\sqrt{x}$ is $(n + 1)$-exact (and, in particular, rational) or for some $\epsilon > 0$, $\mathcal{R}(y, n) = \mathcal{R}(\sqrt{x}, n)$ for all y satisfying $|y - \sqrt{x}| < \epsilon$.

© The Author(s), under exclusive license to Springer Nature Switzerland AG 2022
D. M. Russinoff, *Formal Verification of Floating-Point Hardware Design*,
https://doi.org/10.1007/978-3-030-87181-9_7

In either case, there exist $\ell \in \mathbb{Q}$ and $h \in \mathbb{Q}$ such that $\ell \leq \sqrt{x} \leq h$ and $\mathcal{R}(\ell, n) = \mathcal{R}(\sqrt{x}, n) = \mathcal{R}(h, n)$. Since $\ell^2 \leq x \leq h^2$,

$$\mathcal{R}(\sqrt{x}, n) = \mathcal{R}(\ell, n) \leq r \leq \mathcal{R}(h, n) = \mathcal{R}(\sqrt{x}, n),$$

and hence $r = \mathcal{R}(\sqrt{x}, n)$.

Thus, we would like to prove formally that (7.2) is satisfied by the value r computed by a square root module of interest. For this purpose, it will be useful to have a function that computes, for given x and n, a rational number q that satisfies

$$\mathcal{R}(q, n) = \mathcal{R}(\sqrt{x}, n). \tag{7.3}$$

We shall define a conceptually simple (albeit computationally horrendous) rational function $\sqrt[(k)]{x}$ that serves this need. The definition is motivated by Lemma 6.92, which guarantees that if we are able to arrange that

$$\sqrt[(k)]{x} = RTO(\sqrt{x}, k), \tag{7.4}$$

where $k \geq n + 2$, then (7.3) holds for $q = \sqrt[(k)]{x}$. Of course, (7.4) will not be our formal definition of $\sqrt[(k)]{x}$, nor shall we prove any instance of (7.3). However, after formulating the definition, we shall prove the following (Lemma 7.17):

For all positive rationals ℓ and h and positive integers k and n, if $\ell^2 \leq x \leq h^2$ and $k \geq n + 2$, then

$$\mathcal{R}(\ell, n) \leq \mathcal{R}(\sqrt[(k)]{x}, n) \leq \mathcal{R}(h, n). \tag{7.5}$$

Thus, in order to prove that a computed value r satisfies (7.2), it will suffice to show that $r = \mathcal{R}(\sqrt[(k)]{x}, n)$ for some $k \geq n + 2$. This is the strategy followed in the correctness proof of Chap. 20.

7.1 Truncated Square Root

The first step toward the definition of $\sqrt[(n)]{x}$ is the following recursive function, the name of which is motivated by the unproven observation that for $\frac{1}{4} \leq x < 1$,

$$rtz\text{-}sqrt(x, n) = RTZ(\sqrt{x}, n).$$

Definition 7.1 Let $x \in \mathbb{R}$ and $n \in \mathbb{N}$. If $n = 0$, then $rtz\text{-}sqrt(x, n) = 0$, and if $n > 0$ and $z = rtz\text{-}sqrt(x, n - 1)$, then

$$rtz\text{-}sqrt(x, n) = \begin{cases} z & \text{if } (z + 2^{-n})^2 > x \\ z + 2^{-n} & \text{if } (z + 2^{-n})^2 \leq x. \end{cases}$$

Lemma 7.1 *Let $x \in \mathbb{Q}$ and $n \in \mathbb{Z}^+$. If $x \geq \frac{1}{4}$, then*

$$\frac{1}{2} \leq \textit{rtz-sqrt}(x, n) \leq 1 - 2^{-n}.$$

Proof If $n = 1$, then $\textit{rtz-sqrt}(x, n) = \frac{1}{2}$ and the claim is trivial. Proceeding by induction, let $n > 1$, $z = \textit{rtz-sqrt}(x, n - 1)$, and $w = \textit{rtz-sqrt}((x, n)$, and assume that $\frac{1}{2} \leq z \leq 1 - 2^{1-n}$. If $w = z$, the claim follows trivially; otherwise, $w = z + 2^{-n}$, and

$$\frac{1}{2} \leq z < w = z + 2^{-n} \leq (1 - 2^{1-n}) + 2^{-n} = 1 - 2^{-n}. \qquad \square$$

Corollary 7.2 *Let $x \in \mathbb{Q}$ $n \in \mathbb{Z}^+$. If $x \geq \frac{1}{4}$, then $\textit{expo}(\textit{rtz-sqrt}(x, n)) = -1$.*

Lemma 7.3 *Let $x \in \mathbb{Q}$ and $n \in \mathbb{Z}^+$. If $x \geq \frac{1}{4}$, then $\textit{rtz-sqrt}(x, n)$ is n-exact.*

Proof The claim is trivial for $n = 0$. Let $n > 1$, $z = \textit{rtz-sqrt}(x, n - 1)$, and $w = \textit{rtz}(x, n)$, and assume that z is $(n - 1)$-exact, i.e., $2^{n-1}z \in \mathbb{Z}$. Then either $w = z$ and $2^n w = 2(2^{n-1}z) \in \mathbb{Z}$ or

$$2^n w = 2^n(z + 2^{-n}) = 2(2^{n-1}z) + 1 \in \mathbb{Z}. \qquad \square$$

Lemma 7.4 *Let $x \in \mathbb{Q}$ and $n \in \mathbb{N}$. Assume that $\frac{1}{4} \leq x < 1$ and let $w = \textit{rtz-sqrt}(x, n)$. Then $w^2 \leq x < (w + 2^{-n})^2$.*

Proof The claim is trivial for $n = 0$. Let $n > 0$ and $z = \textit{rtz-sqrt}(x, n - 1)$, and assume that $z^2 \leq x < (z + 2^{1-n})^2$. If $x < (z + 2^{-n})^2$ and $w = z$, the claim is trivial. Otherwise, $x \geq (z + 2^{-n})^2$, $w = z + 2^{-n}$, and

$$w^2 = (z + 2^{-n}) \leq x \leq (z + 2^{1-n})^2 = (w + 2^{-n})^2. \qquad \square$$

According to the next lemma, $\textit{rtz-sqrt}(x, n)$ is uniquely determined by the above properties.

Lemma 7.5 *Let $x \in \mathbb{Q}$, $a \in \mathbb{Q}$, and $n \in \mathbb{Z}^+$. Assume that $\frac{1}{4} \leq x < 1$ and $a \geq \frac{1}{2}$. If a is n-exact and $a^2 \leq x < (a + 2^{-n})^2$, then $a = \textit{rtz-sqrt}(x, n)$.*

Proof Let $w = \textit{rtz-sqrt}(x, n)$. If $a < w$, then by Lemma 4.18.

$$w \geq \textit{fp}^+(a, n) = a + 2^{\textit{expo}(a)+1-n} \geq a + 2^{-n},$$

which implies $w^2 \geq (a + 2^{-n})^2 > x$, contradicting Lemma 7.4. But if $a > w$, then

$$a \geq \textit{fp}^+(w, n) = w + 2^{-n},$$

and by Lemma 7.4, $a^2 \geq (w + 2^{-n})^2 > x$, contradicting our hypothesis. $\qquad \square$

We have the following variation of Lemma 6.12.

Lemma 7.6 *Let $x \in \mathbb{Q}$, $m \in \mathbb{Z}^+$, and $n \in \mathbb{Z}^+$. If $x \geq \frac{1}{4}$ and $n \geq m$, then*

$$RTZ(\textit{rtz-sqrt}(x, n), m) = \textit{rtz-sqrt}(x, m).$$

Proof The case $m = n$ follows from Lemmas 6.8 and 7.3. We proceed by induction on $n - m$. Let $1 < m \leq n$ and assume that $RTZ(\textit{rtz-sqrt}(x, n), m) = \textit{rtz-sqrt}(x, m)$. Then by Lemma 6.12,

$$RTZ(\textit{rtz-sqrt}(x, n), m - 1) = RTZ(RTZ(\textit{rtz-sqrt}(x, n), m), m - 1)$$

$$= RTZ(\textit{rtz-sqrt}(x, m), m - 1),$$

and we need only show that $RTZ(\textit{rtz-sqrt}(x, m), m - 1) = \textit{rtz-sqrt}(x, m - 1)$. Let $w = \textit{rtz-sqrt}(x, m)$ and $z = \textit{rtz-sqrt}(x, m - 1)$. If $w = z$, then w is $(m - 1)$-exact by Lemma 7.3 and $RTZ(w, m - 1) = w = z$ by Lemma 6.8. But otherwise, $w = z + 2^{-m}$, $2^{m-1}z \in \mathbb{Z}$ by Corollary 7.2, and hence, by Definition 6.2,

$$RTZ(w, m - 1) = 2^{1-m} \lfloor 2^{m-1} w \rfloor$$

$$= 2^{1-m} \lfloor 2^{m-1}(z + 2^{-m}) \rfloor$$

$$= 2^{1-m} \lfloor 2^{m-1} z + 1 \rfloor$$

$$= 2^{1-m} (2^{m-1} z)$$

$$= z. \qquad \square$$

7.2 Odd-Rounded Square Root

The name of the following function is motivated by the (once again unproven) observation that for $\frac{1}{4} \leq x < 1$,

$$\textit{rto-sqrt}(x, n) = RTO(\sqrt{x}, n).$$

Definition 7.2 Let $x \in \mathbb{R}$ and $n \in \mathbb{Z}^+$, and let $z = \textit{rtz-sqrt}(x, n - 1)$. Then

$$\textit{rto-sqrt}(x, n) = \begin{cases} z & \text{if } x \leq z^2 \\ z + 2^{-n} & \text{if } x > z^2. \end{cases}$$

Lemma 7.7 *Let $x \in \mathbb{Q}$ and $n \in \mathbb{Z}^+$. If $x \geq \frac{1}{4}$, then*

$$\frac{1}{2} \leq \textit{rto-sqrt}(x, n) \leq 1 - 2^{-n}.$$

Proof If $n = 1$, then $rto\text{-}sqrt(x, n) = \frac{1}{2}$ and the claim is trivial. Let $n > 1$ and $z = rtz\text{-}sqrt(x, n - 1)$. By Lemma 7.1, $\frac{1}{2} \le z < 1$, which implies

$$\frac{1}{2} \le z \le rto\text{-}sqrt(x, n) \le z + 2^{-n} \le (1 - 2^{1-n}) + 2^{-n} = 1 - 2^{-n}. \qquad \square$$

Corollary 7.8 *Let* $x \in \mathbb{Q}$ $n \in \mathbb{Z}^+$. *If* $x \ge \frac{1}{4}$, *then* $expo(rto\text{-}sqrt(x, n)) = -1$.

Lemma 7.9 *Let* $x \in \mathbb{Q}$ *and* $n \in \mathbb{Z}^+$. *If* $x \ge \frac{1}{4}$, *then* $rto\text{-}sqrt(x, n)$ *is n-exact.*

Proof Let $z = rtz\text{-}sqrt(x, n-1)$ and $w = rto\text{-}sqrt(x, n)$. By Corollaries 7.2 and 7.8, $expo(z) = expo(w) = -1$. By Lemma 7.3, $2^{n-1}z \in \mathbb{Z}$. Consequently, since w is either z or $z + 2^{-n}$, $2^n w \in \mathbb{Z}$, i.e., w is n-exact. $\qquad \square$

Lemma 7.10 *Let* $x \in \mathbb{Q}$, $m \in \mathbb{Z}^+$, *and* $n \in \mathbb{N}$. *Assume that* $\frac{1}{4} \le x < 1$ *and* $2 \le n \le m$. *Then*

$$rto(rto\text{-}sqrt(x, m), n) = rto\text{-}sqrt(x, n).$$

Proof We first consider the case $n = m - 1$. Let $z_1 = rtz\text{-}sqrt(x, m - 2)$, $w_1 = rto\text{-}sqrt(x, m - 1)$, $z_2 = rtz\text{-}sqrt(x, m - 1)$, and $w_2 = rto\text{-}sqrt(x, m)$. We shall show that $rto(w_2, m - 1) = w_1$. Note that by Lemmas 7.2, 7.6, and 7.4, $\frac{1}{2} \le z_1^2 \le z_2^2 \le x$.

Case 1: $z_1 = z_2$ and $z_2^2 < x$. *Case 1*: $z_1 = z_2$ and $z_2^2 = x$.
   $z_1 = w_1 = z_2 = w_2$. Since w_2 is $(m - 1)$-exact, Lemma 6.70 implies

$$rto(w_2, m - 1) = w_2 = w_1.$$

Case 2: $z_1 = z_2$ and $z_2^2 < x$.
   Since w_1 is $(m - 2)$-exact, Lemma 4.18 implies that $w_1 = z_1 + 2^{1-n}$ is not $(m - 2)$-exact; similarly, since w_2 is $(m - 1)$-exact, $w_2 = z_2 + 2^{-m}$ is not $(m - 1)$-exact. Therefore,

$$RTO(w_2, m - 1) = RTZ(w_2, m - 2) + 2^{1-n} = z_1 + 2^{1-m} = w_1.$$

Case 3: $z_1 < z_2$ and $z_2^2 = x$.
   By Lemma 7.3, z_1 is $(m - 1)$-exact and z_2 is $(m - 2)$-exact. By Lemma 7.6, $z_1 = RTZ(z_2, m - 2) < z_2$, and it follows from Lemma 6.9 that $z_2 = z_1 + 2^{1-m}$. Thus, $w_1 = z_2 = w_2$ and by Lemma 6.70, $rto(w_2, m - 1) = w_2 = w_1$.

Case 4: $z_1 < z_2$ and $z_2^2 < x$.
   In this case, $w_1 = z_2 = z_1 + 2^{1-m}$ and $w_2 = z_2 + 2^{-m} = z_1 + 2^{1-m} + 2^{-m}$, which is not $(m - 2)$-exact. Thus,

$$rto(w_2, m - 1) = RTZ(w_2, m - 2) + 2^{1-m} = z_1 + 2^{1-m} = w_1.$$

The proof is completed by induction on m. If $m > n$, then by Lemma 6.76,

$$rto(rto\text{-}sqrt(x, m), n) = rto(rto(rto\text{-}sqrt(x, m), m - 1), n)$$

$$= rto(rto\text{-}sqrt(x, m - 1), n)$$

$$= rto\text{-}sqrt(x, n). \qquad \square$$

Lemma 7.11 Let $x \in \mathbb{Q}$, $\ell \in \mathbb{Q}$, $h \in \mathbb{Q}$, and $n \in \mathbb{Z}^+$. Assume that $\frac{1}{4} \le x < 1$, $h > 0$, and $\ell^2 \le x \le h^2$. Then

$$rto(\ell, n) \le rto\text{-}sqrt(x, n) \le rto(h, n).$$

Proof Let $z = rtz\text{-}sqrt(x, n - 1)$ and $w = rto\text{-}sqrt(x, n)$. Suppose $z^2 = x$. Then $w = z$, $\ell^2 \le x = w^2$, and hence $\ell \le w$. By Lemmas 6.74, 6.70, and 7.3,

$$rto(\ell, n) \le rto(w, n) = w.$$

Thus, we may assume $z^2 < x$ and $w = z + 2^{-n}$. By Lemma 7.4, $x < (z + 2^{1-n})^2$, and hence $\ell < z + 2^{1-n} = fp^+(z, n - 1)$. It follows from Lemmas 6.3, 6.7, and 4.18 that $RTZ(\ell, n - 1) \le z$. Therefore,

$$rto(\ell, n) \le RTZ(\ell, n - 1) + 2^{1+expo(\ell)-n} \le z + 2^{-n} = w.$$

To prove the second inequality, we note that if $h \ge w$, then by Lemmas 6.74, 6.70, and 7.3,

$$rto(h, n) \ge rto(w, n) = w.$$

Therefore, we may assume that $h < w$. If $z^2 = x$, then $w = z$, $h^2 \ge x = w^2$, and $h \ge w$. Thus, by Lemma 7.4, $z^2 < x$, which implies $z < h$. It follows from Lemma 4.18 that $RTZ(h, n - 1) = z \ne h$. Thus, h is not $(n - 1)$-exact, and

$$rto(h, n) = RTZ(h, n - 1) + 2^{1+expo(h)-n} = z + 2^{-n} = w. \qquad \square$$

Lemma 7.12 Let $x \in \mathbb{Q}$, $q \in \mathbb{Q}$, and $n \in \mathbb{Z}^+$. Assume that $\frac{1}{4} \le x < 1$, $q > 0$, and q is $(n - 1)$-exact. Then

(a) $q^2 < x \Leftrightarrow q < rto\text{-}sqrt(x, n)$;
(b) $q^2 > x \Leftrightarrow q > rto\text{-}sqrt(x, n)$.

Proof Let $z = rtz\text{-}sqrt(x, n - 1)$ and $w = rto\text{-}sqrt(x, n)$. If $q^2 > x$, then by Lemma 7.4, $q^2 > z^2$, so that $q > z$ and by Lemma 4.18,

$$q \ge z + 2^{1-n} > z + 2^{-n} \ge w.$$

We may assume, therefore, that $q^2 \le x < (z + 2^{1-n})^2$, and hence $q < z + 2^{1-n}$. By Lemma 4.18, $q \le z \le w$. We must show that $q^2 < x$ iff $q < w$. If $q < z$, then $q^2 < z^2 \le x$ and $q < w$. If $q = z = x^2$, then $q = z = w$. Finally, if $q = z < x^2$, then $q = z < z + 2^{-n} = w$. $\square$

7.3   IEEE-Rounded Square Root

The desired approximation function is a simple generalization of *rto-sqrt* to arbitrary positive rationals:

Definition 7.3 Let $x \in \mathbb{Q}$ and $n \in \mathbb{Z}^+$ with $x > 0$. Let $e = \left\lfloor \frac{expo(x)}{2} \right\rfloor + 1$. Then

$$\sqrt[(n)]{x} = 2^e \, rto\text{-}sqrt(2^{-2e}x, n).$$

Lemma 7.13 Let $x \in \mathbb{Q}$, $x > 0$, $e = \left\lfloor \frac{expo(x)}{2} \right\rfloor + 1$, and $x' = 2^{-2e}x$. Then $\frac{1}{4} \le x' < 1$.

Proof Since

$$\frac{expo(x)}{2} - 1 < \left\lfloor \frac{expo(x)}{2} \right\rfloor \le \frac{expo(x)}{2},$$

we have

$$expo(x) < 2\left\lfloor \frac{expo(x)}{2} \right\rfloor + 2 = 2e$$

and

$$expo(x) \ge 2\left\lfloor \frac{expo(x)}{2} \right\rfloor = 2e - 2.$$

By Lemma 4.6, $-2 \le expo(x') < 0$ and the lemma follows. $\square$

Lemma 7.14 Let $x \in \mathbb{Q}$ and $n \in \mathbb{Z}^+$. If $\frac{1}{4} \le x < 1$, then

$$\sqrt[(n)]{x} = rto\text{-}sqrt(x, n).$$

Proof Since $expo(x) \in \{-2, -1\}$, $\left\lfloor \frac{expo(x)}{2} \right\rfloor = -1$ and $e = 0$. $\square$

Lemma 7.15 Let $x \in \mathbb{Q}$ and $n \in \mathbb{Z}^+$ with $x > 0$. For all $k \in \mathbb{Z}$,

$$\sqrt[(n)]{2^{2k}x} = 2^k \sqrt[(n)]{x}.$$

Proof Let $x' = 2^{2k}x$, $e = \left\lfloor \frac{expo(x)}{2} \right\rfloor + 1$, and

$$e' = \left\lfloor \frac{expo(x')}{2} \right\rfloor + 1 = \left\lfloor \frac{expo(x)}{2} + k \right\rfloor + 1 = e + k.$$

Then

$$\begin{aligned}
\sqrt[(n)]{x'} &= 2^{e'} \textit{rto-sqrt}(2^{-2e'}x', n) \\
&= 2^{e+k} \textit{rto-sqrt}(2^{-2(e+k)}2^{2k}x, n) \\
&= 2^k \left(2^e \textit{rto-sqrt}(2^{2e}x, n) \right) \\
&= 2^k \sqrt[(n)]{x}.
\end{aligned}$$

$\square$

Lemma 7.16 *Let* $x \in \mathbb{Q}$, $k \in \mathbb{N}$, $m_1 \in \mathbb{N}$, *and* $n_2 \in \mathbb{N}$ *with* $x > 0$ *and* $2 < k+2 \leq m \leq n$, *and let* $\mathcal{R}$ *be a common rounding mode. Then*

$$\mathcal{R}(\sqrt[(m)]{x}, k) = \mathcal{R}(\sqrt[(n)]{x}, k).$$

Proof Let $e = \left\lfloor \frac{expo(x)}{2} \right\rfloor + 1$. By Definition 7.3 and Lemmas 6.89, 7.10, and 6.92,

$$\begin{aligned}
\mathcal{R}(\sqrt[(m)]{x}, k) &= \mathcal{R}(2^e \textit{rto-sqrt}(2^{-2e}x, m), k) \\
&= 2^e \mathcal{R}(\textit{rto-sqrt}(2^{-2e}x, m), k) \\
&= 2^e \mathcal{R}(\textit{rto}(\textit{rto-sqrt}(2^{-2e}x, n), m), k) \\
&= 2^e \mathcal{R}(\textit{rto-sqrt}(2^{-2e}x, n), k) \\
&= \mathcal{R}(2^e \textit{rto-sqrt}(2^{-2e}x, n), k) \\
&= \mathcal{R}(\sqrt[(n)]{x}, k).
\end{aligned}$$

$\square$

The next lemma establishes the critical property of $\sqrt[(k)]{x}$ discussed at the beginning of this chapter.

Lemma 7.17 *Let* $x \in \mathbb{Q}$, $\ell \in \mathbb{Q}$, $h \in \mathbb{Q}$ $n \in \mathbb{Z}^+$, *and* $k \in \mathbb{Z}^+$. *Assume that* $x > 0$ $h > 0$, $k \geq n + 2$, *and* $\ell^2 \leq x \leq h^2$. *Let* $\mathcal{R}$ *be a common rounding mode. Then*

$$\mathcal{R}(\ell, n) \leq \mathcal{R}(\sqrt[(k)]{x}, n) \leq \mathcal{R}(h, n).$$

Proof Let $e = \left\lfloor \frac{expo(x)}{2} \right\rfloor + 1$, $x' = 2^{-2e}x$, $\ell' = 2^{-e}\ell$, and $h' = 2^{-e}h$. By Lemmas 7.13 and 7.11,

$$RTO(\ell', k) \leq \textit{rto-sqrt}(x', k) \leq RTO(h', k),$$

or

$$RTO(2^{-k}\ell, k) \leq 2^{-k} \; {}^{(k)}\!\sqrt{x} \leq RTO(2^{-k}h, k).$$

By Lemma 6.75,

$$RTO(\ell, k) \leq \; {}^{(k)}\!\sqrt{x} \leq RTO(h, k),$$

and by Lemma 6.92,

$$\mathcal{R}(\ell, n) = \mathcal{R}(RTO(\ell, k), n) \leq \mathcal{R}({}^{(k)}\!\sqrt{x}, n) \leq \mathcal{R}(RTO(h, k), n) = \mathcal{R}(h, n). \quad \square$$

Our final lemma, which is also required for the proof of Chap. 20, warrants some motivation. In practice, a typical implementation of a subtractive square root algorithm produces a final truncated approximation q of the square root and a remainder that provides a comparison between q^2 and the radicand x. A final rounded result r is derived from this approximation in accordance with a given rounding mode $\mathcal{R}$ and precision n. In order to apply (7.5), we would like to show that $r = \mathcal{R}({}^{(k)}\!\sqrt{x}, n)$ for some appropriate k. This may be done, for example, by invoking Lemma 6.97 with q and ${}^{(k)}\!\sqrt{x}$ substituted for x and z, respectively. But this requires showing that $q = RTZ({}^{(k)}\!\sqrt{x}, n)$ and determining whether $q = {}^{(k)}\!\sqrt{x}$. Thus, we require a means of converting inequalities relating q^2 and x to inequalities relating q and ${}^{(k)}\!\sqrt{x}$. This is achieved by the following:

Lemma 7.18 *Let $x \in \mathbb{Q}$, $q \in \mathbb{Q}$, and $n \in \mathbb{N}$. Assume that $x > 0$, $q > 0$, $n > 1$, and q is $(n-1)$-exact. Then*

(a) $q^2 < x \Leftrightarrow q < {}^{(n)}\!\sqrt{x}$;
(b) $q^2 > x \Leftrightarrow q > {}^{(n)}\!\sqrt{x}$;
(c) $q^2 = x \Leftrightarrow q = {}^{(n)}\!\sqrt{x}$.

Proof Let $e = \left\lfloor \frac{expo(x)}{2} \right\rfloor + 1$, $x' = 2^{-2e}x$, and $q' = 2^{-e}q$. Then $\frac{1}{4} \leq x' < 1$ and ${}^{(n)}\!\sqrt{x} = 2^e rto\text{-}sqrt(x', n)$. By Lemma 7.12,

$$q^2 < x \Leftrightarrow q'^2 < x'$$
$$\Leftrightarrow q' < rto\text{-}sqrt(x', n)$$
$$\Leftrightarrow 2^{-e}q < 2^{-e} \; {}^{(n)}\!\sqrt{x}$$
$$\Leftrightarrow q < {}^{(n)}\!\sqrt{x}.$$

The proof of (b) is similar, and (c) follows. $\square$

Corollary 7.19 *Let $x \in \mathbb{Q}$ and $n \in \mathbb{N}$ with $x > 0$ and $n > 1$. If ${}^{(n)}\!\sqrt{x}$ is $(n-1)$-exact, then $({}^{(n)}\!\sqrt{x})^2 = x$.*

Proof Instantiate Lemma 7.18 with $q = {}^{(n)}\!\sqrt{x}$. $\square$

Corollary 7.20 *Let* $x \in \mathbb{Q}$, $k \in \mathbb{N}$, $n \in \mathbb{N}$, *and* $m \in \mathbb{N}$ *with* $x > 0$, $k > 1$, $n > k$, *and* $m > k$. *If* $\sqrt[(n)]{x}$ *is* $(n-1)$-*exact, then* $\sqrt[(m)]{x} = \sqrt[(n)]{x}$.

Proof By Corollary 7.19, $(\sqrt[(n)]{x})^2 = x$. The corollary again follows from Lemma 7.18.   □

Lemma 7.18 is also critical in the detection of floating-point precision exceptions. As described more fully in Sects. 12.5, 13.5, and 14.3, this exception is signaled when an instruction returns a rounded result r that differs from the precise mathematical value u of an operation. But in the case of the square root, the ACL2 formalization compares r to $\sqrt[(p+2)]{x}$ rather than $u = \sqrt{x}$, where p is the target precision. This is justified by (c) above, from which it follows that $r = \sqrt{x}$ iff $r = \sqrt[(p+2)]{x}$.

Part III
Implementation of Elementary Operations

The marvel of efficient and accurate arithmetic performed in silicon is the culmination of several layers of technology and artistry. In the preceding chapters, we discussed the primitive operations and schemes on which implementations of arithmetic are based: bit manipulation, floating-point representation, and rounding. The following chapters collect some of the commonly used algorithms and methods that employ these primitives in implementing the elementary operations of addition, multiplication, division, and square root extraction. The application of these methods in the design of commercial floating-point units will be illustrated in Part V.

Addition, the most basic of the arithmetic operations, is modeled at the level of bit vectors in Chap. 8, applying the foregoing general theory. We address two distinct and equally challenging tasks: (1) the design of an integer adder and (2) its application to the addition of floating-point numbers. In contrast, the design of an integer multiplier is a more interesting problem than its application to floating-point multiplication. Thus, in Chap. 9, we limit our attention to an analysis of the former, based on several variants of the ubiquitous Booth encoding algorithm [2].

As division and square root extraction are implemented as sequences of these simpler operations, we model them at the more abstract level of rational numbers, treating addition and multiplication as primitives. Two general approaches form the basis of most implementations of division: (1) digit recurrence, which generates a fixed number of quotient bits on each iteration and subtracts a corresponding multiple of the divisor from the current remainder and (2) multiplicative methods based on a convergent sequence of approximations of the reciprocal of the divisor derived by Newton's iterative method. Both approaches involve recurrence relations that may be adapted to the computation of the square root, and each has been used in a wide range of variations driven by application requirements and technological constraints. In Chaps. 10 and 11, we present and analyze representative instances of both.

Part III
Implementation of Elementary Operations

Chapter 8
Addition

Integer adders are central elements of processor designs, as they are used by a wide variety of floating-point and integer arithmetic instructions as well as memory addressing and program counting. Since it is important that the operation be performed for common bit-widths within a single cycle, the execution speed of an adder is often a dominant factor in determining the processor's clock speed and is therefore generally a more important consideration than chip area or power consumption. Integer addition is often treated by the RTL designer as a primitive operation to be implemented by a logic synthesis tool (see, for example, the adder of Chap. 17). This amounts to a selection from a library of predefined adder modules, based on width and timing requirements. If other considerations are more important, however, as in the context of a GPU or a small-core CPU, the designer may prefer to exercise more control and hand-code a suitable solution (as in the GPU adder of Chap. 22).

Section 8.1 is a brief introduction to the gate-level implementation of integer addition. In Sect. 8.2, we present an important class of integer adders that may be tailored according to the efficiency requirements of a design.

Floating-point addition presents additional challenges, mainly due to the required shifting: an alignment shift prior to the addition determined by the exponent difference and a normalizing shift afterward in the event of cancellation. Section 8.3 deals with *leading zero anticipation*, an optimization technique that is commonly used in the normalization of floating-point sums.

8.1 Bit Vector Addition

As a first step, we consider the simple 2-gate module of Fig. 8.1, known as a 1-bit *half adder*. According to the first lemma below, the outputs of this module may be interpreted as the 2-bit sum of its 1-bit inputs.

© The Author(s), under exclusive license to Springer Nature Switzerland AG 2022
D. M. Russinoff, *Formal Verification of Floating-Point Hardware Design*,
https://doi.org/10.1007/978-3-030-87181-9_8

Lemma 8.1 *If u and v are 1-bit vectors, then*

$$u + v = \{1\,'(u \ \& \ v),\, 1\,'(u \ \hat{} \ v)\} = u \ \hat{} \ v + 2(u \ \& \ v).$$

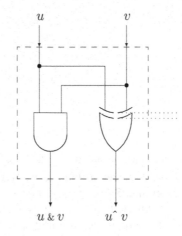

Fig. 8.1 Half adder

Proof The equation may be checked exhaustively, i.e., for all four possible combinations of u and v. □

The *propagate* and *generate* vectors of two integers x and y are defined as

$$p = x \ \hat{} \ y$$

and

$$g = x \ \& \ y,$$

respectively. Obviously, the least significant n bits of these vectors may be computed by the n-bit half adder shown in Fig. 8.2. The following lemma gives a simple reformulation of $x + y$ in terms of p and g.

Lemma 8.2 *Given integers x and y, let $p = x \ \hat{} \ y$ and $g = x \ \& \ y$. Then*

$$x + y = p + 2g.$$

Proof Let $x' = \lfloor x/2 \rfloor$, $y' = \lfloor y/2 \rfloor$, $p' = \lfloor p/2 \rfloor$, and $g' = \lfloor g/2 \rfloor$. By Lemma 3.4(b), $p' = x' \ \hat{} \ y'$ and $g' = x' \ \& \ y'$. By Definition 1.3, Lemma 8.1, and induction,

$$x + y = 2(x' + y') + x[0] + y[0]$$
$$= 2(p' + 2g') + p[0] + 2g[0]$$

$$= (2p' + p[0]) + 2(2g' + g[0])$$

$$= p + 2g. \qquad \qquad \square$$

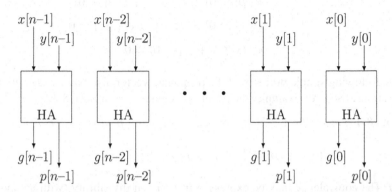

Fig. 8.2 Propagate and generate vectors

Clearly, the propagate and generate vectors for a pair of integers may be computed much more quickly than their sum, but they reveal a variety of properties of the sum [17]. Lemma 8.2 is used, for example, in the proof of the following lemma, which is relevant to the rounding of a floating-point sum. For this purpose, it is useful to predict the least index of the sum at which a 1 occurs. This result provides a method of computing, in constant time, an integer that has the same number of trailing zeroes as the sum of two given integers and a carry-in bit:

Lemma 8.3 *Let* $x \in \mathbb{Z}$, $y \in \mathbb{Z}$, *and* $c \in \{0, 1\}$. *Let*

$$\tau = x \wedge y \wedge (2(x \mid y) \mid c).$$

Then for all $k \in \mathbb{N}$,

$$(x + y + c)[k : 0] = 0 \Leftrightarrow \tau[k : 0] = 0.$$

Proof Let $p = x \wedge y$ and $g = x \& y$. By Lemmas 3.3 and 8.2

$$2(x \mid y) \mid c = 2(x \mid y) + c$$

$$= 2(p + g) + c$$

$$= (2g + p) + c + p$$

$$= x + y + c + p.$$

Thus,

$$\tau[k:0] = 0 \Leftrightarrow (p \mathbin{\char`\^} (2(x \mid y) \mid c))[k:0] = 0$$
$$\Leftrightarrow p[k:0] = (2(x \mid y) \mid c)[k:0]$$
$$\Leftrightarrow p[k:0] = (x + y + c + p)[k:0]$$
$$\Leftrightarrow (x + y + c)[k:0] = 0. \qquad \square$$

The following simple property of the propagate vector also has a wide variety of applications. (See, for example, the proof of Lemma 8.27 in Sect. 8.3.)

Lemma 8.4 *Let $x \in \mathbb{N}$, $y \in \mathbb{N}$, $n \in \mathbb{N}$, $s = x + y$, and $p = x \mathbin{\char`\^} y$. Then*

$$s[n-1:0] = 2^n - 1 \Leftrightarrow p[n-1:0] = 2^n - 1.$$

Proof The equivalence may be expressed in terms of divisibility (with an apology for overloading the symbol "$\mid$"):

$$2^n \mid (x + y + 1) \Leftrightarrow 2^n \mid (p + 1).$$

This holds trivially for $n = 0$. Let $n > 0$, and assume that

$$2^{n-1} \mid \left(\left\lfloor \frac{x}{2} \right\rfloor + \left\lfloor \frac{y}{2} \right\rfloor + 1 \right) \Leftrightarrow 2^{n-1} \mid \left(\left\lfloor \frac{x}{2} \right\rfloor \mathbin{\char`\^} \left\lfloor \frac{y}{2} \right\rfloor + 1 \right).$$

According to Lemma 3.4(b),

$$\left\lfloor \frac{x}{2} \right\rfloor \mathbin{\char`\^} \left\lfloor \frac{y}{2} \right\rfloor = \left\lfloor \frac{p}{2} \right\rfloor.$$

We may assume that $p[0] = 1$, for otherwise $x + y + 1$ and $p + 1$ are both odd. Thus,

$$x + y + 1 = 2 \left\lfloor \frac{x}{2} \right\rfloor + x[0] + 2 \left\lfloor \frac{y}{2} \right\rfloor + y[0] + 1 = 2 \left(\left\lfloor \frac{x}{2} \right\rfloor + \left\lfloor \frac{y}{2} \right\rfloor + 1 \right),$$

$$p + 1 = 2 \left\lfloor \frac{p}{2} \right\rfloor + 1 + 1 = 2 \left(\left\lfloor \frac{p}{2} \right\rfloor + 1 \right),$$

and

$$2^n \mid (x + y + 1) \Leftrightarrow 2^{n-1} \mid \left(\left\lfloor \frac{x}{2} \right\rfloor + \left\lfloor \frac{y}{2} \right\rfloor + 1 \right) \Leftrightarrow 2^{n-1} \mid \left(\left\lfloor \frac{p}{2} \right\rfloor + 1 \right) \Leftrightarrow 2^n \mid (p + 1).$$

$$\square$$

The following variation of Lemma 8.4 simplifies the computation of Lemma 8.3 when the carry-in is known to be set:

Lemma 8.5 *Let x and y be n-bit vectors, where $n \in \mathbb{N}$. For all $k \in \mathbb{N}$, if $k < n$, then*

$$(x + y + 1)[k : 0] = 0 \Leftrightarrow \sim (x \; \char94 \; y)[k : 0] = 0.$$

Proof By Lemma 3.18, $\sim (x \; \char94 \; y)[k : 0] = 2^{k+1} - (x \; \char94 \; y)[k : 0] - 1$, and hence, by Lemma 8.4,

$$\sim (x \; \char94 \; y)[k : 0] = 0 \Leftrightarrow (x \; \char94 \; y)[k : 0] = 2^{k+1} - 1$$
$$\Leftrightarrow (x + y)[k : 0] = 2^{k+1} - 1$$
$$\Leftrightarrow (x + y + 1)[k : 0] = 0. \qquad \square$$

A *full adder* is a module composed of two half adders and an or-gate, as shown in Fig. 8.3. The outputs of the full adder may be readily computed as

$$u \; \char94 \; v \; \char94 \; w$$

and

$$u \; \& \; v \; | \; (u \; \char94 \; v) \; \& \; w = u \; \& \; v \; | \; u \; \& \; w \; | \; v \; \& \; w.$$

Its arithmetic functionality is characterized as follows.

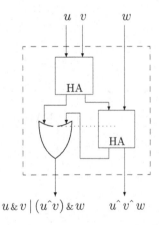

Fig. 8.3 Full adder

Lemma 8.6 *If u, v, and w are 1-bit vectors, then*

$$u + v + w = \{1' \, (u \; \& \; v \; | \; u \; \& \; w \; | \; v \; \& \; v), 1' \, (u \; \char94 \; v \; \char94 \; w)\}.$$

Proof The equation may be checked exhaustively, i.e., for all eight possible combinations of u, v, and w.   □

Thus, a full adder computes the 2-bit sum of three 1-bit inputs. Replacing the half adders of Fig. 8.2 with full adders results in the *3:2 compressor* of Fig. 8.4, also known as a *carry-save adder*, which reduces a sum of three vectors to a sum of two.

Lemma 8.7 *Given integers x, y, and z, let*

$$a = x \wedge y \wedge z$$

and

$$b = x \,\&\, y \mid x \,\&\, z \mid y \,\&\, z.$$

Then

$$x + y + z = a + 2b.$$

Proof The proof is similar to that of Lemma 8.2.   □

The following consequence of Lemma 8.7 reduces the problem of detecting a zero sum to that of testing equality of two vectors:

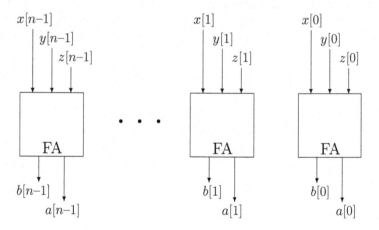

Fig. 8.4 3:2 compressor

Lemma 8.8 *For integers x and y, $x + y = 2(x \mid y) - x \wedge y$.*

Proof By Lemma 8.7 together with Lemma 3.5 and Definition 3.2,

$$x + y = (x + y + (-1)) + 1$$
$$= x \wedge y \wedge (-1) + 2(x \,\&\, y \mid x \,\&\, (-1) \mid y \,\&\, (-1)) + 1$$

$$= \sim(x \text{ ^ } y) + 2(x \text{ \& } y \mid x \mid y) + 1$$
$$= -(x \text{ ^ } y) - 1 + 2(x \mid y) + 1$$
$$= 2(x \mid y) - (x \text{ ^ } y). \qquad \square$$

The module displayed in Fig. 8.5 is constructed from the same hardware as the 3:2 compressor, but the third input $z[k]$ of each adder is eliminated and replaced by c_k, the carry bit generated at index $k - 1$. The resulting circuit, known as a *ripple-carry adder* (RCA), produces the sum $x + y + c_0$ of two n-bit vectors x and y and a carry-in bit c_0, represented as a single *n*-bit vector s and a carry-out c_n. This is established by the following result.

Lemma 8.9 *Let $s = x + y + c_0$, where $x \in \mathbb{N}$, $y \in \mathbb{N}$, and $c_0 = \{0, 1\}$. For $i > 0$, let*

$$c_i = x[i - 1] \text{ \& } y[i - 1] \mid x[i - 1] \text{ \& } c_{i-1} \mid y[i - 1] \text{ \& } c_{i-1}.$$

Then for all $i \in \mathbb{N}$,

$$s[i] = x[i] \text{ ^ } y[i] \text{ ^ } c_i.$$

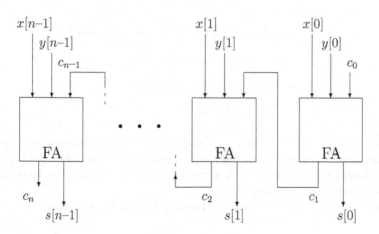

Fig. 8.5 Ripple-carry adder

Proof For $i \in \mathbb{Z}$, let $\sigma_i = x[i : 0] + y[i : 0] + c_0$. Then for $i < 0$, $\sigma_i = c_0$, and for $i \geq 0$,

$$\sigma_i = (2^i x[i] + x[i - 1 : 0]) + (2^i y[i] + y[i - 1 : 0]) + c_0 = 2^i (x[i] + y[i]) + \sigma_{i-1}$$

and

$$0 \leq \sigma_i \leq (2^{1+1} - 1) + (2^{1+1} - 1) + 1 < 2^{i+2}.$$

We shall show that for $i \geq 0$, $c_i = 1$ iff $\sigma_{i-1} \geq 2^i$: the claim is trivial for $i = 0$, and for $i > 0$, by induction,

$$c_i = 1 \Leftrightarrow x[i-1]=y[i-1]=1 \text{ or } ((x[i-1]=1 \text{ or } y[i-1]=1) \text{ and } c_{i-1}=1)$$

$$\Leftrightarrow x[i-1]=y[i-1]=1 \text{ or } \left((x[i-1]=1 \text{ or } y[i-1]=1) \text{ and } \sigma_{i-2} \geq 2^{i-1}\right)$$

$$\Leftrightarrow \sigma_{i-1} \geq 2^i.$$

Since $0 \leq \sigma_{i-1} < 2^{i+1} = 2 \cdot 2^i$, it follows that $c_i = \lfloor 2^{-i}\sigma_{i-1} \rfloor$.

By Lemmas 2.1(c) and 1.15,

$$\sigma_i \bmod 2^{i+1} = (x \bmod 2^{i+1} + y \bmod 2^{i+1} + c_0) \bmod 2^{i+1}$$

$$= (x + y + c_0) \bmod 2^{i+1}$$

$$= s \bmod 2^{i+1},$$

and by Lemma 2.16 (l) and (c),

$$s[i] = \sigma_i[i]$$

$$= \lfloor 2^{-i}\sigma_i \rfloor \bmod 2$$

$$= \left\lfloor 2^{-i}\left(2^i(x[i]+y[i]) + \sigma_{i-1}\right) \right\rfloor \bmod 2$$

$$= (x[i] + y[i] + \lfloor 2^{-i}\sigma_{i-1} \rfloor) \bmod 2$$

$$= (x[i] + y[i] + c_i) \bmod 2$$

$$= x[i] \text{ \textasciicircum } y[i] \text{ \textasciicircum } c_i. \qquad \square$$

Since the preceding carry bit c_{i-1} is required for the computation of c_i and $s[i]$, the execution time of an RCA of width n is a linear function of n. In order to improve the efficiency of bit vector addition, some degree of parallelism must be introduced in the computation of the carry bits.

Note that the recurrence formula of Lemma 8.9 may be written as

$$c_{i+1} = g_i \mid p_i \text{ \& } c_i, \qquad (8.1)$$

where

$$g_i = g[i] = x[i] \text{ \& } y[i]$$

and

$$p_i = p[i] = x[i] \text{ \textasciicircum } y[i].$$

Assuming that some c_i has been computed, successive carry bits may thus be computed as follows:

$$c_{i+2} = g_{i+1} \mid p_{i+1} \ \& \ c_{i+1}$$
$$= g_{i+1} \mid p_{i+1} \ \& \ g_i \mid p_{i+1} \ \& \ p_i \ \& \ c_i,$$
$$c_{i+3} = g_{i+2} \mid p_{i+2} \ \& \ c_{i+2}$$
$$= g_{i+2} \mid p_{i+2} \ \& \ g_{i+1} \mid p_{i+2} \ \& \ p_{i+1} \ \& \ g_i \mid p_{i+2} \ \& \ p_{i+1} \ \& \ p_i \ \& \ c_i,$$
$$c_{i+4} = g_{i+3} \mid p_{i+3} \ \& \ c_{i+3}$$
$$= g_{i+3} \mid p_{i+3} \ \& \ g_{i+2} \mid p_{i+3} \ \& \ p_{i+2} \ \& \ g_{i+1} \mid p_{i+3} \ \& \ p_{i+2} \ \& \ p_{i+1} \ \& \ g_i \mid$$
$$p_{i+3} \ \& \ p_{i+2} \ \& \ p_{i+1} \ \& \ p_i \ \& \ c_i,$$

$\ldots$

If we take $i = 0$, then all carry bits $c_1, \ldots, c_n$ may thus be computed in parallel, thereby reducing overall execution time. The resulting circuit is known as a *carry-lookahead* adder (CLA). Unfortunately, this design is impractical for all but very small values of n, since the number of gates required increases quadratically with n and the fan-in (maximum number of inputs to a gate) increases linearly. In practice, this technique is used only in very narrow adders, which are connected in series to add wider vectors. The optimal width of a link in such a chain is a function of the propagation delays and other characteristics of the underlying technology, but the most common width is 4, as illustrated by the adder of width $4m$ shown in Fig. 8.6, comprising m 4-bit CLAs.

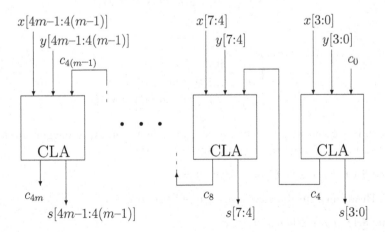

Fig. 8.6 A chain of carry-lookahead adders

In the next section, we present a general framework for employing parallelism in the computation of carry bits, which is applicable to adders of arbitrary width and gives rise to a variety of designs that exhibit measurable trade-offs with respect to the objectives of execution speed, area efficiency, and power consumption.

8.2 Parallel Prefix Adders

Let x and y be 2^n-bit vectors, where $n \in \mathbb{N}$, and let $c_0 \in \{0, 1\}$. These values will be fixed throughout this section. Our objective is to compute the sum $s = x + y + c_0$.

Once again, let $g = x$ & y and $p = x$ ^ y, the generate and propagate vectors, and for $i \in \mathbb{N}$, let $g_i = g[i]$ and $p_i = p[i]$. As noted in Eq. (8.1), the carry bits of Lemma 8.9 may be formulated as

$$c_i = g_{i-1} \mid p_{i-1} \ \& \ c_{i-1},$$

providing a straightforward sequential computation. As we have observed, once we have computed g, p, and the c_i, the sum may be derived according to Lemma 8.9. In the present section, we shall investigate a class of algorithms based on this observation that improve efficiency by employing parallelism in the computation of the carry bits.

The generate and propagate bits g_i and p_i are generalized as follows:

Definition 8.1 Given $i \in \mathbb{Z}$ and $j \in \mathbb{Z}$,

$$G^i_j = \begin{cases} 0 & \text{if } i < j \text{ or } i < 0 \\ x[i] & \text{if } i \geq j \text{ and } x[i] = y[i] \\ G^{i-1}_j & \text{if } i \geq j \text{ and } x[i] \neq y[i] \end{cases}$$

and

$$P^i_j = \begin{cases} 1 & \text{if } i < j \text{ or } i < 0 \\ 0 & \text{if } i \geq j \text{ and } x[i] = y[i] \\ P^{i-1}_j & \text{if } i \geq j \text{ and } x[i] \neq y[i]. \end{cases}$$

Although it is convenient to define these functions for arbitrary integer arguments, our interest in them is limited to the case $i \geq j \geq 0$.

Lemma 8.10 *For* $i \in \mathbb{N}$, $G^i_i = g_i$ *and* $P^i_i = p_i$.

Proof This is an immediate consequence of Definition 8.1. $\square$

Lemma 8.11 *For* $i \in \mathbb{N}$ *and* $j \in \mathbb{N}$,

$$G^i_j = \begin{cases} 1 \ \textit{if } i \geq j \textit{ and } x[i : j] + y[i : j] \geq 2^{i+1-j} \\ 0 \ \textit{otherwise} \end{cases}$$

and

$$P_j^i = \begin{cases} 1 & \text{if } i < j \text{ or } x[i:j] + y[i:j] = 2^{i+1-j} - 1 \\ 0 & \text{otherwise.} \end{cases}$$

Proof We shall prove the first claim; the second is similar. The proof is by induction on $i - j$. The case $i \le j$ is trivial. In the remaining case, if $x[i] = y[i] = 1$, then $G_j^i = 1$ and

$$x[i:j] + y[i:j] = (2^{i-j} + x[i-1:j]) + (2^{i-j} + y[i-1:j]) \ge 2^{i-j} + 2^{i-j} = 2^{i+1-j};$$

if $x[i] = y[i] = 0$, then $G_j^i = 0$ and

$$x[i:j] + y[i:j] = x[i-1:j] + y[i-1:j] < 2^{i-j} + 2^{i-j} = 2^{i+1-j};$$

finally, if $x[i] \ne y[i]$, then

$$x[i:j] + y[i:j] = 2^{i-j} + x[i-1:j] + y[i-1:j],$$

and by induction,

$$G_j^i = 1 \Leftrightarrow G_j^{i-1} = 1$$
$$\Leftrightarrow x[i-1:j] + y[i-1:j] \ge 2^{i-j}$$
$$\Leftrightarrow x[i:j] + y[i:j] \ge 2^{i-j} + 2^{i-j} = 2^{i+1-j}. \qquad \square$$

The following ordered pairs are the basis of parallel prefix addition:

Definition 8.2 Given $i \in \mathbb{N}$ and $j \in \mathbb{N}$,

$$GP_j^i = (G_j^i, P_j^i)$$

and

$$GP_i = GP_i^i = (g_i, p_i).$$

The *fundamental carry operator* is a binary operation on ordered pairs of booleans, defined as follows:

Definition 8.3 Given boolean values G', P', G'', and P'',

$$(G', P') \oplus (G'', P'') = (G' \mid P' \,\&\, G'', P' \,\&\, P'').$$

Associativity is a critical property of this operation:

Lemma 8.12 *For booleans* G', P', G'', P'', G''', *and* P''',

$$((G', P') \oplus (G'', P'')) \oplus (G''', P''') = (G', P') \oplus ((G'', P'') \oplus (G''', P''')).$$

Proof Both expressions are readily reduced to

$$(G' \mid P' \& G'' \mid P' \& P'' \& G''', P' \& P'' \& P''').$$ □

The following related result is relevant to the problem at hand:

Lemma 8.13 *If* $x \in \mathbb{N}$, $y \in \mathbb{N}$, $i \in \mathbb{N}$, $j \in \mathbb{N}$, $k \in \mathbb{N}$, *and* $j \leq k < i$, *then*

$$GP_{k+1}^i \oplus GP_j^k = GP_j^i.$$

Proof This follows from Lemma 8.11 by a case analysis using Lemma 2.10, which yields the identities

$$x[i : j] = 2^{k+1-j} x[i : k+1] + x[k : j]$$

and

$$y[i : j] = 2^{k+1-j} y[i : k+1] + y[k : j].$$ □

Lemma 8.13 provides the following expression for the carry bits:

Lemma 8.14 *For* $i \in \mathbb{N}$, $c_{i+1} = G_0^i \mid P_0^i \& c_0$.

Proof For $i = 0$, we have

$$c_1 = g_0 \mid p_0 \& c_0 = G_0^0 \mid P_0^0 \& c_0.$$

For the inductive step, we invoke Lemma 8.13 with $k = i - 1$ and $j = 0$, which yields

$$\begin{aligned}
(G_0^i, P_0^i) &= GP_0^i \\
&= GP_i^i \oplus GP_0^{i-1} \\
&= (g_i, p_i) \oplus (G_0^{i-1}, P_0^{i-1}) \\
&= (g_i \mid p_i \& G_0^{i-1}, p_i \& P_0^{i-1}).
\end{aligned}$$

Thus, $G_0^i = g_1 \mid p_i \& G_0^{i-1}$, $P_0^i = p_i \& P_0^{i-1}$, and

$$\begin{aligned}
c_{i+1} &= g_i \mid p_i \& c_i \\
&= g_i \mid p_i \& (G_0^{i-1} \mid P_0^{i-1} \& c_0)
\end{aligned}$$

$$= g_i \mid (p_i \And G_0^{i-1} \mid p_i \And P_0^{i-1} \And c_0)$$

$$= (g_i \mid p_i \And G_0^{i-1}) \mid p_i \And P_0^{i-1} \And c_0$$

$$= G_0^i \mid P_0^i \And c_0. \qquad\qquad \square$$

Our objective, therefore, is to compute GP_0^i for $i = 0, \ldots, 2^n$. By Lemmas 8.12 and 8.13, we may unambiguously write

$$GP_0^i = GP_i \oplus GP_{i-1} \oplus \ldots \oplus GP_0. \qquad (8.2)$$

This is an example of a *prefix sum*, i.e., a recurrent application of an associative binary operation, which is susceptible to efficient computation through parallelism. A *parallel prefix adder* of width 2^n computes the sum

$$s = x + y + c_0 = 2^{2^n} c_{2^n} + s[2^n - 1 : 0]$$

as follows:

(1) For $0 < i < 2^n$, $g_i = x[i] \And y[i]$, $p_i = x[i] \mathbin{\char`\^} y[i]$, and $GP_i = (g_i, p_i)$ are computed;
(2) For $0 < i < 2^n$, GP_0^i is computed according to (8.2) by one of the various algorithms presented below, all of which are based on Lemma 8.13;
(3) For $0 < i \le 2^n$, c_i is computed according to Lemma 8.14;
(4) For $0 < i < 2^n$, $s[i]$ is computed as $p_i \mathbin{\char`\^} c_i$.

The computation of the set $\{GP_0^i\}$ in Step (2) consists of a set of instantiations of the fundamental carry operator, which may be viewed as the nodes of an acyclic directed graph. Each arc of the graph is labeled with a pair of integers (i, j), where $0 \le j \le i < 2^n$. Each node has two incoming arcs, labeled (i_1, j_1) and (i_2, j_2), where $j_1 = i_2 + 1$, and any positive number of outgoing arcs, each labeled (i_1, j_2). Such a node represents the computation of

$$GP_{j_2}^{i_1} = GP_{j_1}^{i_1} \oplus GP_{j_2}^{i_2}.$$

For each $i < 2^n$, there is exactly one input arc (i.e., one that is not an outgoing arc of any node), labeled (i, i), and exactly one output arc (one that is not an incoming arc of any node), labeled $(i, 0)$.

The graph is naturally partitioned into *levels*. The level of an arc is 0 if the arc is an input and otherwise that of its source node; the level of a node is $\max(d_1, d_2) + 1$, where d_1 and d_2 are the levels of its incoming arcs.

The complexity of a parallel prefix adder may be characterized by three parameters:

Adder	Depth	Area	Fan-out
Ripple-carry	$2^n - 1$	$2^n - 1$	2
Ladner-Fischer	n	$2^{n-1}n$	$2^{n-1} + 1$
Kogge-Stone	n	$2^n(n - 1) + 1$	$n + 1$
Brent-Kung	$2n - 2$	$2^{n+1} - n + 2$	$n + 1$
Han-Carlson	$\min(n + k, 2n - 2)$	$2^{n+1} + 2^{n-k}(n-k-3) - k + 1$	$n + 1$

Fig. 8.7 Complexity of parallel prefix adders of width 2^n

- Depth: the maximum level of a node;
- Area: the total number of nodes;
- Fan-out: the maximum number of outgoing arcs of a node.

Note that depth and fan-out both contribute to execution time.

Since the graph must contain at least one node for each i, $0 < i < 2^n$, the area must be at least $2^n - 1$. If we define the *span* of an arc labeled (i, j) to be $i - j + 1$, then it is clear that the span of an outgoing arc of a node is the sum of the spans of its incoming arcs. It follows that the span of an arc at level d is at least $d + 1$ and at most 2^d. Since the span of the output labeled $(2^n - 1, 0)$ is 2^n, the depth of the graph is at least n; since the span of an arc cannot exceed 2^n, the depth is at most $2^n - 1$.

The structural variety of these graphs allows a parallel prefix adder to be selected according to the relative priorities of these complexity parameters as dictated by the design goals of an implementation. We shall examine five such adders, the complexities of which are compared in Fig. 8.7. In each case, we shall define a recursive function based on the fundamental carry operator, to be implemented in hardware, that performs the corresponding computation of GP_0^i for given i. The implementation of each of these functions as a provably equivalent iterative computation at the register-transfer level is straightforward, as illustrated by the Han-Carlson adder found in the FMA of Chap. 22, the pseudocode for which appears online at https://go.sn.pub/fma32 (see the function HC64).

We shall diagram the graph of each of these adders for the case $n = 4$. The diagrams employ a representation scheme that has been established in the literature. Each graph is organized into 2^n columns. For each $i < 2^n$, Column i contains all nodes that compute GP_j^i for some $j < i$.

The simplest of all parallel prefix adders is represented by the graph of Fig. 8.8. This is a degenerate case that employs no parallelism whatsoever, with minimum

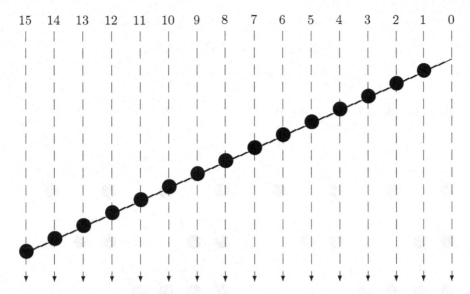

Fig. 8.8 Parallel prefix representation of a ripple-carry adder of width 16

area and fan-out and maximum depth. For $0 < i < 2^n$, the output GP_0^i is computed at level i as follows:

Definition 8.4 For $i \in \mathbb{N}$,

$$RC(i) = \begin{cases} GP_0 & \text{if } i = 0 \\ GP_i \oplus RC(i-1) & \text{if } i = 0. \end{cases}$$

The correctness of this computation is trivial:

Lemma 8.15 *For $i \in \mathbb{N}$, $RC(i) = GP_0^i$.*

Proof By Lemma 8.13 and induction, for $i > 0$,

$$RC(i) = GP_i \oplus RC(i-1) = GP_i \oplus GP_0^{i-1} = GP_0^i. \qquad \square$$

Thus, the outputs of the graph are computed recursively as

$$GP_0^i = GP_i \oplus GP_0^{i-1} = (g_i, p_i) \oplus (G_0^{i-1}, P_0^{i-1}) = (g_i \mid p_i \,\&\, G_0^{i-1}, \, p_i \,\& P_0^{i-1}).$$

Upon examining the computation of the carry bits by this adder according to Lemma 8.14, it becomes clear that we have essentially replicated the logic of the ripple-carry adder of width 2^n:

$$c_{i+1} = G_0^i \mid P_0^i \,\&\, c_0$$
$$= (g_i \mid p_i \,\&\, G_0^{i-1}) \mid (p_i \,\&\, P_0^{i-1}) \,\&\, c_0$$

$$= g_i \mid p_i \And G_0^{i-1} \mid p_i \And P_0^{i-1} \And c_0$$

$$= g_i \mid p_i \And (G_0^{i-1} \mid P_0^{i-1} \And c_0)$$

$$= g_i \mid p_i \And c_i.$$

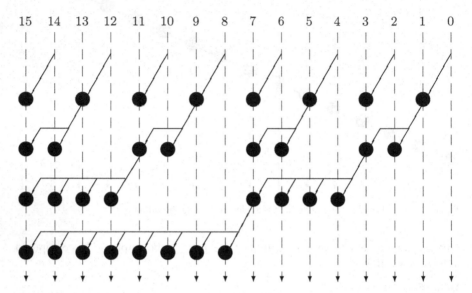

Fig. 8.9 Ladner-Fischer adder of width 16

The *Ladner-Fischer* adder (Fig. 8.9) has minimum depth n but an exceedingly large fan-out. For $0 \le i < 2^n$, the computation of G_0^i is performed by $LF(i, n)$, defined below. More generally, for $d \le n$, $LF(i, d)$ is the value GP_j^i computed by the node in Column i of maximum level not exceeding d, or GP_i if no such node exists. Note that the definition refers to the truncation operator of Sect. 1.3:

Definition 8.5 For $i \in \mathbb{N}$ and $d \in \mathbb{N}$,

$$LF(i, d) = \begin{cases} GP_i & \text{if } d = 0 \\ LF(i, d - 1) & \text{if } d > 0 \text{ and } i \bmod 2^d < 2^{d-1} \\ LF(i, d - 1) \oplus LF(i', d - 1) & \text{if } d > 0 \text{ and } i \bmod 2^d \ge 2^{d-1}, \end{cases}$$

where $i' = i^{(-d)} + 2^{d-1} - 1$.

Note that $i^{(-d)} = 2^d \lfloor 2^{-d} i \rfloor \in \mathbb{N}$, and therefore $i' \in \mathbb{N}$.

The correctness of *LF* is a special case of the following more general result:

Lemma 8.16 *For* $i \in \mathbb{N}$ *and* $d \in \mathbb{N}$, $LF(i, d) = GP_{i^{(-d)}}^i$.

Proof If $d = 0$, then by Definition 1.5, $i^{(-d)} = i$ and the claim is trivial. Assume $d > 0$. By Definitions 1.5 and 1.3,

$$i^{(-d)} = 2^d \lfloor 2^{-d} i \rfloor = i - i \bmod 2^d.$$

Suppose $i \bmod 2^d < 2^{d-1}$. We invoke Lemma 1.19(a) with $m = i$, $n = 2^{d-1}$, and $a = 2\lfloor 2^{-d} i \rfloor$: since $an \le m < (a + 1)n$, i.e.,

$$2^d \lfloor 2^{-d} i \rfloor \le i = 2^d \lfloor 2^{-d} i \rfloor + i \bmod 2^d < 2^d \lfloor 2^{-d} i \rfloor + 2^{d-1},$$

the lemma yields $m \bmod n = m - an$, i.e.,

$$i \bmod 2^{d-1} = i - 2^d \lfloor 2^{-d} i \rfloor = i \bmod 2^d$$

and by induction,

$$LF(i, d) = LF(i, d - 1) = GP^i_{i^{(1-d)}} = GP^i_{i^{(-d)}}.$$

Now suppose $i \bmod 2^d \ge 2^{d-1}$. Then

$$i = 2^d \lfloor 2^{-d} i \rfloor + i \bmod 2^d = 2^{d-1} (2\lfloor 2^{-d} i \rfloor + 1) + (i \bmod 2^d - 2^{d-1}),$$

where $0 \le i \bmod 2^d - 2^{d-1} < 2^{d-1}$. Thus,

$$2^{d-1} (2\lfloor 2^{-d} i \rfloor + 1) \le i < 2^{d-1} (2\lfloor 2^{-d} i \rfloor + 2)$$

and, by Lemma 1.19(a) with $m = i$, $n = 2^{d-1}$, and $a = 2\lfloor 2^{-d} i \rfloor + 1$,

$$i \bmod 2^{d-1} = i - 2^{d-1} (2\lfloor 2^{-d} i \rfloor + 1) = i - 2^d \lfloor 2^{-d} i \rfloor - 2^{d-1} = i \bmod 2^d - 2^{d-1}.$$

Thus,

$$i^{(1-d)} = i - i \bmod 2^{d-1} = i - i \bmod 2^d + 2^{d-1} = i^{(-d)} + 2^{d-1}$$

and

$$
\begin{aligned}
\left(i^{(1-d)} - 1 \right)^{(1-d)} &= 2^{d-1} \lfloor 2^{1-d} (i^{(1-d)} - 1) \rfloor \\
&= 2^{d-1} \lfloor 2^{1-d} (2^{d-1} \lfloor 2^{1-d} i \rfloor - 1) \rfloor \\
&= 2^{d-1} \lfloor \lfloor 2^{1-d} i \rfloor - 2^{1-d} \rfloor \\
&= 2^{d-1} (\lfloor 2^{1-d} i \rfloor + \lfloor -2^{1-d} \rfloor) \\
&= 2^{d-1} (\lfloor 2^{1-d} i \rfloor - 1)
\end{aligned}
$$

$$= i^{(1-d)} - 2^{d-1}$$
$$= i^{(-d)}.$$

Combining these equations with induction, we have

$$LF(i^{(-d)} + 2^{d-1} - 1, d - 1) = LF(i^{(1-d)} - 1, d - 1)$$
$$= GP^{i^{(1-d)}-1}_{(i^{(1-d)}-1)^{(1-d)}}$$
$$= GP^{i^{(1-d)}-1}_{i^{(-d)}}$$

and

$$LF(i, d) = LF(i, d - 1) \oplus LF(i^{(-d)} + 2^{d-1} - 1, d - 1) = GP^i_{i^{(1-d)}} \oplus GP^{i^{(1-d)}-1}_{i^{(-d)}}.$$

By Lemma 1.24, $i^{(-d)} \leq i^{(1-d)}$. If $i^{(-d)} \leq i^{(1-d)} - 1$, then the lemma follows from Lemma 8.13. Otherwise, $i^{(-d)} = i^{(1-d)}$, and applying Definitions 8.1, 8.2, and 8.3, the above equation reduces to

$$LF(i, d) = GP^i_{i^{(-d)}} \oplus (0, 1) = GP^i_{i^{(-d)}}. \qquad \square$$

Corollary 8.17 *If $i \in \mathbb{N}$ and $i < 2^n$, then $LF(i, n) = GP^i_0$.*

Proof This follows from Lemma 8.16 and the observation that

$$i^{(-n)} = 2^n \lfloor 2^{-n} i \rfloor = 2^n \cdot 0 = 0. \qquad \square$$

The *Kogge-Stone* adder (Fig. 8.10) has minimum depth and roughly twice the area of the Ladner-Fischer (for large n) but a more manageable fan-out. The computation of GP^i_0 is represented by $KS(i, n)$, defined as follows:

Definition 8.6 For $i \in \mathbb{N}$ and $d \in \mathbb{N}$,

$$KS(i, d) = \begin{cases} GP_i & \text{if } d = 0 \\ KS(i, d - 1) & \text{if } d > 0 \text{ and } i < 2^{d-1} \\ KS(i, d - 1) \oplus KS(i - 2^{d-1}, d - 1) & \text{if } d > 0 \text{ and } i \geq 2^{d-1}. \end{cases}$$

Lemma 8.18 *If $i \in \mathbb{N}$, $d \in \mathbb{N}$, and $m = \max(0, i - 2^d + 1)$, then*

$$KS(i, d) = GP^i_m.$$

Proof If $d = 0$, then $m = i$ and

$$KS(i, d) = GP_i = GP^i_i = GP^i_m.$$

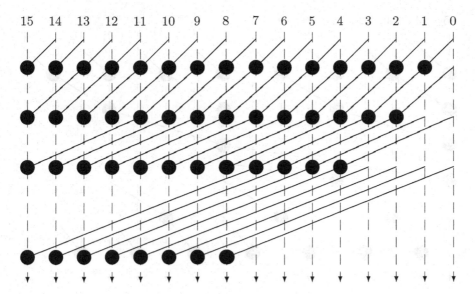

Fig. 8.10 Kogge-Stone adder of width 16

Assume $d > 0$. Let $m' = \max(0, i - 2^{d-1} + 1)$. If $i < 2^{d-1}$, then $m = m' = 0$ and

$$KS(i, d) = KS(i, d - 1) = GP^i_{m'} = GP^i_m.$$

On the other hand, suppose $i \geq 2^{d-1}$. Let

$$i' = i - 2^{d-1}$$

and

$$m'' = \max(0, i' - 2^{d-1} + 1) = \max(0, i - 2^d + 1) = m.$$

Then

$$m' = i - 2^{d-1} + 1 = i' + 1$$

and

$$
\begin{aligned}
KS(i, d) &= KS(i, d - 1) \oplus KS(i', d - 1) \\
&= GP^i_{m'} \oplus GP^{i'}_{m''} \\
&= GP^i_{i'+1} \oplus GP^{i'}_{m} \\
&= GP^i_m.
\end{aligned}
$$

$\square$

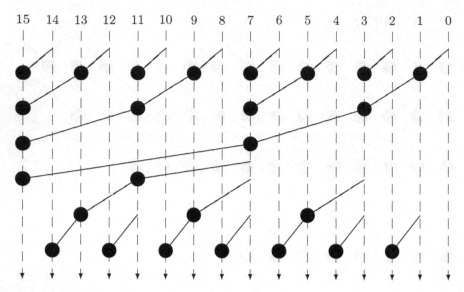

Fig. 8.11 Brent-Kung adder of width 16

Corollary 8.19 *If $i \in \mathbb{N}$ and $i < 2^n$, then $KS(i, n) = GP_0^i$.*

Proof This follows from Lemma 8.18 and the observation that $i - 2^n + 1 \leq 0$. □

The *Brent-Kung* adder (Fig. 8.11) has nearly twice the depth of Kogge-Stone but a smaller area. The computation of GP_0^i is represented by $BK(i, n)$, defined below. It consists of two stages. The first, computed by $BK_0(i, n)$, contains n levels. The second contains $n - 1$ levels, but these depend only on the first $n - 1$ levels of the first stage, resulting in a combined depth of $2n - 2$.

The definition is based on the following: for an integer $k \neq 0$, we define $\Pi(k)$ to be the maximum integer p such that k is divisible by 2^p. Note that the graph of BK extends that of BK_0 with a single node in Column i for each i such that $i + 1$ is not a power of 2.

Definition 8.7 Let $i \in \mathbb{N}$, $d \in \mathbb{N}$, and $p = \Pi(i + 1)$. Then

$$
BK_0(i, d) = \begin{cases} GP_i & \text{if } d = 0 \\ BK_0(i, d - 1) & \text{if } d > 0 \text{ and } p < d \\ BK_0(i, d - 1) \oplus BK_0(i - 2^{d-1}, d - 1) & \text{if } d > 0 \text{ and } p \geq d \end{cases}
$$

and

$$
BK(i, n) = \begin{cases} BK_0(i, n) & \text{if } i = 2^p - 1 \\ BK_0(i, n) \oplus BK(i - 2^p, n) & \text{if } i \neq 2^p - 1. \end{cases}
$$

Lemma 8.20 *Let $i \in \mathbb{N}$, $d \in \mathbb{N}$, $p = \Pi(i + 1)$, and $m = \min(p, d)$. Then*

$$BK_0(i, d) = GP^i_{i-2^m+1}.$$

Proof If $d = 0$, then $m = 0$ and

$$BK_0(i, d) = GP_i = GP^i_i = GP^i_{i-2^m+1}.$$

Assume $d > 0$ and let $m' = \min(p, d - 1)$. If $p \leq d - 1$, then $m = m' = p$ and

$$BK_0(i, d) = BK_0(i, d - 1) = GP^i_{i-2^{m'}+1} = GP^i_{i-2^m+1}.$$

On the other hand, suppose $p \geq d$. Then $m = d$ and $m' = d - 1$. Let $i' = i - 2^{d-1}$, $p' = \Pi(i' + 1)$, and $m'' = \min(p', d - 1)$. Then

$$p' = d - 1 = m''$$

and

$$i' - 2^{m''} = i - 2^{d-1} - 2^{d-1} = i - 2^d = i - 2^m.$$

Since $i + 1$ is divisible by 2^p, $i + 1 \geq 2^p > 2^{d-1}$, i.e., $i - 2^{d-1} \geq 0$. Thus, by induction and Lemma 8.13,

$$
\begin{aligned}
BK_o(i, d) &= BK_o(i, d - 1) \oplus BK_o(i', d - 1) \\
&= GP^i_{i-2^{m'}+1} \oplus GP^{i'}_{i'-2^{m''}+1} \\
&= GP^i_{i'+1} \oplus GP^{i'}_{i-2^m+1} \\
&= GP^i_{i-2^m+1}.
\end{aligned}
$$
□

Corollary 8.21 *If $i \in \mathbb{N}$, $i < 2^n$, and $p = \Pi(i + 1)$, then*

$$BK_0(i, n) = GP^i_{i-2^p+1}.$$

Proof This follows from Lemma 8.20 and the observation that $p \leq n$. □

Lemma 8.22 *If $i \in \mathbb{N}$ and $i < 2^n$, then $BK(i, n) = GP^i_0$.*

Proof The proof is by induction on i. Let $p = \Pi(i + 1)$. If $i = 2^p - 1$, then

$$BK(i, n) = BK_0(i, n) = GP^i_{i-2^p+1} = GP^i_0.$$

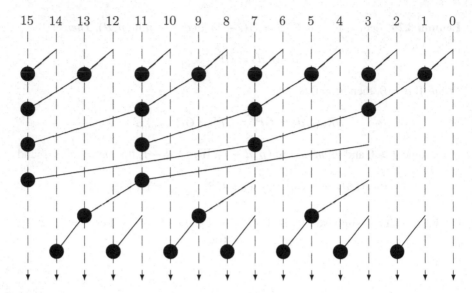

Fig. 8.12 Han-Carlson adder of width 16 with $k = 2$

Otherwise, $i \geq 2^p$, and by Corollary 8.21, Lemma 8.13, and induction,

$$BK(i, n) = BK_0(i, n) \oplus BK(i - 2^p, n)$$
$$= GP^i_{i-2^p+1} \oplus GP^{i-2^p}_0$$
$$= GP^i_0. \qquad \square$$

The *Han-Carlson* (Fig. 8.12) is a hybrid adder, combining elements of the Kogge-Stone and the Brent-Kung. It consists of three stages and is constructed according to an additional parameter k, $0 \leq k \leq n$, The first stage consists of the first k levels of the Brent-Kung; the second consists of $n - k$ levels similar to those of the Kogge-Stone; the third is similar to the second Brent-Kung stage and contains $\min(k, n-2)$ additional levels.

As shown in Fig. 8.7, the Han-Carlson is a compromise between the two preceding adders with respect to depth and area when $k \approx \frac{n}{2}$; this is a popular choice. In the extreme cases $k = n$ and $k = 0$, it reduces to the Brent-Kung and Kogge-Stone, respectively.

The function HC_0 computes the first two stages, calling BK_0 to compute the first; HC computes the third stage:

Definition 8.8 If $i \in \mathbb{N}$, $k \in \mathbb{N}$, $d \in \mathbb{N}$, and $p = \Pi(i + 1)$, then

$$HC_0(i, k, d) = \begin{cases} BK_0(i, k) & \text{if } d = 0 \text{ or } p < k \\ HC_0(i, k, d-1) & \text{if } d > 0, \ p \geq k, \text{ and } i < 2^{k+d-1} \\ HC_0(i, k, d-1) & \\ \quad \oplus HC_0(i - 2^{k+d-1}, k, d-1) & \text{otherwise} \end{cases}$$

and if $k \leq n$, then

$$HC(i, k, n) = \begin{cases} HC_0(i, k, n-k) & \text{if } p \geq k \text{ or } i = 2^p - 1 \\ HC_0(i, k, n-k) \oplus HC(i - 2^p, k, n) & \text{otherwise.} \end{cases}$$

Lemma 8.23 *Let $i \in \mathbb{N}$, $k \in \mathbb{N}$, $d \in \mathbb{N}$, and $p = \Pi(i+1)$. If $p \geq k$, then*

$$HC_0(i, k, d) = GP_m^i$$

where $m = \max(0, i - 2^{k+d} + 1)$.

Proof The proof is by induction on d. Consider the case $d = 0$. Since $i + 1 \geq 2^p \geq 2^k$, $m = i - 2^k + 1$. We shall invoke Lemma 8.20, substituting k for d. Since $\min(p, k) = k$,

$$HC_0(i, k, d) = BK_0(i, k) = GP_{i-2^k+1}^i = GP_m^i.$$

Now suppose $d > 0$. Let $m' = \max(0, i - 2^{k+d-1} + 1)$. If $i < 2^{k+d-1}$, then $m = m' = 0$, and by induction,

$$HC_0(i, k, d) = HC_0(i, k, d-1) = GP_{m'}^i = GP_m^i.$$

Finally, suppose $i \geq 2^{k+d-1}$. Let $i' = i - 2^{k+d-1}$ and

$$m'' = \max(0, i' - 2^{k+d-1} + 1).$$

Then $m' = i - 2^{k+d-1} + 1 = i' + 1$, and since $i' - 2^{k+d-1} = i - 2^{k+d}$, $m'' = m$ and

$$\frac{i' + 1}{2^k} = \frac{i + 1}{2^k} - 2^d \in \mathbb{Z}.$$

Thus, by induction,

$$HC_0(i, k, d) = HC_0(i, k, d-1) \oplus HC_0(i', k, d-1)$$
$$= GP_{m'}^i \oplus GP_{m''}^{i'}$$
$$= GP_{i'+1}^i \oplus GP_m^{i'}$$
$$= GP_m^i. \qquad \square$$

Corollary 8.24 *Let $i \in \mathbb{N}$, $k \in \mathbb{N}$, and $p = \Pi(i+1)$. If $i < 2^n$ and $p \leq k \leq n$, then*

$$HC_0(i, k, n-k) = GP_0^i.$$

Proof Substituting $n - k$ for d in Lemma 8.23, we have $HC_0(i, k, n - k) = GP_0^m$, where

$$m = \max(0, i - 2^{k+(n-k)} + 1) = \max(0, i - 2^n + 1) = 0. \qquad \square$$

Lemma 8.25 *If* $i \in \mathbb{N}$, $k \in \mathbb{N}$, $k \leq n$, *and* $i < 2^n$, *then* $HC(i, k, n) = GP_0^i$.

Proof Let $p = \Pi(i + 1)$. If $p \geq k$, then $HC(i, k, n) = HC_0(i, k, n - k)$ and the claim follows from Corollary 8.24. Thus, we may assume $p < k$. By Lemma 8.20,

$$HC_0(i, k, n - k) = BK_0(i, k) = GP_{i-2^p+1}^i.$$

If $i = 2^p - 1$, then

$$HC(i, k, n) = HC_0(i, k, n - k) = GP_{i-2^p+1}^i = GP_0^i.$$

In the remaining case, by induction and Lemma 8.13,

$$HC(i, k, n) = HC_0(i, k, n - k) \oplus HC(i - 2^p, k, n)$$
$$= GP_{i-2^p+1}^i \oplus GP_0^{i-2^p}$$
$$= GP_0^i. \qquad \square$$

8.3   Leading Zero Anticipation

We turn now to the problem of adding two numbers that are encoded in a floating-point format. The following procedure represents a naive approach to the design of a floating-point adder:

(1) Compare the exponent fields of the summands to determine the right shift necessary to align the significands;
(2) Perform the required right shift on the significand field that corresponds to the lesser exponent;
(3) Add (or subtract) the aligned significands, together with the appropriate rounding constant;
(4) Determine the left shift required to normalize the result;
(5) Perform the left shift and adjust the exponent accordingly;
(6) Compute the final result by assembling the sign, exponent, and significand fields.

It is possible to reduce the latency of floating-point addition by executing some of these operations in parallel. While a large left shift may be required (in the case of subtraction, if massive cancellation occurs), and a large right shift may be required (if the exponents are vastly different), only one of these possibilities will be realized for any given pair of inputs. Thus, an efficient adder typically includes two data

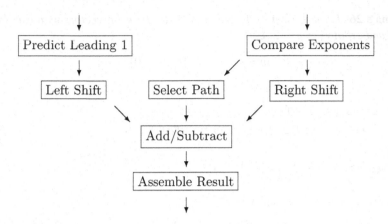

Fig. 8.13 Split-path adder

paths, called the *near* and *far* paths. On the near path, computation proceeds under the assumption that an effective subtraction is to be performed and the exponents differ by at most 1. Thus, the summands are aligned quickly, but time is allocated to Steps (4) and (5). On the far path, which handles the remaining case, Steps (1) and (2) are time-consuming, but the sum is easily normalized, requiring at most a 1-bit shift. A concurrent analysis of the exponents determines which of these paths is actually used to produce the final result.

A critical optimization on the near path is the technique of *leading zero anticipation* (LZA), which allows the normalizing left shift to be determined, and perhaps even performed, in advance of the subtraction. While the precise determination of this shift is a problem of roughly the same complexity as the subtraction itself, a typical LZA allows a 1-bit error in the predicted leading one of the difference (some allow an underestimate, others an overestimate), which may be quickly detected upon sampling a single bit once the difference becomes available. This results in a much faster execution. First, a vector that exhibits the predicted leading one is computed in constant time. A leading zero counter is then applied to this vector to determine the predicted normalizing left shift; this may be performed in logarithmic time with respect to the width of the adder.

The point at which the paths merge is implementation-dependent. Figure 8.13 depicts a design in which Steps (4) and (5) of the near path are executed concurrently with Steps (1) and (2) of the far path and the merging occurs before the addition is performed, thereby avoiding duplication of the adder hardware. On the other hand, many designs use separate adders for the two paths, and some use separate adders to handle various subcases within the near path.

The case in which the operand exponents differ by 1 (and their ordering is known) admits a relatively simple solution. The significand of the smaller operand is shifted by 1 bit, and the following result applies:

Lemma 8.26 *Let $n \in \mathbb{N}$, $n > 1$, and let x and y be n-bit vectors with $expo(x) = n - 1$ and $expo(y) = n - 2$. Let*

$$w = (x \mid \sim y)[n - 2 : 0].$$

Then

(a) $w = 0 \Leftrightarrow x - y = 1$;
(b) $w \neq 0 \Rightarrow expo(w) \leq expo(x - y) \leq expo(w) + 1$.

Proof
(a) Since $x \geq 2^{n-1}$ and $y < 2^{n-1}$,

$$x - y = 1 \Leftrightarrow x = 2^{n-1} \text{ and } y = 2^{n-1} - 1$$

$$\Leftrightarrow x[n-2 : 0] = 0 \text{ and } \sim y[n-2 : 0] = 2^{n-1} - y[n-2 : 0] - 1 = 0$$

$$\Leftrightarrow w = 0.$$

(b) Suppose $w \neq 0$ and let $expo(w) = k$. Then $x[n - 2 : k + 1] = 0$, which implies

$$x = 2^{n-1} + x[k : 0],$$

and

$$\sim y[n - 2 : k + 1] = 2^{n-k-2} - y[n - 2 : k + 1] - 1 = 0,$$

which implies

$$y[n - 2 : k + 1] = 2^{n-k-2} - 1$$

and

$$y = 2^{k+1}(2^{n-k-2} - 1) + y[k : 0] = 2^{n-1} - 2^{k+1} + y[k : 0].$$

Thus,

$$x - y = 2^{k+1} + x[k : 0] - y[k : 0] < 2^{k+1} + 2^{k+1} - 0 = 2^{k+2}.$$

Since $w[k] = 1$, either $x[k] = 1$ or $y[k] = 0$, which implies

$$x - y = 2^{k+1} + x[k : 0] - y[k : 0]$$

$$= 2^{k+1} + 2^k(x[k] - y[k]) + x[k - 1 : 0] - y[k - 1 : 0]$$

$$\geq 2^{k+1} + 0 + 0 - y[k - 1 : 0]$$

$$> 2^{k+1} - 2^k$$

$$= 2^k. \qquad \square$$

The case of equal exponents is more complex. We shall present two separate LZA algorithms that are both generally applicable, imposing no restrictions on the exponent difference, and geared for different requirements with respect to execution speed and resource allocation.

Σ	P	P	P	P	P	P	G	K	K	K	P	K	G	P	P
a	1	0	1	1	0	1	1	0	0	0	1	0	1	0	0
b	0	1	0	0	1	0	1	0	0	0	0	0	1	1	1
s	0	0	0	0	0	0	0	0	0	1	1	0	1	1	
p	1	1	1	1	1	1	0	0	0	0	1	0	0	1	1
g	0	0	0	0	0	0	1	0	0	0	0	0	1	0	0
k	0	0	0	0	0	0	0	1	1	1	0	1	0	0	0

Fig. 8.14 Leading zero anticipation

The first algorithm requires the identification of the larger of the two operands, which simplifies the computation, resulting in faster execution. When the exponents are equal, however, the operand comparison is usually based on the output of the adder, well after the LZA has been initiated. It is necessary, therefore, to replicate the hardware that performs this operation on separate paths corresponding to the two possible orderings and later to select the output of the appropriate path. This scheme, which naturally increases the hardware area requirement, is implemented in the "big core" adder of Chap. 17, for which execution speed is a higher priority. In this design, the least significant bits of the exponents are compared, and a 1-bit significand shift, if required, is performed in advance of the LZA computation. On the other hand, an implementation may further reduce execution time by using four LZA paths—two assuming different exponents, based on Lemma 8.26, and two assuming equal exponents—and selecting the appropriate output once the exponent and significand comparisons have completed.

The second algorithm produces correct results regardless of operand ordering. The computation is more complicated and executes more slowly, but need not be duplicated. Thus, only a single LZA path is required, but the exponent comparison and significand alignment must be performed in advance. This is the basis of the GPU adder of Chap. 22.

Beginning with the simpler algorithm, we note that subtraction of bit vectors x and y with $0 < y < x < 2^n$ is naturally implemented as an n-bit addition that is guaranteed to overflow:

$$x + (\sim y[n-1:0] + 1) = x + 2^n - y - 1 + 1 = 2^n + x - y. \tag{8.3}$$

Let a and b be n-bit vectors with $s = a + b > 2^n$. Our objective is to predict the location of the leading one of the sum, i.e., the greatest $i < n$ such that $s[i] = 1$, or $expo(s[n-1:0])$. We shall compute, in constant time (independent of a, b, and n), a positive integer w such that $expo(s[n-1:0])$ is either $expo(w)$ or $expo(w) - 1$.

We begin with an informal motivating discussion; a formal solution is given by Lemma 8.27. The technique is based on the propagate and generate vectors, $p = a \wedge b$ and $g = a \ \& \ b$. We also define the *kill* vector, $k =\sim a \ \& \ \sim b$. As illustrated in Fig. 8.14, for each index $i < n$, exactly one of $p[i]$, $g[i]$, and $k[i]$ is asserted, and we associate i with one of the symbols P, G, and K accordingly, creating a string of symbols $\Sigma = \sigma_{n-1} \ldots \sigma_1 \sigma_0$, where

$$\sigma_i = \begin{cases} P \text{ if } p[i] = 1 \\ G \text{ if } g[i] = 1 \\ K \text{ if } k[i] = 1. \end{cases}$$

We shall identify an index j such that the leading one of $s[n-1 : 0]$ occurs at either j or $j - 1$. While the computation is actually performed in constant time, it is instructive to view the search as a left-to-right traversal of Σ. Since the hypothesis $s > 2^n$ implies $a[n-1] + b[n-1] \geq 1$, we must have $k[n-1] = 0$, and σ_{n-1} is either P or G. If $\sigma_{n-1} = P$, then there must be a carry into index $n - 1$, resulting in $s[n-1] = 0$, and we may ignore this index and move to index $n - 2$. We continue in this manner until we reach the first occurrence of the symbol G, which we find at some index i. Thus, $\sigma_{n-1} = \sigma_{n-2} = \ldots = \sigma_{i+1} = P$ and $\sigma_i = G$. If σ_{i-1} is either P or G, then the leading one of s must occur at either i or $i - 1$ depending on whether a carry is produced at $i - 1$, and our search is concluded with index $j = i$. But if $\sigma_{i-1} = K$, then the traversal continues until we reach the index j at which the final K occurs, i.e., $\sigma_{i-1} = \ldots = \sigma_j = K$ and $\sigma_{j-1} \neq K$. In this case, the leading one must occur at either j or $j - 1$, depending on whether a carry is produced at $j - 1$.

Thus, we select j as the terminal index of the maximal prefix of the string of the form P^*GK^* (0 or more Ps, followed by a single G, followed by 0 or more Ks). Clearly, this is the maximal $i < n$ for which $\sigma_i \sigma_{i-1}$ is not one of the following combinations: PP, PG, GK, or KK. But this may also be characterized as the maximal $i < n$ such that $w[i] = 1$, where

$$w[i] = \sim\big((p[i] \ \& \ p[i-1]) \mid (p[i] \ \& \ g[i-1]) \mid (g[i] \ \& \ k[i-1]) \mid (k[i] \ \& \ k[i-1])\big)$$

$$= \sim\big((p[i] \ \& \ (p[i-1] \mid g[i-1])) \mid ((g[i] \mid k[i]) \ \& \ k[i-1])\big)$$

$$= \sim\big((p[i] \ \& \ \sim k[i-1]) \mid (\sim p[i] \ \& \ k[i-1])\big)$$

$$= \sim(p[i] \wedge k[i-1])$$

$$= \sim(p[i] \wedge (2k)[i])$$

$$= \sim(p \wedge 2k)[i]),$$

that is, the leading one of the vector $w =\sim (p \wedge 2k)[n-1 : 0]$. Our conclusion, stated in terms of exponents, is that if $j = expo(w)$, then $expo(s[n - 1 : 0])$ is either j or $j - 1$.

A more rigorous version of this derivation is given in the proof of Lemma 8.27, which includes the additional observation that the same conclusion holds when the sum $s = a + b$ is replaced by $s' = a + b + 1$. The significance of this somewhat surprising variation (which we shall derive as a consequence of Lemma 8.4) is that the prediction of the exponent of a positive difference $x - y$ may be based on the vectors $a = x$ and $b =\sim y[n - 1 : 0]$ instead of $b =\sim y[n - 1 : 0] + 1$, thereby avoiding the expense of the increment, since this yields

$$s'[n - 1 : 0] = (a + b + 1) \bmod 2^n = \left(x + (2^n - y - 1) + 1\right) \bmod 2^n = x - y.$$

Note that for this choice of a and b, the assumption $a+b > 2^n$, which is necessary to ensure that the pattern P^*KG^* is broken (i.e., $w \neq 0$), is violated when $x - y = 1$. This case is handled specially by the lemma.

On the other hand, suppose we are computing $\lfloor x - y \rfloor$, where $x \in \mathbb{Z}$ and $y \notin \mathbb{Z}$, and that $a = x$ and b is the complement of the integer truncation of y, i.e.,

$$b = 2^n - \lfloor y \rfloor - 1.$$

Then

$$s = a + b = x + (2^n - \lfloor y \rfloor - 1) = 2^n + \lfloor x - y \rfloor$$

and

$$s[n - 1 : 0] = \lfloor x - y \rfloor.$$

In this case, we are interested in s rather than s'. Thus, (b) and (c) of the lemma are of equal interest.

Lemma 8.27 *Let* $n \in \mathbb{Z}^+$ *and let* a *and* b *be* n-*bit vectors. Let* $p = a$ `^` b, $k =\sim a[n-1 : 0]$ `&` $\sim b[n-1 : 0]$, *and*

$$w =\sim (p \text{ `^` } 2k)[n-1 : 0].$$

Let $s = a + b$ *and* $s' = a + b + 1$.

(1) If $s = 2^n$, *then* $w = s'[n - 1 : 0] = 1$.
(2) If $s > 2^n$, *then*

    (a) $w \geq 2$;
    (b) $expo(w) - 1 \leq expo(s[n - 1 : 0]) \leq expo(w)$;
    (c) $expo(w) - 1 \leq expo(s'[n - 1 : 0]) \leq expo(w)$.

Proof
(1) First note that

$$(p \text{ `^` } 2k)[0] = p[0] \text{ `^` } 2k[0] = s[0] \text{ `^` } 0 = 0 \text{ `^` } 0 = 0,$$

and therefore $w[0] = 1$. We must show that for $0 < i < n$, $w[i] = 0$. Using the notation of the preceding discussion, it suffices to show that $\sigma_i \sigma_{i-1} \in \{PP, PG, GK, KK\}$ because for each element of that set, it may be shown directly that $w[i] = 0$.

Note that by Lemmas 8.9, 8.11, and 8.14, for $0 \le j < n$,

$$p[j] = s[j] \Leftrightarrow a[j-1:0] + b[j-1:0] < 2^j.$$

If $\sigma_i = P$, then since $s[i] = 0 \neq p[i]$, $a[i-1:0] + b[i-1:0] \ge 2^i$, which implies $\sigma_{i-1} \neq K$. Thus, $\sigma_i \sigma_{i-1} \in \{PP, PG\}$.

If $\sigma_i = K$, then $a[i-1:0] + b[i-1:0] = a[i:0] + b[i:0] = s[i:0] = 0$, which implies $\sigma_{i-1} = K$ and $\sigma_i \sigma_{i-1} = KK$.

Finally, if $\sigma_i = G$, then since

$$s[i] = 0, a[i-1:0] + b[i-1:0] = s[i-1:0] = 0$$

which implies $\sigma_{i-1} = K$ and $\sigma_i \sigma_{i-1} = GK$.

(2) First, we shall prove by induction that for $0 \le j < n$, if $w[n-1:j+1] = 0$, then

(i) $a[j:0] + b[j:0] \ge 2^{j+1} \Leftrightarrow k[j] = 0$ and
(ii) $s[n-1:j+1] = 0$.

Clearly, (i) and (ii) hold for $j = n-1$. Suppose the claim is true for some j, $0 < j < n$, and that $w[n-1:j] = 0$. Then $w[n-1:j+1] = 0$, which implies (i) and (ii). We shall show that both conditions hold with $j-1$ substituted for j. Note that since $w[j] = \sim (p \ \hat{} \ 2k)[j] = 0$, $p[j] \neq (2k)[j] = k[j-1]$.

(i) We may assume $k[j-1] = 0$ and therefore $p[j] = 1$. Since $k[j] = 0$, we must have

$$2^{j+1} \le a[j:0] + b[j:0]$$
$$= 2^j (a[j] + b[j]) + a[j-1:0] + b[j-1:0]$$
$$= 2^j + a[j-1:0] + b[j-1:0],$$

and hence $a[j-1:0] + b[j-1:0] \ge 2^j$.

(ii) Since $s[n-1:j+1] = 0$, it will suffice to show that $s[j] = 0$, i.e., $s[j:0] < 2^j$. If $p[j] = 1$ and $k[j-1] = 0$, then, as we have shown, $a[j-1:0] + b[j-1:0] \ge 2^j$, and therefore,

$$s[j:0] = (a[j:0] + b[j:0]) \bmod 2^{j+1}$$
$$= (2^j + a[j-1:0] + b[j-1:0]) \bmod 2^{j+1}$$
$$= (2^{j+1} + (a[j-1:0] + b[j-1:0] - 2^j)) \bmod 2^{j+1}$$
$$= a[j-1:0] + b[j-1:0] - 2^j < 2^j.$$

But if $p[j] = 0$ and $k[j-1] = 1$, then either $k[j] = 1$ and

$$s[j : 0] = a[j : 0] + b[j : 0] = a[j-2 : 0] + b[j-2 : 0] < 2^{j-1} + 2^{j-1} = 2^j$$

or $a[j] = b[j] = 1$ and

$$\begin{aligned}
s[j : 0] &= (a[j : 0] + b[j : 0]) \bmod 2^{j+1} \\
&= (2^j + a[j-2 : 0] + 2^j + b[j-2 : 0]) \bmod 2^{j+1} \\
&= a[j-2 : 0] + b[j-2 : 0] \\
&< 2^j.
\end{aligned}$$

The conclusions of the lemma now follow:

(a) If $w < 2$, i.e., $w[n - 1 : 1] = 0$, then (ii) implies $s[0] = s[n - 1 : 0] \neq 0$, and therefore $a[0] + b[0] = s[0] = 1$, contradicting (i).

(b) Let $j = expo(w)$. Then $w[n - 1 : j + 1] = 0$, which implies (i) and (ii), and $w[j] = 1$, which implies $p[j] = k[j - 1]$. If $p[j] = k[j - 1] = 1$, then

$$\begin{aligned}
a[j : 0] &+ b[j : 0] \\
&= 2^j(a[j] + b[j]) + 2^{j-1}(a[j - 1] + b[j - 1]) + a[j - 2] + b[j - 2] \\
&= 2^j + a[j - 2] + b[j - 2] \\
&< 2^j + 2^{j-1} + 2^{j-1} \\
&= 2^{j+1},
\end{aligned}$$

contradicting (i). Therefore, we must have $p[j] = k[j - 1] = 0$. Since $p[j] = 0$, either $a[j] = b[j] = 0$, in which case

$$s[n - 1 : 0] = s[j : 0] = a[j - 1] + b[j - 1],$$

or $a[j] = b[j] = 1$, which implies

$$\begin{aligned}
s[n - 1 : 0] = s[j : 0] &= (a[j : 0] + b[j : 0]) \bmod 2^{j+1} \\
&= (2^j + a[j - 1 : 0] + b[j - 1 : 0]) \bmod 2^{j+1} \\
&= a[j - 1 : 0] + b[j - 1 : 0].
\end{aligned}$$

Since $k[j - 1] = 0$,

$$2^{j-1} \leq a[j - 1 : 0] + b[j - 1 : 0] < 2^{j+1},$$

i.e., $2^{j-1} \leq s[n - 1 : 0] < 2^{j+1}$ and

$$j - 1 \leq expo(s[n - 1 : 0]) \leq j.$$

(c) We note first that since $2^n < s < s' < 2^{n+1}$,

$$s'[n - 1 : 0] = s' - 2^n = s - 2^n + 1 = s[n - 1 : 0] + 1.$$

We may assume $expo(s[n - 1 : 0]) \neq expo(s'[n - 1 : 0])$; otherwise the claim follows trivially from (b). Let $e = expo(s'[n - 1 : 0])$. Then $s'[n - 1 : 0] = 2^e$ and $s[n - 1 : 0] = 2^e - 1$. Thus,

$$s[e : 0] = s[e - 1 : 0] = 2^e - 1 \neq 2^{e+1} - 1,$$

and by Lemma 8.4, $p[e - 1 : 0] = 2^e - 1$ and $p[e : 0] \neq 2^{e+1} - 1$. It follows that $p[e] = 0$ and $p[e - 1] = 1$, which implies $w[e] = 1$. Therefore, $expo(w) \geq e = 1 + expo(s[n - 1 : 0])$, and according to (b), we must have $expo(w) = e = expo(s'[n - 1 : 0])$.   □

Now suppose we are computing the sum $s = a + b$, where $a = x$ and $b = \sim y[n - 1 : 0]$, and that the ordering of x and y is unknown. Since $s = 2^n + (x - y) - 1$, $s[n] = 1$ iff $x > y$, and thus the two cases are distinguished by examining that bit. If $x < y$, then we are interested in counting the leading zeroes of

$$\sim s[n - 1 : 0] = 2^n - s - 1 = 2^n - (2^n - (x - y) - 1) - 1 = y - x$$

and hence the leading ones, rather than leading zeroes, of $s[n - 1 : 0]$. In the context of the discussion preceding Lemma 8.27, this may be achieved by locating the terminal index of the maximal prefix of the string $\Sigma = \sigma_{n-1} \ldots \sigma_0$ of the form $P^* K G^*$. Thus, without knowing which is the case, we seek the maximal prefix of either form $P^* G K^*$ or $P^* K G^*$. We shall assume that $\sigma_{n-1} = P$, so that the length of this prefix is at least 2. In practice, this is ensured by padding each of x and y with a leading one.

To achieve this objective, we examine adjacent triples rather than pairs of Σ. In the case $x > y$, the pattern is maintained as long as

$$\sigma_{i+1}\sigma_i\sigma_{i-1} \in S_1 = \{PPP, PPG, PGK, GKK, KKK\}.$$

It is easily shown by exhaustive computation that this condition is equivalent to $z_1[i] = 1$, where

$$z_1[i] = (p[i + 1] \; \hat{} \; k[i]) \; \& \; (p[i] \; \hat{} \; k[i - 1]) = ((p[n : 1] \; \hat{} \; k) \; \& \; (p \; \hat{} \; 2k))[i].$$

Similarly, if $x < y$, the condition is

$$\sigma_{i+1}\sigma_i\sigma_{i-1} \in S_2 = \{PPP, PPK, PKG, KGG, GGG\},$$

or $z_2[i] = 1$, where

$$z_2[i] = (p[i + 1] \verb|^| g[i]) \ \& \ (p[i] \verb|^| g[i - 1]) = ((p[n : 1] \verb|^| g) \ \& \ (p \verb|^| 2g))[i].$$

This is the motivation for examining triples rather than pairs: If $x > y$ and $\sigma_{n-1} \ldots \sigma_{i-1}$ is the maximal prefix of the form $P^* G K^*$, then $\sigma_{i+1} \sigma_i \sigma_{i-1} \in S_1 - S_2$, which implies that the next triple $\sigma_i \sigma_{i-1} \sigma_{i-2} \notin S_2 - S_1$, and thus the pattern cannot be broken by an element of S_2. Similarly, for $x < y$, the pattern cannot be broken by an element of S_1. Consequently, we seek the maximal index i such that $\sigma_{i+1} \sigma_i \sigma_{i-1} \notin S_1 \cup S_2$, which is the index of the leading one of the vector $w[n - 2 : 0]$, where

$$w =\sim (z_1 \mid z_2).$$

This analysis, however, assumes that the pattern is broken at some point in the string Σ. This is not the case when Σ is of the form P^*, $P^* K G^*$, or $P^* G K^*$. The values of s that correspond to these patterns are $2^n - 1$, $2^n - 2$, and 2^n, respectively, which are handled specially by the following lemma.

Lemma 8.28 *Let $n \in \mathbb{Z}^+$ and let a and b be n-bit vectors. Let $p = a \verb|^| b$, $g = a \ \& \ b$, $k =\sim a[n-1 : 0] \ \& \ \sim b[n-1 : 0]$,*

$$z = (p[n : 1] \verb|^| k) \ \& \ (p \verb|^| 2k) \mid (p[n : 1] \verb|^| g) \ \& \ (p \verb|^| 2g),$$

and

$$w =\sim z[n - 2 : 0].$$

Let $s = a + b$ and $s' = a + b + 1$.

(1) If $s = 2^n - 1$, then $w = 0$.
(2) If $n > 1$ and either $s = 2^n$ or $s = 2^n - 2$, then $w = 1$.
(3) If $s > 2^n$ and $p[n - 1] = 1$, then

    (a) $w \geq 2$;
    (b) $expo(w) - 1 \leq expo(s[n - 1 : 0]) \leq expo(w)$;
    (c) $expo(w) - 1 \leq expo(s'[n - 1 : 0]) \leq expo(w)$.

(4) If $s < 2^n - 2$ and $p[n - 1] = 1$, then

    (a) $w \geq 2$;
    (b) $expo(w) - 1 \leq expo(\sim s[n - 1 : 0]) \leq expo(w)$;
    (c) $expo(w) - 1 \leq expo(\sim s'[n - 1 : 0]) \leq expo(w)$.

Proof
(1) By Lemma 8.4, $p[i] = 1$ for $0 \leq i < n$. Thus, for $0 \leq i < n - 1$,

$$w[i] = \sim\big((p[i+1] \hat{~} k[i]) \text{ \& } (p[i] \hat{~} k[i-1]) \mid (p[i+1] \hat{~} g[i]) \text{ \& } (p[i] \hat{~} g[i-1])\big)$$

$$= \sim(p[i+1] \text{ \& } p[i])$$

$$= 0.$$

(2) First note that

$$w[0] = \sim\big((p[1] \hat{~} k[0]) \text{ \& } (p[0] \hat{~} 0) \mid (p[1] \hat{~} g[0]) \text{ \& } (p[0] \hat{~} 0)\big) = \sim(p[0] \text{ \& } p[1]).$$

Thus, in both cases, $w[0] = 1$, for otherwise $s[1 : 0] = s \bmod 4 = 3$. We must show that $w[i] = 0$ for $0 < i < n-1$. It suffices to show that $\sigma_{i+1}\sigma_i\sigma_{i-1} \in S_1 \cup S_2$, because for each element of that set it may be directly verified that $w[i] = 0$. We shall prove this for the case $s = 2^n$; the case $2^n - 2$ is similar. The proof is an extension of that of Lemma 8.27 (1).

If $\sigma_i = P$, as we have observed in the preceding proof, $\sigma_{i-1} \neq K$. Since $s[i+1] = 0$, $\sigma_{i+1} = P$. Thus,

$$\sigma_{i+1}\sigma_i\sigma_{i-1} \in \{PPP, PPG\} \subset S_1.$$

If $\sigma_i = K$, then once again $\sigma_{i-1} = K$, and since $s[i+1] = 0$, $\sigma_{i+1} \in \{G, K\}$ and

$$\sigma_{i+1}\sigma_i\sigma_{i-1} \in \{KKK, GKK\} \subset S_1.$$

Finally, if $\sigma_i = G$, then once again $\sigma_{i-1} = K$, and since $s[i+1] = 0$, $\sigma_{i+1} = P$ and

$$\sigma_{i+1}\sigma_i\sigma_{i-1} = PGK \in S_1.$$

(3) The proof for this case, which is similar to that of Lemma 8.27, is based on the following claim: for $0 \leq j < n-1$, if $w[n-2 : j+1] = 0$, then

(i) $a[j : 0] + b[j : 0] \geq 2^{j+1} \Leftrightarrow k[j] = 0$, and
(ii) $s[n-1 : j+1] = 0$.

First we show that for $0 < j < n-1$, if $w[n-2 : j+1] = 0$ and (i) and (ii) hold, then

$$w[j] = 1 \Leftrightarrow p[j] = k[j-1]. \tag{8.4}$$

We have

$$z[j] = (p[j+1] \hat{~} k[j]) \text{ \& } (p[j] \hat{~} k[j-1]) \mid (p[j+1] \hat{~} g[j]) \text{ \& } (p[j] \hat{~} g[j-1]).$$

We shall show that this may be simplified to

$$z[j] = p[j] \hat{~} k[j-1],$$

which is equivalent to (8.4). By (i), since $s[j+1] = 0$,

$$p[j+1] = 1 \Leftrightarrow a[j:0] + b[j:0] \geq 2^{j+1} \Leftrightarrow k[j] = 0. \tag{8.5}$$

Thus, $p[j+1] \wedge k[j] = 1$, and

$$z[j] = (p[j] \wedge k[j-1]) \mid (p[j+1] \wedge g[j]) \, \& \, (p[j] \wedge g[j-1]).$$

If $p[j] \wedge k[j-1] = 1$, then clearly $z[j] = 1$, and therefore we may assume $p[j] = k[j-1]$, and we need only show

$$(p[j+1] \wedge g[j]) \, \& \, (p[j] \wedge g[j-1]) = 0,$$

i.e., either $p[j+1] = g[j]$ or $p[j] = g[j-1]$. In fact, we shall show that $p[j+1] = g[j]$.

If $p[j] = 1$, then it follows from (i) that $a[j:0]+b[j:0] \geq 2^{j+1}$, which implies $k[j-1] = 0$, contradicting our assumption that $p[j] = k[j-1]$. We are left with the case $p[j] = k[j-1] = 0$. Now either $g[j] = 1$ or $k[j] = 1$. We appeal to the equivalence (8.5). If $g[j] = 1$, then $k[j] = 0$, which implies $p[j+1] = 1 = g[j]$. On the other hand, if $k[j] = 1$, then $p[j+1] = 0 = g[j]$. This completes the proof of (8.4).

Next, we show that the above claim holds for $j = n - 1$ (the base case of the induction). Since $p[n-1] = 1$, $a[n-1] + b[n-1] = 1$, and (i) and (ii) follow easily:

$$\begin{aligned}
a[j:0] + b[j:0] &= a[n-2:0] + b[n-2:0] \\
&= a + b - 2^{n-1}(a[n-1] + b[n-1]) \\
&= a + b - 2^{n-1} \\
&> 2^n - 2^{n-1} \\
&= 2^{n-1} \\
&= 2^{j+1}
\end{aligned}$$

and

$$2^n < s = 2^{n-1} + a[n-2:0] + b[n-2:0] < 2^{n-1} + 2^{n-1} + 2^{n-1} = 2^n + 2^{n-1},$$

and therefore $s[n-1:0] = s \bmod 2^n < 2^{n-1}$, which implies $s[n-1:j+1] = s[n-1] = 0$.

The induction may be completed as in the proof of Lemma 8.27, invoking (8.4). The proofs of (a), (b), and (c) are also the same.

(4) We shall derive this result as a consequence of (3). Let $\hat{a} =\sim a[n-1:0]$, $\hat{b} =\sim b[n-1:0]$, $\hat{s} = \hat{a} + \hat{b}$, and $\hat{s}' = \hat{a} + \hat{b} + 1$. Then

$$\hat{s} = (2^n - a - 1) + (2^n - b - 1)$$
$$= 2^{n+1} - (a+b) - 2$$
$$> 2^{n+1} - (2^n - 2) - 2$$
$$= 2^n.$$

Let $\hat{p} = \hat{a} \,\hat{}\, \hat{b}$, $\hat{g} = \hat{a} \,\&\, \hat{b}$, $\hat{k} =\sim\hat{a}[n-1:0] \,\&\, \sim\hat{b}[n-1:0]$,

$$\hat{z} = (\hat{p}[n:1] \,\hat{}\, \hat{k}) \,\&\, (\hat{p} \,\hat{}\, 2\hat{k}) \mid (\hat{p}[n:1] \,\hat{}\, \hat{g}) \,\&\, (\hat{p} \,\hat{}\, 2\hat{g}),$$

and

$$\hat{w} =\sim\hat{z}[n-2:0].$$

Then the conclusions of (3) apply to $\hat{s}$, $\hat{s}'$, and $\hat{w}$. But $\hat{p} = p$, $\hat{g} = k$, and $\hat{k} = g$. Thus, $\hat{z} = z$, $\hat{w} = w$, and we have

$$w \geq 2,$$

$$expo(w) - 1 \leq expo(\hat{s}[n-1:0]) \leq expo(w),$$

and

$$expo(w) - 1 \leq expo(\hat{s}'[n-1:0]) \leq expo(w).$$

But

$$\hat{s} = 2^{n+1} - (a+b+1) - 1 = 2^n + (2^n - s' - 1) = 2^n + \sim s'[n-1:0],$$

and hence

$$\hat{s}[n-1:0] =\sim s'[n-1:0].$$

Similarly,

$$\hat{s}' = \hat{s} + 1 = 2^{n+1} - (a+b) - 1 = 2^n + (2^n - s - 1) = 2^n + \sim s[n-1:0]$$

and

$$\hat{s}'[n-1:0] =\sim s[n-1:0]. \qquad\qquad \square$$

8.4 Counting Leading Zeroes

Once the vector w with predicted leading one has been constructed by any of the preceding algorithms, its leading zeroes remain to be counted. The algorithm of Definition 8.9 below is a logarithmic-time recursive solution to this problem due to Oklobdzija [24]. It operates on a 2^n-bit nonzero vector x, so that w generally must be padded with zeroes on the right to increase its width to a power of 2. At the kth stage of the computation, where $k = 0, \ldots, n$, x is conceptually partitioned into 2^k-bit segments, $S_k(i) = x[2^k(i + 1) - 1 : 2^k i], 0 \le i < 2^{n-k}$. Two values associated with each segment are recursively computed: (a) $Z_k(i)$ is a boolean that is asserted iff $S_k(i) = 0$, and (b) $C_k(i)$ is the leading zero count of $S_k(i)$, valid if $Z_k(i) = 0$. Since the lone segment of the final stage is $S_n(0) = x$, the desired result is $CLZ(x, n) = C_n(0)$.

Definition 8.9 Let x be a 2^n-bit vector, where $n \in \mathbb{Z}^+$. For $0 \le k \le n$ and $0 \le i < 2^{n-k}$, let

$$Z_k(i) = \begin{cases} 1 & \text{if } k = 0 \text{ and } x[i] = 0 \\ 0 & \text{if } k = 0 \text{ and } x[i] = 1 \\ 1 & \text{if } k > 0 \text{ and } Z_{k-1}(2i + 1) = Z_{k-1}(2i) = 1 \\ 0 & \text{otherwise} \end{cases}$$

and

$$C_k(i) = \begin{cases} 0 & \text{if } k = 0 \\ 2^{k-1} \mid C_{k-1}(2i) & \text{if } Z_{k-1}(2i + 1) = 1 \\ C_{k-1}(2i + 1) & \text{if } Z_{k-1}(2i + 1) = 0. \end{cases}$$

Then $CLZ(x, n) = C_n(0)$.

Lemma 8.29 *Let x be a 2^n-bit vector, where $n \in \mathbb{N}$. If $x \ne 0$, then*

$$CLZ(x, n) = 2^n - 1 - expo(x).$$

Proof Note that for $1 \le k \le n$, $S_k(i)$ is the concatenation of

$$S_{k-1}(2i + 1) = x[2^{k-1}((2i + 1) + 1) - 1 : 2^{k-1}(2i + 1)]$$
$$= x[2^k(i + 1) - 1 : 2^k i + 2^{k-1}]$$

and

$$S_{k-1}(2i) = x[2^{k-1}(2i + 1) - 1 : 2^{k-1}(2i)]$$
$$= x[2^k i + 2^{k-1} - 1 : 2^k i],$$

i.e.,

$$S_k(i) = 2^{2^{k-1}} S_{k-1}(2i + 1) + S_{k-1}(2i).$$

We shall prove the following by induction on k:

(a) $Z_k(i) = 1 \Leftrightarrow S_k(i) = 0$;
(b) $Z_k(i) = 0 \Rightarrow C_k(i) = 2^k - 1 - expo(S_k(i))$.

By hypothesis, $S_n(0) = x[2^n - 1 : 0] = x \neq 0$, and hence it follows from the claim that $Z_n(0) = 0$ and $CLZ(x, n) = C_n(0) = 2^n - 1 - expo(x)$.

The claim holds trivially for $k = 0$. Assume that it holds for $k - 1$, where $0 < k \leq n$. Let $0 \leq i < 2^{n-k}$. Then

$$Z_k(i) = 1 \Leftrightarrow Z_{k-1}(2i + 1) = Z_{k-1}(2i) = 1$$

$$\Leftrightarrow S_{k-1}(2i + 1) = S_{k-1}(2i) = 0$$

$$\Leftrightarrow S_k(i) = 0.$$

Assume $Z_k(i) = 0$. If $Z_{k-1}(2i + 1) = 1$, then $Z_{k-1}(2i) = 0$, $S_{k-1}(2i + 1) = 0$, $S_k(i) = S_{k-1}(2i)$, and by Corollary 3.11,

$$C_k(i) = 2^{k-1} \mid C_{k-1}(2i)$$

$$= 2^{k-1} + C_{k-1}(2i)$$

$$= 2^{k-1} + 2^{k-1} - 1 - expo(S_{k-1}(2i))$$

$$= 2^k - 1 - expo(S_k(i)).$$

Thus, we may also assume $Z_{k-1}(2i + 1) = 0$, and hence $S_{k-1}(2i + 1) \neq 0$. Let $e = expo(S_{k-1}(2i + 1))$. Then $2^e \leq S_{k-1}(2i + 1) \leq 2^{e+1} - 1$. Since

$$S_k(i) \geq 2^{2^{k-1}} S_{k-1}(2i + 1) \geq 2^{2^{k-1}} 2^e = 2^{2^{k-1}+e},$$

$expo(S_k(i)) \geq 2^{k-1} + e$. Since

$$S_k(i) < 2^{2^{k-1}} S_{k-1}(2i + 1) + 2^{2^{k-1}}$$

$$= 2^{2^{k-1}}(S_{k-1}(2i + 1) + 1)$$

$$\leq 2^{2^{k-1}} 2^{e+1}$$

$$= 2^{2^{k-1}+e+1},$$

$expo(S_k(i)) < 2^{k-1} + e + 1$, and hence $expo(S_k(i)) = 2^{k-1} + e$. Thus,

$$C_k(i) = C_{k-1}(2i + 1)$$

$$= 2^{k-1} - 1 - e$$

$$= 2^{k-1} - 1 - (expo(S_k(i)) - 2^{k-1})$$

$$= 2^k - 1 - expo(S_k(i)). \qquad \square$$

Chapter 9
Multiplication

While the RTL implementation of integer multiplication is a challenging problem, its extension to the floating-point domain, unlike addition, does not present any significant difficulties that have not already been addressed. The focus of this chapter, therefore, is the multiplication of integers.

Let $x \in \mathbb{N}$ and $y \in \mathbb{N}$. We shall refer to x and y as the *multiplicand* and the *multiplier*, respectively. A natural approach to the computation of the product xy begins with the bit-wise decomposition of the multiplier provided by Lemma 2.20:

$$y = (\beta_{w-1} \cdots \beta_0)_2 = \sum_{i=0}^{w-1} 2^i \beta_i, \tag{9.1}$$

where we assume $y < 2^w$ and for $i = 0, \ldots, w - 1$, $\beta_i = y[i]$. The product may then be computed as

$$xy = \sum_{i=0}^{w-1} 2^i \beta_i x.$$

Thus, the computation is reduced to the summation of at most w nonzero terms, called *partial products*, each of which is derived by an appropriate shift of x. In practice, this summation is performed with the use of a tree of compressors similar to the 3:2 compressor shown in Fig. 8.4, which reduces the number of addends to 2 so that only a single carry-propagate addition is required. It is clear that two 3:2 compressors may be combined to form a 4:2 compressor and that 2^{k-2} 4:2 compressors may be used to reduce a sum of 2^k terms to 2^{k-1} in constant time. Consequently, the hardware needed to compress w terms to two grows linearly with w, and the run time grows logarithmically.

Naturally, any reduction in the number of partial products generated in a multiplication would tend to reduce the latency of the operation. Most modern

© The Author(s), under exclusive license to Springer Nature Switzerland AG 2022
D. M. Russinoff, *Formal Verification of Floating-Point Hardware Design*,
https://doi.org/10.1007/978-3-030-87181-9_9

i	11	10	9	8	7	6	5	4	3	2	1	0
$y[i]$	0	1	1	1	0	0	1	1	1	1	1	0
β_i	1	0	0	-1	0	1	0	0	0	0	-1	0

Fig. 9.1 Radix-2 Booth encoding of $y = (011100111110)_2$

multipliers achieve this objective through some version of a technique discovered by A. D. Booth in the early days of computer arithmetic [2]. After a brief discussion of Booth's original algorithm (Sect. 9.1), we shall present a popular refinement known as the *radix-4 modified Booth algorithm* [15] (Sect. 9.2), which limits the number of partial products to half the multiplier width. We begin with a signed integer version, which is applicable to integer multiplication instructions. We then describe a restriction of this design to unsigned integers, which results in a slightly simpler computation and is suitable for floating-point operations. Each of the three subsequent sections contains a variant of this algorithm.

9.1 Radix-2 Booth Encoding

Booth encoding is based on the observation that if we allow -1, along with 0 and 1, as a value of the digit β_i in (9.1), then the representation is no longer unique. Thus, we may seek to minimize the number of nonzero digits and consequently the number of partial products in the expression for xy, at the expense of introducing a negation along with the shift of x in the case $\beta_i = -1$. Moreover, this scheme extends naturally to the encoding of signed integers.

In fact, as illustrated for a 12-bit vector y in Fig. 9.1, any maximal uninterrupted sequence of 1s in the binary expansion of y,

$$y[k] = y[k+1] = \ldots = y[\ell-1] = 1,$$

where $\ell > k$, may be replaced with as few as two nonzero entries,

$$\beta_k = -1, \beta_{k+1} = \ldots = \beta_{\ell-1} = 0, \beta_\ell = 1,$$

using the identity

$$\sum_{i=k}^{\ell-1} 2^i = 2^\ell - 2^k.$$

Thus, the decomposition $\beta_i = y[i]$ is replaced by the formula

$$\beta_i = \begin{cases} 1 \text{ if } y[i] = 0 \text{ and } y[i-1] = 1 \\ -1 \text{ if } y[i] = 1 \text{ and } y[i-1] = 0 \\ 0 \text{ if } y[i] = y[i-1], \end{cases}$$

or more succinctly,

$$\beta_i = y[i-1] - y[i].$$

The correctness of this encoding for the general case of a signed integer is easily established: if y is an n-bit vector, then since $y[-1] = 0$,

$$\sum_{i=0}^{n-1} 2^i \beta_i = \sum_{i=0}^{n-1} 2^i y[i-1] - \sum_{i=0}^{n} 2^i y[i]$$

$$= \sum_{i=0}^{n-2} 2^{i+1} y[i] - \sum_{i=0}^{n-1} 2^i y[i]$$

$$= \sum_{i=0}^{n-1} 2^i y[i] - 2^n y[n-1]$$

$$= y - 2^n y[n-1]$$

$$= si(y, n).$$

Clearly, the reduction in partial products afforded by this scheme is greatest for arguments containing long strings of 1s, while in the worst case (alternating 0s and 1s), all digits are nonzero, and of course, the number of such digits cannot be predicted. Consequently, any benefit is limited to designs that generate and accumulate partial products sequentially, as was common in early computing, when execution speed was readily sacrificed to conserve die area. Modern multiplier architectures, however, employ compression trees that combine the partial products in a single cycle. An effective multiplier encoding scheme today must reduce the depth of such a tree in a consistent and predictable way.

9.2   Radix-4 Booth Encoding

As a notational convenience, we shall assume that x and y are signed integer encodings of widths n and $2m$, respectively, with $\hat{x} = si(x, n)$ and $\hat{y} = si(y, 2m)$. Our objective is an efficient computation of the product $\hat{x}\hat{y}$ as a sum of m partial products. Conceptually, the multiplier y is partitioned into m 2-bit slices, $y[2i + 1 :$

$2i]$, $i = 0, \ldots, m - 1$. We seek to define an integer digit θ_i corresponding to each slice such that

$$\hat{y} = \sum_{i=0}^{m-1} 2^{2i} \theta_i.$$

The standard radix-4 representation,

$$\theta_i = 2y[2i+1] + y[2i] = y[2i + 1 : 2i],$$

offers no advantage over the ordinary binary representation, since the case $\theta_i = 3$ effectively involves two partial products rather than one. Instead, we define θ_i as a combination of two successive radix-2 Booth digits:

$$\begin{aligned}
\theta_i &= 2\beta_{2i+1} + \beta_{2i} \\
&= 2(y[2i] - y[2i+1]) + (y[2i-1] - y[2i]) \\
&= y[2i-1] + y[2i] - 2y[2i+1].
\end{aligned}$$

$y[2i+1:2i-1]$	$\beta_{2i+1}\beta_{2i}$		θ_i
000	0	0	0
001	0	1	1
010	1	-1	1
011	1	0	2
100	-1	0	-2
101	-1	1	-1
110	0	-1	-1
111	0	0	0

Fig. 9.2 Radix-4 Booth encoding

Since β_{2i+1} and β_{2i} cannot be nonzero and equal, these digits are confined to the range $-2 \leq \theta_i \leq 2$, and we have an expression for y as a sum of at most m nonzero terms, each with absolute value a power of 2, as summarized in Fig. 9.2.

Definition 9.1 For $y \in \mathbb{N}$ and $i \in \mathbb{N}$,

$$\theta_i(y) = y[2i-1] + y[2i] - 2y[2i+1].$$

Lemma 9.1 *If y is a bit vector of positive width $2m$ and $\hat{y} = si(y, 2m)$, then*

$$\hat{y} = \sum_{i=0}^{m-1} 2^{2i} \theta_i(y).$$

Proof We shall prove, by induction, that for $0 \leq k \leq m$,

$$y[2k-1:0] = \sum_{i=0}^{k-1} 2^{2i}\theta_i + 2^{2k}y[2k-1],$$

where $\theta_i = \theta_i(y)$. The claim is trivial for $k = 0$. Assuming that it holds for some $k < m$, we have

$$y[2k+1:0] = 2^{2k}y[2k+1:2k] + y[2k-1:0]$$

$$= 2^{2k}(2y[2k+1] + y[2k]) + \sum_{i=0}^{k-1} 2^{2i}\theta_i + 2^{2k}y[2k-1]$$

$$= 2^{2k}(2y[2k+1] + y[2k] + y[2k-1]) + \sum_{i=0}^{k-1} 2^{2i}\theta_i$$

$$= 2^{2k}(4y[2k+1] + \theta_k) + \sum_{i=0}^{k-1} 2^{2i}\theta_i$$

$$= \sum_{i=0}^{k} 2^{2i}\theta_i + 2^{2(k+1)}y[2(k+1)-1],$$

which completes the induction. In particular, substituting m for k, we have

$$\hat{y} = y[2m-1:0] - 2^{2m}y[2m-1]$$

$$= \left(\sum_{i=0}^{m-1} 2^{2i}\theta_i + 2^{2m}y[2m-1]\right) - 2^{2m}y[2m-1]$$

$$= \sum_{i=0}^{m-1} 2^{2i}\theta_i. \qquad \qquad \square$$

Our goal is to compute

$$\hat{x} \cdot \hat{y} = \hat{x} \sum_{i=0}^{m-1} 2^{2i}\theta_i = \sum_{i=0}^{m-1} 2^{2i}\theta_i\hat{x}. \qquad (9.2)$$

Each term of this sum will correspond to a row in an array of m partial products of width $2m+n$. These are constructed by means of a 5-to-1 multiplexer, which selects an n-bit vector B_i, representing the multiple of x determined by θ_i, as given by the following lemma:

Lemma 9.2 *Let x be an n-bit vector with $\hat{x} = si(x, n)$. Let $\theta \in \{-2, -1, 0, 1, 2\}$,*

$$\sigma = \begin{cases} 0 \ if \theta \geq 0 \\ 1 \ if \theta < 0, \end{cases}$$

$$\tau = \begin{cases} \sigma \ \hat{} \ x[n-1] \ if \theta \neq 0 \\ 0 \qquad\qquad if \theta = 0, \end{cases}$$

and

$$B = \begin{cases} (x\theta)[n-1:0] \ if \sigma = 0 \\ \sim(x|\theta|)[n-1:0] \ if \sigma = 1. \end{cases}$$

Then $B = \hat{x}\theta + 2^n\tau - \sigma$.

Proof First suppose $x[n-1] = 0$ so that $\hat{x} = x$. If $\sigma = 0$, then $\tau = 0$ and

$$B = (x\theta)[n-1:0] = x\theta,$$

and if $\sigma = 1$, then $\tau = 1$ and

$$B = \sim(x|\theta|)[n-1:0] = 2^n - (x|\theta|)[n-1:0] - 1 = 2^n - x|\theta| - 1 = 2^n + x\theta - 1.$$

On the other hand, suppose $x[n-1] = 1$. Then $\hat{x} = x - 2^n$. The case $\theta = 0$ is trivial. Note that if $|\theta| = 2$, then we have

$$(x|\theta|)[n-1:0] = (2x)[n-1:0] = 2x - 2^n = 2(\hat{x} + 2^n) - 2^n = \hat{x}|\theta| + 2^n,$$

and if $|\theta| = 1$, then the same relation holds:

$$(x|\theta|)[n-1:0] = x[n-1:0] = x = \hat{x} + 2^n = \hat{x}|\theta| + 2^n.$$

```
0 0 0 0 0 0 0 0 0 0 0 0 0 0 1 τ̄₀ • • • • • •
0 0 0 0 0 0 0 0 0 0 0 0 1 τ̄₁ • • • • • • 0 σ₀
0 0 0 0 0 0 0 0 0 0 1 τ̄₂ • • • • • • 0 σ₁ 0 0
0 0 0 0 0 0 0 0 1 τ̄₃ • • • • • • 0 σ₂ 0 0 0 0
0 0 0 0 0 0 1 τ̄₄ • • • • • • 0 σ₃ 0 0 0 0 0 0
0 0 0 0 1 τ̄₅ • • • • • • 0 σ₄ 0 0 0 0 0 0 0 0
0 0 1 τ̄₆ • • • • • • 0 σ₅ 0 0 0 0 0 0 0 0 0 0
1 τ̄₇ • • • • • • 0 σ₆ 0 0 0 0 0 0 0 0 0 0 0 0
```

Fig. 9.3 Radix-4 partial product array

Thus, if $\theta > 0$, then $\sigma = 0$, $\tau = 1$, and

$$B = (x\theta)[n-1:0] = (x|\theta|)[n-1:0] = \hat{x}|\theta| + 2^n = \hat{x}\theta + 2^n,$$

and if $\theta < 0$, then $\sigma = 1$, $\tau = 0$, and

$$B = \sim (x|\theta|)[n-1:0] = 2^n - (x|\theta|)[n-1:0] - 1 = 2^n - (\hat{x}|\theta| + 2^n) - 1 = \hat{x}\theta - 1. \quad \square$$

We shall construct an array of m partial products pp_i, $0 \le i < m$, corresponding to the terms of (9.2). As suggested by Lemma 9.2, pp_i will essentially be the n-bit vector

$$B_i = \hat{x}\theta_i + 2^n\tau_i - \sigma_i$$

determined by $\theta_i = \theta_i(y)$, shifted by $2i$ bits. As a visual aid, the array is depicted for the case $m = 8$, $n = 6$ in Fig. 9.3. In order to understand its construction, note that we must correct for the discrepancy between $2^{2i}B_i$ and $2^{2i}\hat{x}\theta_i$ by adding $2^{2i}\sigma_i$ and subtracting $2^{2i+n}\tau_i$. The addition is achieved simply by the insertion of σ_i at index $2i$ of pp_{i+1} (although the term corresponding to $i = m-1$ must be handled otherwise). The subtraction may be viewed as a two-step process. First, we insert 1s at indices $2i+n$ and $2i+n+1$ of pp_i. If we include an additional 1 at index n, then the cumulative effect of this is merely a carry-out to index $2m + n$:

$$2^n + \sum_{i=0}^{m-1}(2^{n+1+2i} + 2^{n+2i}) = 2^n\left(1 + \sum_{i=0}^{m-1}(2^{2i+1} + 2^{2i})\right) \tag{9.3}$$

$$= 2^n\left(1 + \sum_{j=0}^{2m-1} 2^j\right)$$

$$= 2^{n+2m}.$$

But the motivation for this step is that the 1 at index $2i + n$ may now be subtracted off in the case $\tau_i = 1$. Thus, in the second step, we replace that 1 with $\bar{\tau}_i = 1 - \tau_i$.

Note that we still must account for 2 missing bits: the 2s complement completion term $2^{2(m-1)}\sigma_{m-1}$ and the term 2^n required to complete the carry-out. These may be handled by an additional array entry.

This strategy is formalized by the following:

Lemma 9.3 *Let x and y be bit vectors of positive widths n and $2m$, respectively, with $\hat{x} = si(x, n)$ and $\hat{y} = si(y, 2m)$. For $i = 0, \ldots, m-1$, let $\theta_i = \theta_i(y)$,*

$$\sigma_i = \begin{cases} 0 \ \text{if } \theta_i \ge 0 \\ 1 \ \text{if } \theta_i < 0, \end{cases}$$

$$\tau_i = \begin{cases} \sigma_i \mathbin{\char`\^} x[n-1] & \text{if } \theta_i \neq 0 \\ 0 & \text{if } \theta_i = 0, \end{cases}$$

$$\bar{\tau}_i = 1 - \tau_i,$$

$$B_i = \begin{cases} (x\theta_i)[n-1:0] & \text{if } \sigma_i = 0 \\ \sim(x|\theta_i|)[n-1:0] & \text{if } \sigma_i = 1, \end{cases}$$

and

$$pp_i = \begin{cases} 2^{n+1} + 2^n \bar{\tau}_0 + B_0 & \text{if } i = 0 \\ 2^{n+1+2i} + 2^{n+2i} \bar{\tau}_i + 2^{2i} B_i + 2^{2(i-1)} \sigma_{i-1} & \text{if } i > 0. \end{cases}$$

Then

$$2^n + 2^{2(m-1)} \sigma_{m-1} + \sum_{i=0}^{m-1} pp_i \equiv \hat{x}\hat{y} \pmod{2^{n+2m}}.$$

Proof The left-hand side of the conclusion of the lemma is

$$2^n + 2^{2(m-1)} \sigma_{m-1} + \sum_{i=0}^{m-1} \left(2^{n+1+2i} + 2^{n+2i}(1 - \tau_i) + 2^{2i} B_i \right) + \sum_{i=1}^{m-1} 2^{2(i-1)} \sigma_{i-1}.$$

The constant terms of this sum contribute 2^{2m+n} as shown in (9.3). The final term may be rewritten as

$$\sum_{i=1}^{m-1} 2^{2(i-1)} \sigma_{i-1} = \sum_{i=0}^{m-2} 2^{2i} \sigma_i = \sum_{i=0}^{m-1} 2^{2i} \sigma_i - 2^{2(m-1)} \sigma_{m-1}.$$

Thus, the second term is canceled, and Lemma 9.1 yields

$$2^n + 2^{2(m-1)} \sigma_{m-1} + \sum_{i=0}^{m-1} pp_i = 2^{n+2m} + \sum_{i=0}^{m-1} (-2^{n+2i} \tau_i + 2^{2i} B_i) + \sum_{i=0}^{m-1} 2^{2i} \sigma_i$$

$$= 2^{n+2m} + \sum_{i=0}^{m-1} 2^{2i} \left(B_i - 2^n \tau_i + \sigma_i \right)$$

$$= 2^{n+2m} + \sum_{i=0}^{m-1} 2^{2i} \hat{x}\theta_i$$

$$= 2^{n+2m} + \hat{x}\hat{y}. \qquad \qquad \square$$

$$
\begin{array}{cccccccccccccccccccccccc}
0 & 0 & 0 & 0 & 0 & 0 & 0 & 0 & 0 & 0 & 0 & 0 & 0 & \bar{\sigma}_0 & \sigma_0 & \sigma_0 & \bullet & \bullet & \bullet & \bullet & \bullet & \bullet \\
0 & 0 & 0 & 0 & 0 & 0 & 0 & 0 & 0 & 0 & 0 & 0 & 1 & \bar{\sigma}_1 & \bullet & \bullet & \bullet & \bullet & \bullet & \bullet & 0 & \sigma_0 \\
0 & 0 & 0 & 0 & 0 & 0 & 0 & 0 & 0 & 0 & 1 & \bar{\sigma}_2 & \bullet & \bullet & \bullet & \bullet & \bullet & \bullet & 0 & \sigma_1 & 0 & 0 \\
0 & 0 & 0 & 0 & 0 & 0 & 0 & 0 & 1 & \bar{\sigma}_3 & \bullet & \bullet & \bullet & \bullet & \bullet & \bullet & 0 & \sigma_2 & 0 & 0 & 0 & 0 \\
0 & 0 & 0 & 0 & 0 & 0 & 1 & \bar{\sigma}_4 & \bullet & \bullet & \bullet & \bullet & \bullet & \bullet & 0 & \sigma_3 & 0 & 0 & 0 & 0 & 0 & 0 \\
0 & 0 & 0 & 0 & 1 & \bar{\sigma}_5 & \bullet & \bullet & \bullet & \bullet & \bullet & \bullet & 0 & \sigma_4 & 0 & 0 & 0 & 0 & 0 & 0 & 0 & 0 \\
0 & 0 & 1 & \bar{\sigma}_6 & \bullet & \bullet & \bullet & \bullet & \bullet & \bullet & 0 & \sigma_5 & 0 & 0 & 0 & 0 & 0 & 0 & 0 & 0 & 0 & 0 \\
1 & \bar{\sigma}_7 & \bullet & \bullet & \bullet & \bullet & \bullet & \bullet & 0 & \sigma_6 & 0 & 0 & 0 & 0 & 0 & 0 & 0 & 0 & 0 & 0 & 0 & 0 \\
\end{array}
$$

Fig. 9.4 Radix-4 partial product array (second version)

In the design of a floating-point multiplier, we enjoy the luxury of avoiding negative operands. One consequence of this is the minor simplification that since $x[n-1] = 0$, $\sigma_i = \tau_i$ for every i. Another is that since $y[2m-1] = 0$, $\theta_{m-1} = y[2m-2] + y[2m-1] \geq 0$ and $\sigma_{m-1} = 0$, eliminating one of the two terms injected into the sum:

Lemma 9.4 *Let x and y be bit vectors of positive widths $n-1$ and $2m-1$, respectively. For $i = 0, \ldots, m-1$, let $\theta_i = \theta_i(y)$,*

$$
\sigma_i = \begin{cases} 0 \ \text{if}\ \theta_i \geq 0 \\ 1 \ \text{if}\ \theta_i < 0, \end{cases}
$$

$$
\bar{\sigma}_i = 1 - \sigma_i,
$$

$$
B_i = \begin{cases} x\theta_i & \text{if}\ \sigma_i = 0 \\ \sim(x|\theta_i|)[n-1:0] & \text{if}\ \sigma_i = 1, \end{cases}
$$

and

$$
pp_i = \begin{cases} 2^{n+1} + 2^n\bar{\sigma}_0 + B_0 & \text{if}\ i = 0 \\ 2^{n+1+2i} + 2^{n+2i}\bar{\sigma}_i + 2^{2i}B_i + 2^{2(i-1)}\sigma_{i-1} & \text{if}\ i > 0. \end{cases}
$$

Then

$$
2^n + \sum_{i=0}^{m-1} pp_i \equiv xy \ (\mathrm{mod}\ 2^{n+2m}).
$$

The remaining injected term, 2^n, may also be eliminated through a minor modification of the low-order entry pp_0 as shown in Fig. 9.4.

Lemma 9.5 *Let x and y be bit vectors of positive widths $n - 1$ and $2m - 1$, respectively. For $i = 0, \ldots, m - 1$, let θ_i, σ_i, $\bar{\sigma}_i$, B_i, and pp_i be as defined as in Lemma 9.4, except that*

$$pp_0 = 2^{n+2}\bar{\sigma}_0 + 2^{n+1}\sigma_0 + 2^n\sigma_0 + B_0.$$

Then

$$\sum_{i=0}^{m-1} pp_i \equiv xy \pmod{2^{n+2m}}.$$

Proof The difference between the two definitions of pp_0 is

$$(2^{n+2}\bar{\sigma}_0 + 2^{n+1}\sigma_0 + 2^n\sigma_0) - (2^{n+1} + 2^n\bar{\sigma}_0)$$

$$= 2^n \left(4(1 - \sigma_0) + 2\sigma_0 + \sigma_0 - 2 - (1 - \sigma_0)\right)$$

$$= 2^n. \qquad \qquad \qquad \qquad \qquad \qquad \qquad \qquad \square$$

Another common minor optimization is in the determination of the sign bit σ_i. Note that for $i > 0$, both σ_i and σ_{i-1} are required in the construction of pp_i. If we define $\sigma_i = y[2i + 1]$, then both of these required bits may be easily extracted from the current slice $y[2i+1 : 2i-1]$. This may produce a different result in the case $\theta_i = 0$, but the overall sum is unchanged. This is the variation of the algorithm that is used in the multiplier of Chap. 16:

Lemma 9.6 *Let x and y be bit vectors of positive widths $n - 1$ and $2m - 1$, respectively. For $i = 0, \ldots, m - 1$, let θ_i, σ_i, $\bar{\sigma}_i$, B_i, and pp_i be as defined as in Lemma 9.5, except that*

$$\sigma_i = y[2i + 1].$$

Then

$$\sum_{i=0}^{m-1} pp_i \equiv xy \pmod{2^{n+2m}}.$$

Proof Of the eight possible values of the slice $y[2i+1 : 2i-1]$, the modification produces the same σ_i, except when $y[2i+1 : 2i-1] = (111)_2$, in which case we have $\sigma_i = 1$ and $B_i = 2^n - 1$, whereas the corresponding values computed by the definitions of Lemma 9.4 are $\sigma_i' = B_i' = 0$. When this occurs, we must have $0 \leq i < 2m - 1$, and the affected partial products are pp_i and pp_{i+1}. If $i > 0$, then the resulting change in pp_i is

$$2^{2i} \left(2^n\bar{\sigma}_i + B_i\right) - 2^{2i} \left(2^n\bar{\sigma}_i' + B_i'\right) = 2^{2i} \left(0 + (2^n - 1)\right) - 2^{2i} \left(2^n + 0\right)$$

$$= -2^{2i},$$

and if $i = 0$, the result is the same:

$$\left(2^{n+2}\bar{\sigma}_0' + 2^{n+1}\sigma_0' + 2^n\sigma_0' + B_0'\right) - \left(2^{n+2}\bar{\sigma}_0 + 2^{n+1}\sigma_0 + 2^n\sigma_0 + B_0\right)$$

$$= \left(0 + 2^{n+1} + 2^n + (2^n - 1)\right) - \left(2^{n+2} + 0 + 0 + 0\right)$$

$$= -1$$

$$= -2^{2i}.$$

On the other hand, the change in pp_{i+1} is

$$2^{2i}(\sigma_i - \sigma_i') = 2^{2i}(1 - 0) = 2^{2i}.$$

Thus, the net change in the sum is 0. $\qquad\square$

9.3 Encoding Carry-Save Sums

In the context of an iterative multiplication-based algorithm (such as the division and square root algorithms of Chap. 10), it often occurs that the result of a multiplication is used as the multiplier y in the next iteration. In this case, if the Booth encoding of y can be derived directly from the carry-save representation produced by the compression tree, then y need not be computed explicitly and the expensive final step of carry-propagate addition may be avoided.

In this section, we shall assume once again that x is an $(n - 1)$-bit vector, but now the multiplier to be encoded is expressed as a sum

$$y = a + b,$$

where a and b are bit vectors of width $2m - 2$. As a consequence,

$$y \le (2^{2m-2} - 1) + (2^{2m-2} - 1) = 2^{2m-1} - 2 < 2^{2m-1}.$$

Our objective is to encode y as a sequence of digits $\psi_0, \ldots, \psi_{m-1}$ and derive a modified version of Lemma 9.5 with θ_i replaced by ψ_i. Examining the proofs of Lemmas 9.4 and 9.5, we see that the relevant properties of θ_i are as follows:

(1) $y = \sum_{i=0}^{m-1} 2^{2i}\theta_i$;
(2) $\theta_i \in \{-2, -1, 0, 1, 2\}$ for $i = 0, \ldots, m - 1$;
(3) $\theta_{m-1} \ge 0$.

We shall show that the same properties hold for the digits ψ_i, defined as follows.

Definition 9.2 Let $a \in \mathbb{N}$, $b \in \mathbb{N}$, $c \in \mathbb{N}$, and $d \in \mathbb{N}$. For all $i \in \mathbb{N}$, let

$$a_i = a[2i + 1 : 2i]$$

and

$$b_i = b[2i + 1 : 2i],$$

and let γ_i and δ_i be defined recursively as follows: $\gamma_0 = 0$, $\delta_0 = 0$, and for $i > 0$,

$$\gamma_i = a_{i-1}[1] \mid b_{i-1}[1]$$

and

$$\delta_i = (a_{i-1}[0] \,\&\, b_{i-1}[0] \mid a_{i-1}[0] \,\&\, \gamma_{i-1}[0] \mid b_{i-1}[0] \,\&\, \gamma_{i-1}[0]) \,\&\, \sim(a_{i-1}[1] \,\hat{}\, b_{i-1}[1]).$$

Then for all $i \in \mathbb{N}$,

$$\psi_i(a, b) = a_i + b_i + \gamma_i + \delta_i - 4(\gamma_{i+1} + \delta_{i+1}).$$

Lemma 9.7 *Let $m \in \mathbb{Z}^+$ and $y = a + b$, where a and b are $(2m - 2)$-bit vectors. Then*

$$y = \sum_{i=0}^{m-1} 2^{2i} \psi_i(a, b).$$

Proof Let $\psi_i = \psi_i(a, b, c, d)$, and let a_i, b_i, γ_i, and δ_i be as specified in Definition 9.2. We shall prove, by induction, that for $0 \leq k \leq m$,

$$a[2k-1 : 0] + b[2k-1 : 0] = \sum_{i=0}^{k-1} 2^{2i} \psi_i + 2^k(\gamma_k + \delta_k).$$

Assume that the statement holds for some $k < m$. Then

$$a[2k + 1 : 0] + b[2k + 1 : 0] = 2^k a_k + a[2k-1 : 0] + 2^k b_k + b[2k-1 : 0]$$

$$= 2^k(a_k + b_k) + \sum_{i=0}^{k-1} 2^{2i} \psi_i + 2^k(\gamma_k + \delta_k)$$

$$= 2^k(a_k + b_k + \gamma_k + \delta_k) + \sum_{i=0}^{k-1} 2^{2i} \psi_i$$

$$= 2^k(\psi_k + 4(\gamma_{k+1} + \delta_{k+1})) + \sum_{i=0}^{k-1} 2^{2i} \psi_i$$

$$= \sum_{i=0}^{k} 2^{2i} \psi_i + 2^{k+1}(\gamma_{k+1} + \delta_{k+1}).$$

Note that $a_{m-1} = b_{m-1} = 0$ and therefore, as a consequence of Definition 9.2, $\gamma_m = \delta_m = 0$. Thus,

$$a + b = a[2m-1 : 0] + b[2m-1 : 0]$$

$$= \sum_{i=0}^{m-1} 2^{2i} \psi_i + 2^m(\gamma_m + \delta_m)$$

$$= \sum_{i=0}^{m-1} 2^{2i} \psi_i. \qquad \square$$

It is not obvious that the ψ_i lie within the prescribed range.

Lemma 9.8 *Let a and b be $(2m - 2)$-bit vectors, where $m \in \mathbb{Z}^+$. Then for $i = 0, \ldots, m - 1$,*

$$|\psi_i(a, b)| \leq 2.$$

Proof Let $\psi_i = \psi_i(a, b)$ and let $a_i, b_i, \gamma_i,$ and δ_i be as specified in Definition 9.2. Then

$$\psi_i = a_i + b_i + \gamma_i + \delta_i - 4(\gamma_{i+1} + \delta_{i+1}),$$

where γ_{i+1} and δ_{i+1} are functions of $a_i, b_i,$ and γ_i. Thus, we may express ψ_l as a function of $a_i, b_i, \gamma_i,$ and δ_i. The inequality may then be trivially verified for each of the $4 \cdot 4 \cdot 2 \cdot 2 = 64$ possible sets of values of these arguments. $\qquad \square$

The remaining required property is trivial:

Lemma 9.9 *If a and b be $(2m - 2)$-bit vectors, where $m \in \mathbb{Z}^+$, then*

$$\psi_{m-1}(a, b) \geq 0.$$

Proof According to Definition 9.2, $a_{m-1} = b_{m-1} = 0$, $\gamma_m = \delta_m = 0$, and hence $\psi_{m-1} \geq 0$. $\qquad \square$

By the same arguments used in the proofs of Lemmas 9.4 and 9.5, we have the following:

Lemma 9.10 *Let* $m \in \mathbb{Z}^+$ *and* $n \in \mathbb{Z}^+$. *Let* x *be an* $(n-1)$-*bit vector, and let* $y = a + b$, *where* a *and* b *are* $(2m-2)$-*bit vectors. For* $i = 0, \ldots, m-1$, *let* pp_i *be as defined in Lemma 9.5 with* $\theta_i(y)$ *replaced by* $\psi_i(a, b)$. *Then*

$$\sum_{i=0}^{m-1} pp_i \equiv xy \pmod{2^{n+2m}}.$$

9.4 Statically Encoded Multiplier Arrays

In a practical implementation of the algorithm of Sect. 9.2, although the Booth digits θ_i are not actually computed arithmetically as suggested by the formula of Lemma 9.1, some combinational logic is required to derive an encoding of each θ_i from the corresponding multiplier bits $y[2i+1 : 2i-1]$. If the value of the multiplier is known in advance, i.e., at design time, then these encoded values may be stored instead of the multiplier itself, thereby saving the time and hardware associated with the encoding logic. However, since the range of θ_i consists of five values, 3 bits are required for each encoding, and therefore $3m$ bits in total, as compared to $2m$ bits for unencoded multiplier. For a single multiplier, this is a negligible expense, but if the multiplier is to be selected from an array of vectors, then the penalty incurred by such static encoding could be a 50% increase in the size of a large ROM.

In this section, we present an alternative encoding scheme that involves four rather than five encoded values, which allows 2-bit encodings and thereby eliminates any increase in space incurred by statically encoded arrays. Again we assume that the multiplicand x is an $(n-1)$-bit vector, but the bound on the multiplier is somewhat weaker than that of the preceding section, requiring only that

$$y \le \sum_{i=0}^{m-1} 2^{2i+1} = \frac{2}{3}(2^{2m} - 1).$$

Under this scheme, the coefficients θ_i are replaced with the values ϕ_i defined below. Note that the recursive nature of this definition precludes parallel computation of the m values. Consequently, this technique is not suitable for designs that require dynamic encoding.

The definition of ϕ_i involves a pair of mutually recursive auxiliary functions.

Definition 9.3 For all $y \in \mathbb{N}$ and $i \in \mathbb{N}$,

(a) $\mu_i(y) = y[2i + 1 : 2i] + \chi_i(y)$;

(b) $\chi_i(y) = \begin{cases} 1 \text{ if } i > 0 \text{ and } \mu_{i-1}(y) \ge 3 \\ 0 \text{ otherwise}; \end{cases}$

(c) $\phi_i(y) = \begin{cases} -1 & \text{if } \mu_i[1 : 0] = 3 \\ \mu_i[1 : 0] \text{ if } \mu_i[1 : 0] \ne 3, \end{cases}$ where $\mu_i = \mu_i(y)$.

Thus, ϕ_i is limited to a set of four values, $\{-1, 0, 1, 2\}$, and in particular, the second property of θ_l listed in Sect. 9.3 is satisfied. We shall establish the other two properties as well. The proof that $\phi_{m-1} \geq 0$ involves a nontrivial induction.

Lemma 9.11 *Let $y \in \mathbb{N}$ and $m \in \mathbb{N}$. If $y \leq \sum_{i=0}^{m-1} 2^{2i+1}$, then $\chi_m(y)=0$.*

Proof Let $\chi_i = \chi_i(y)$ and $\mu_i = \mu_i(y)$. More generally, we shall prove that for $0 \leq k \leq m$, if $y[2k-1 : 0] \leq \sum_{i=0}^{k-1} 2^{2i+1}$, then $\chi_k = 0$. Assuming that this claim holds for some $k < m$ and proceeding by induction, we must show that if $y[2k+1 : 0] \leq \sum_{i=0}^{k} 2^{2i+1}$, then $\chi_{k+1} = 0$.

First, suppose that $y[2k + 1 : 2k] \leq 1$. Then

$$\mu_k = y[2k+1 : 2k] + \chi_k \leq 1 + 1 < 3,$$

and hence $\chi_{k+1} = 0$. Thus, we may assume $y[2k + 1 : 2k] \geq 2$. Since

$$2^{2k} y[2k+1 : 2k] + y[2k-1 : 0] = y[2k+1 : 0]$$

$$\leq \sum_{i=0}^{k} 2^{2i+1}$$

$$= 2^{2k+1} + \sum_{i=0}^{k-1} 2^{2i+1}$$

$$< 2^{2k+1} + 2^{2k}$$

$$= 2^{2k} \cdot 3,$$

we must have $y[2k+1 : 2k] = 2$ and $y[2k-1 : 0] \leq \sum_{i=0}^{k-1} 2^{2i+1}$. Now, the inductive hypothesis yields $\chi_k = 0$. Hence,

$$\mu_k = y[2k+1 : 2k] + \chi_k = 2 + 0 < 3,$$

and once again, $\chi_{k+1} = 0$. □

The desired result now follows from Lemma 9.11 and Definition 9.3.

Corollary 9.12 *If $y \leq \sum_{i=0}^{m-1} 2^{2i+1}$, then $\phi_{m-1} \geq 0$.*

Lemma 9.11 is also needed for the remaining required property of ϕ_i:

Lemma 9.13 *Let $y \in \mathbb{N}$ and $m \in \mathbb{N}$. If $y \leq \sum_{i=0}^{m-1} 2^{2i+1}$, then*

$$y = \sum_{i=0}^{m-1} 2^{2i} \phi_i(y).$$

Proof Let $\chi_i = \chi_i(y)$, $\mu_i = \mu_i(y)$, and $\phi_i = \phi_i(y)$. First note that $4\chi_{i+1}+\phi_i = \mu_i$, for if $\mu_i = 3$, then

$$4\chi_{i+1} + \phi_i = 4 - 1 = 3,$$

and in all other cases,

$$4\chi_{i+1} + \phi_i = 4\mu_i[2] + \mu[1:0] = \mu_i.$$

We shall show that for $k = 0, \ldots, m$,

$$y[2k-1:0] = \sum_{i=0}^{k-1} 2^{2i}\phi_i + 2^{2k}\chi_k.$$

The claim is trivial for $k = 0$. For $0 \le k < m$, by induction, we have

$$y[2k+1:0] = 2^{2k}y[2k+1:2k] + y[2k-1:0]$$

$$= 2^{2k}y[2k+1:2k] + \sum_{i=0}^{k-1} 2^{2i}\phi_i + 2^{2k}\chi_k$$

$$= 2^{2k}(y[2k+1:2k] + \chi_k) + \sum_{i=0}^{k-1} 2^{2i}\phi_i$$

$$= 2^{2k}\mu_k + \sum_{i=0}^{k-1} 2^{2i}\phi_i$$

$$= 2^{2k}(4\chi_{k+1} + \phi_k) + \sum_{i=0}^{k-1} 2^{2i}\phi_i$$

$$= \sum_{i=0}^{k} 2^{2i}\phi_i + 2^{2(k+1)}\chi_{k+1}.$$

In particular, substituting $m - 1$ for k, we have

$$y = y[2m-1:0] = \sum_{i=0}^{m-1} 2^{2i}\phi_i + 2^{2m}\chi_m = \sum_{i=0}^{m-1} 2^{2i}\phi_i. \qquad \square$$

Thus, our multiplier y may be statically encoded as a vector of width $2m$,

$$z = \{\mu_{m-1}[1:0], \ldots, \mu_0[1:0]\},$$

from which each ϕ_i may be readily recovered as

$$\phi_i = \begin{cases} -1 & \text{if } z[2i{+}1 : 2i] = 3 \\ \mu_i[1 : 0] & \text{if } z[2i{+}1 : 2i] \neq 3. \end{cases}$$

We may now conclude the following result. Note that as a further optimization, the 5-to-1 multiplexer that produces the B_i of Lemma 9.4 is replaced with a 4-to-1 multiplexer.

Lemma 9.14 *Let $m \in \mathbb{Z}^+$ and $n \in \mathbb{Z}^+$. Let x be a bit vector of width $n - 1$ and let $y \in \mathbb{N}$ satisfy $y \leq \sum_{i=0}^{m-1} 2^{2i+1}$. For $i = 0, \ldots, m - 1$, let pp_i be as defined in Lemma 9.5 with $\theta_i(y)$ replaced by $\phi_i(y)$. Then*

$$\sum_{i=0}^{m-1} pp_i \equiv xy \pmod{2^{n+2m}}.$$

9.5 Radix-8 Booth Encoding

A partition of the multiplier y into slices of 3 bits instead of 2 leads to a decomposition

$$y = \sum 2^{3i} \eta_i,$$

where

$$\eta_i = 4\beta_{3i+2} + 2\beta_{3i+1} + \beta_{3i}$$
$$= 4(y[3i{+}1] - y[3i{+}2]) + 2(y[3i] - y[3i{+}1]) + (y[3i{-}1] - y[3i])$$
$$= y[3i{-}1] + y[3i] + 2y[3i{+}1] - 4y[3i{+}2].$$

While the number of terms of this sum is only 1/3 (rather than 1/2) of the width of y, the range of digits is now $-4 \leq \eta_i \leq 4$, and hence the value of each term is no longer guaranteed to be a power of 2 in absolute value. Consequently, a multiplier based on this radix-8 scheme generates fewer partial products than a radix-4 multiplier, but the computation of each partial product is more complex. In particular, a partial product corresponding to an encoding $\eta_i = \pm 3$ requires the computation of $3x$ and therefore a full addition.

While radix-4 multiplication is more common, radix-8 may offer an advantage, depending on the timing details of a hardware design. In a typical implementation, the partial products are computed in one clock cycle and the compression tree is executed in the next. If there is sufficient time during the first cycle to perform the addition required for radix-8 encoding (which may be the case, for example, for a low-precision operation), then this scheme is feasible. Since most of the silicon area allocated to a multiplier is associated with the compression tree, the resulting reduction in the number of partial products may represent a significant gain in efficiency.

For the purpose of this analysis, which is otherwise quite similar to that of Sect. 9.2, we shall assume that x and y are bit vectors of widths $n - 2$ and $3m - 1$, respectively.

Definition 9.4 For $y \in \mathbb{N}$ and $i \in \mathbb{N}$,

$$\eta_i(y) = y[3i{-}1] + y[3i] + 2y[3i{+}1] - 4y[3i{+}2].$$

Lemma 9.15 *Let y be a bit vector of width 2^{3m-1}, where $m \in \mathbb{Z}^+$. Then*

$$y = \sum_{i=0}^{m-1} 2^{3i} \eta_i(y).$$

Proof The proof is essentially the same as that of Lemma 9.1. We shall show by induction that for $0 \leq k \leq m$,

$$y[3k{-}1 : 0] = \sum_{i=0}^{k-1} 2^{3i} \eta_i + 2^{3k} y[3k{-}1],$$

where $\eta_i = \eta_i(y)$. The claim is trivial for $k = 0$. Assuming that it holds for some $k < m$, we have

$$y[3(k{+}1){-}1 : 0] = y[3k{+}2 : 0]$$

$$= 2^{3k} y[3k{+}2 : 3k] + y[3k{-}1 : 0]$$

$$= 2^{3k}(4y[3k{+}2] + 2y[3k{+}1] + y[3k]) + \sum_{i=0}^{k-1} 2^{3i} \eta_i + 2^{3k} y[3k{-}1]$$

$$= 2^{3k}(4y[3k{+}2] + 2y[3k{+}1] + y[3k] + y[3k{-}1]) + \sum_{i=0}^{k-1} 2^{3i} \eta_i$$

$$= 2^{3k}(8y[3k{+}2] + \eta_k) + \sum_{i=0}^{k-1} 2^{3i} \eta_i$$

$$= \sum_{i=0}^{k} 2^{3i} \eta_i + 2^{3(k+1)} y[3(k+1){-}1],$$

which completes the induction. In particular, substituting m for k, we have

$$y = y[3m{-}1:0] = \sum_{i=0}^{m-1} 2^{3i}\eta_i + 2^{3m}y[3m{-}1] = \sum_{i=0}^{m-1} 2^{3i}\eta_i. \qquad \square$$

The partial product array, as depicted in Fig. 9.5 for the case $n = 6$, $m = 8$, consists of m bit vectors of width $n + 3m$. Its structure is quite similar to that of the radix-4 array of Lemma 9.5, as its proof of correctness:

```
0 0 0 0 0 0 0 0 0 0 0 0 0 0 0 0 0 0 0 0 σ̄₀ σ₀ σ₀ σ₀ • • • • • • •
0 0 0 0 0 0 0 0 0 0 0 0 0 0 0 0 0 1 1 σ̄₁ • • • • • • • 0 0 σ₀
0 0 0 0 0 0 0 0 0 0 0 0 0 0 1 1 σ̄₂ • • • • • • • 0 0 σ₁ 0 0 0
0 0 0 0 0 0 0 0 0 0 0 1 1 σ̄₃ • • • • • • 0 0 σ₂ 0 0 0 0 0 0
0 0 0 0 0 0 0 0 1 1 σ̄₄ • • • • • • 0 0 σ₃ 0 0 0 0 0 0 0 0 0
0 0 0 0 0 0 1 1 σ̄₅ • • • • • • 0 0 σ₄ 0 0 0 0 0 0 0 0 0 0 0 0
0 0 0 1 1 σ̄₆ • • • • • • 0 0 σ₅ 0 0 0 0 0 0 0 0 0 0 0 0 0 0 0
1 1 σ̄₇ • • • • • • 0 0 σ₆ 0 0 0 0 0 0 0 0 0 0 0 0 0 0 0 0 0 0
```

Fig. 9.5 Radix-8 partial product array

Lemma 9.16 *Let x and y be bit vectors of widths $n - 2$ and $3m - 1$, respectively, where $m \in \mathbb{Z}^+$ and $n \in \mathbb{Z}^+$. For $i = 0, \ldots, m - 1$, let $\eta_i = \eta_i(y)$,*

$$\sigma_i = \begin{cases} 0 \text{ if } \eta_i \geq 0 \\ 1 \text{ if } \eta_i < 0, \end{cases}$$

$$\bar{\sigma}_i = 1 - \sigma_i,$$

$$B_i = \begin{cases} \eta_i x & \text{if } \sigma_i = 0 \\ {\sim}(-\eta_i x)\lfloor n{-}1:0\rfloor & \text{if } \sigma_i = 1, \end{cases}$$

and

$$pp_i = \begin{cases} \{(3m{-}4)\,'0, \bar{\sigma}_0, \sigma_0, \sigma_0, \sigma_0, n\,'B_0\} & \text{if } i = 0 \\ \{3(m{-}i{-}1)\,'0, 1, 1, \bar{\sigma}_i, n\,'B_i, 0, 0, \sigma_{i-1}, 3(i{-}1)\,'0\} & \text{if } i > 0. \end{cases}$$

Then

$$\sum_{i=0}^{m-1} pp_i \equiv xy \pmod{2^{n+3m}}.$$

Proof Note that B_i may be expressed as $x\eta_i + (2^n - 1)\sigma_i$. Thus,

$$pp_0 = B_0 + 2^n\sigma_0 + 2^{n+1}\sigma_0 + 2^{n+2}\sigma_0 + 2^{n+3}(1-\sigma_0)$$
$$= \left(x\eta_0 + (2^n-1)\sigma_0\right) - 2^n\sigma_0 + 2^{n+3}$$
$$= x\eta_0 - \sigma_0 + 2^{n+3}$$

and for $i > 0$,

$$pp_i = 2^{3(i-1)}\sigma_{i-1} + 2^{3i}B_i + 2^{n+3i}(1-\sigma_i) + 2^{n+3i+1} + 2^{n+3i+2}$$
$$= 2^{3(i-1)}\sigma_{i-1} + 2^{3i}\left(x\eta_i + (2^n-1)\sigma_i\right) + 2^{n+3i}(1-\sigma_i) + 2^{n+3i+1} + 2^{n+3i+2}$$
$$= 2^{3(i-1)}\sigma_{i-1} + 2^{3i}x\eta_i - 2^{3i}\sigma_i + 2^{n+3(i+1)} - 2^{n+3i}.$$

Combining these expressions and noting that $\sigma_{m-1} = 0$, we have

$$\sum_{i=0}^{m-1} pp_i$$

$$= \sum_{i=1}^{m-1} 2^{3(i-1)}\sigma_{i-1} + \sum_{i=0}^{m-1} x\eta_i - \sum_{i=0}^{m-1} 2^{3i}\sigma_i + \sum_{i=1}^{m-1}(2^{n+3(i+1)}-2^{n+3i}) + 2^{n+3}$$

$$= \sum_{i=0}^{m-2} 2^{3i}\sigma_i + xy - \sum_{i=0}^{m-1} 2^{3i}\sigma_i + 2^{n+3m}$$

$$= xy + 2^{n+3m}. \qquad\qquad\qquad \square$$

Chapter 10
SRT Division and Square Root

The simplest and most common approach to computer division is *digit recurrence*, an iterative process whereby at each step, a multiple of the divisor is subtracted from the current remainder and the quotient is updated accordingly by appending a fixed number of bits k, determined by the underlying radix, $r = 2^k$. Thus, quotient convergence is linear, resulting in fairly high latencies of high-precision operations for the most common radices, $r = 2$, 4, and 8.

Since division and square root lend themselves to similar recurrence formulas for the remainder, the same methods are generally applicable to both operations. An important class of digit recurrence algorithms are grouped under the name *SRT*, in recognition of the independent contributions of Sweeney [4], Robertson [28], and Tocher [37] in the late 1950s. The common element is a table, indexed by approximations of the divisor or root and the remainder, from which an integer multiplier is extracted on each iteration. SRT dividers are ubiquitous in contemporary microprocessor design and notoriously prone to implementation error. They are, therefore, an important application of formal verification.

Elsewhere [33], we explore the sharing of SRT tables between division and square root as an area-conserving optimization. Here, we take a simpler approach using separate tables, following [5]. The results of this chapter are the basis of the floating-point division and square root designs of Chaps. 18 and 20, as well as the integer division module of Chap. 19.

10.1 SRT Division

Our objective is to compute an approximation of the quotient $\frac{x}{d}$ of given positive rational numbers x and d. In most cases, the operands are shifted as required to ensure that

$$d \leq x < 2d, \tag{10.1}$$

© The Author(s), under exclusive license to Springer Nature Switzerland AG 2022
D. M. Russinoff, *Formal Verification of Floating-Point Hardware Design*,
https://doi.org/10.1007/978-3-030-87181-9_10

thereby confining the quotient to the interval $[1, 2)$, but for the purpose of the present analysis, we shall assume only that

$$0 < x < 2d. \tag{10.2}$$

The computation is governed by a fixed power of 2, $r \geq 2$, the *radix* of the operation, which determines the number of bits contributed to the quotient on each iteration. We shall construct a sequence of *quotient digits* $q_j \in \mathbb{Z}$ and the resulting sequence of *partial quotients*,

$$Q_j = \sum_{i=1}^{j} r^{1-i} q_i, \tag{10.3}$$

which converges to $\frac{x}{d}$. We also define the *partial remainders*,

$$R_j = r^{j-1}(x - dQ_j), \tag{10.4}$$

which may be computed by the following recurrence relation:

Lemma 10.1 $R_0 = \frac{x}{r}$ *and for* $j \geq 0$, $R_{j+1} = rR_j - q_{j+1}d$.

Proof The claim is trivial for $j = 0$, and for $j > 0$,

$$\begin{aligned}
R_{j+1} &= r^j(x - Q_{j+1}d) \\
&= r^j \left(x - (Q_j + r^{-j}q_{j+1})d \right) \\
&= r^j(x - Q_jd) - q_{j+1}d \\
&= rR_j - q_{j+1}d.
\end{aligned}$$
$\square$

The quotient digits are selected from a set of integers $\{-a, \dots, a\}$, where a is chosen so that the *redundancy factor*

$$\rho = \frac{a}{r-1}$$

satisfies

$$\frac{1}{2} < \rho \leq 1. \tag{10.5}$$

The *minimally redundant* case $a = \frac{r}{2}$ minimizes the number of multiples of d that must be computed, while the *maximally redundant* case $a = r - 1$ provides greater flexibility in the selection of digits.

The digits are selected with the goal of preserving the invariant

$$|R_j| \le \rho d, \tag{10.6}$$

or equivalently,

$$\left| \frac{x}{d} - Q_j \right| \le \rho r^{1-j}. \tag{10.7}$$

This choice of bound is motivated by the observation that if

$$\frac{x}{d} = \lim_{j \to \infty} Q_j = \sum_{i=1}^{\infty} r^{1-i} q_i,$$

then

$$\left| \frac{x}{d} - Q_j \right| = \left| \sum_{i=j+1}^{\infty} r^{1-i} q_i \right| \le a \sum_{i=j+1}^{\infty} r^{1-i} = \frac{a r^{1-j}}{r - 1} = \rho r^{1-j}.$$

Thus, (10.7) is equivalent to convergence.

The invariant holds trivially for $j = 0$:

Lemma 10.2 $R_0 \le \rho d$.

Proof This follows from Eqs. (10.2) and (10.5): if $r = 2$, then $a = 1$, $\rho = 1$, and

$$R_0 = \frac{x}{2} < d = \rho d;$$

otherwise, $r \ge 4$ and

$$R_0 = \frac{x}{r} < \frac{2d}{r} \le \frac{d}{2} < \rho d. \qquad \square$$

For $k \in \{-a, \ldots, a\}$, the *selection interval* $[L_k(d), U_k(d)]$, defined by

$$U_k(d) = (k + \rho)d$$

and

$$L_k(d) = (k - \rho)d,$$

is so named because if the shifted partial remainder $r R_j$ lies in this interval, then the invariant (10.6) may be preserved by choosing $q_{j+1} = k$:

Lemma 10.3 *If* $L_k(d) \le r R_j \le U_k(d)$ *and* $q_{j+1} = k$, *then* $|R_{j+1}| \le \rho d$.

Proof By Lemma 10.1,

$$L_k(d) \le rR_j \le U_k(d) \Rightarrow (k - \rho)d \le rR_j \le (k + \rho)d$$
$$\Rightarrow -\rho d \le rR_j - kd \le \rho d$$
$$\Rightarrow -\rho d \le R_{j+1} \le \rho d. \qquad \square$$

Thus, the existence of a quotient digit q_{j+1} that preserves (10.6) is guaranteed if the selection intervals cover the entire range of rR_j, i.e.,

$$[-r\rho d, r\rho d] \subseteq \bigcup_{k=-a}^{a} [L_k(j), U_k(j)].$$

This is ensured by the following two lemmas. Note that the proof of the first accounts for the bounds (10.5) imposed on ρ:

Lemma 10.4 *For all $k \in \mathbb{Z}$,*

(a) $L_k(d) < L_{k+1}(d) < U_k(d) < U_{k+1}(d)$;
(b) $U_k(d) \le L_{k+2}(d)$.

Proof

(a) $L_{k+1}(d) - L_k(d) = U_{k+1}(d) - U_k(d) = d > 0$ and $U_k(d) - L_{k+1}(d) = (2\rho - 1)d > 0$.
(b) $L_{k+2}(d) - U_k(d) = 2(1 - \rho)d \ge 0$.

$$\square$$

Lemma 10.5 $U_a(d) = r\rho d$ *and* $L_{-a}(d) = -r\rho d$.

Proof First note that

$$a + \rho = \frac{a(r - 1) + a}{r - 1} = \frac{ar}{r - 1} = r\rho.$$

Thus, $U_a(d) = (a + \rho)d = r\rho d$ and $L_{-a}(d) = (-a - \rho)d = -r\rho d$. $\qquad \square$

Typically, the selection of a quotient digit that satisfies the hypothesis of Lemma 10.3 is achieved by means of a set of *comparison constants* $m_k(d)$, $-a < k \le a$, which satisfy

$$L_k(d) < m_k(d) \le U_{k-1}(d). \tag{10.8}$$

If the range of possible values of d is partitioned into sufficiently small subintervals, a single set of such constants may be identified for each subinterval.

Note that if $-a < k < a$, then (10.8), along with Lemma 10.4(b), implies that $m_k(d) \le U_{k-1}(d) \le L_{k+1}(d) < m_{k+1}(d)$, i.e., $m_k(d)$ is a strictly increasing function of k. As a matter of convenience, we also define $m_{-a}(d) = -\infty$. If q_{j+1} is selected as the largest k for which $m_k(d) \le rR_j$, then the required bound on $|R_{j+1}|$ follows from Lemmas 10.3 and 10.5.

In practice, however, the partial remainder R_j is represented in a redundant form, i.e., as a sum or difference of two vectors, and therefore, such direct comparisons are not possible. Instead, the comparisons are based on an approximation A_j of rR_j, produced by a narrow adder applied to the leading bits of the component vectors of R_j. Such an implementation is based on an integer parameter t, the number of fractional bits of A_j, which determines the width of the adder and the accuracy of the approximation.

In the radix-8 SRT divider of Chap. 18, for example, $8R_j$ is represented in *borrow-save* form, i.e., as a difference of 2 bit vectors, which are truncated to 8 fractional bits and subtracted to produce an approximation A_j that satisfies

$$|A_j - 8R_j| < 2^{-8}.$$

For a design that represents the shifted remainder in *carry-save* form, i.e., as a sum rather than a difference, a similar error bound may be achieved by incrementing the truncated sum.

Note that a lower redundancy factor, which results in narrower selection intervals with less overlap, requires a more accurate approximation, i.e., a wider adder. This consideration is usually outweighed, however, by the advantage of a smaller set of digits. Minimally redundant radix-4 dividers ($r = 2, a = 2$) are common because all of the required multiples of the divisor may be computed with shifts, complements, and increments.

A stricter version of (10.8) based on the accuracy of the approximation is required to ensure correctness. In order to simplify the comparison with A_j, the constants $m_k(d)$ should be selected to have as few fractional bits as possible. We need only assume, however, that $m_k(d)$ has no more fractional bits than the approximation.

Lemma 10.6 *Let $t \in \mathbb{N}$. Let $m_{-a}(d) = -\infty$, and for $-a < k \le a$, let $m_k(d) \in \mathbb{Q}$, such that $2^t m_k(d) \in \mathbb{Z}$ and*

$$L_k(d) + 2^{-t} \le m_k(d) \le U_{k-1}(d).$$

Let $j \in \mathbb{N}$ and assume that $|R_j| \le \rho d$. Let $A_j \in \mathbb{Q}$ such that $2^t A_j \in \mathbb{Z}$ and

$$|A_j - rR_j| \le 2^{-t}.$$

If q_{j+1} is the greatest $k \in \{-a, \dots, a\}$ such that $m_k(d) \le A_j$, then $|R_{j+1}| \le \rho d$.

Proof Let $q = q_{j+1}$. We shall show that $L_q(d) \le rR_j \le U_q(d)$ and invoke Lemma 10.3. Since Lemma 10.5 ensures that $L_{-a}(d) \le rR_j \le U_a(d)$, the required bounds are reduced to

$$-a < q \le a \Rightarrow rR_j \ge L_q(d)$$

and

$$-a \leq q < a \Rightarrow rR_j \leq U_q(d).$$

But according to hypothesis, if $-a < q \leq a$, then

$$L_q(d) \leq m_q(d) - 2^{-t} \leq A_j - 2^{-t} \leq rR_j$$

and if $-a \leq q < a$, then $m_{q+1}(d) > A_j$, which implies $m_{q+1}(d) \geq A_j + 2^{-t}$, and hence

$$U_q(d) \geq m_{q+1}(d) \geq A_j + 2^{-t} \geq rR_j. \qquad \square$$

The parameter t must be large enough to ensure the existence of comparison constants $m_k(d)$ that satisfy the hypothesis of Lemma 10.6 but should be as small as possible in order to minimize the width of the adder that generates A_j. The number of fractional bits of these constants is at most t (as required by the lemma) and should be further reduced if possible in order to simplify the comparisons.

In the maximally redundant case ($\rho = 1$), the relative weakness of the inequality $|R_j| \leq d$ presents two distinct difficulties that are not present when $\rho < 1$. Both of these are illustrated by the radix-2 divider of Chap. 21. Note that a radix-2 implementation is necessarily both minimally and maximally redundant, since the constraint $\frac{1}{2} < \rho \leq 1$, where $\rho = \frac{a}{r-1}$ and $r = 2$, requires $a = 1 = \frac{r}{2} = r - 1$.

One problem is that if neither of the extremes $\pm d$ can be excluded as the value of the final remainder, then the selection of the final truncated quotient is more complicated. Thus, in Chap. 21, we replace the upper bound with a strict inequality, proving that

$$-d \leq R_j < d.$$

The second issue pertains to the approximation of the partial remainder and the wrap-around problem discussed at the end of Sect. 2.4. When $\rho < 1$, the inequality $|R_j| \leq \rho d < 2\rho$ bounds R_j away from ± 2. As illustrated in the proof of Lemma 18.6, if the representation of R_j includes two implicit integer bits, this allows the use of Lemma 2.37 to guarantee the accuracy of the approximation. For a maximally redundant divider, this is not the case. The most obvious remedy is to widen the representation of the remainder, increasing the number of integer bits to 3 and the width of the adder to 4. The design of Chap. 21, however, provides a more sophisticated solution that circumvents the need for an extra bit.

10.2 Minimally Redundant Radix-4 Division

As an example, we consider the case $r = 4$, $a = 2$, and $\rho = \frac{2}{3}$. This is a particularly common instance of SRT division because multiplication of the divisor by each element of the digit set $\{-2, -1, 0, 1, 2\}$ may be performed as a simple shift of d or its complement.

In general, a table of comparison constants is constructed to satisfy the hypothesis of Lemma 10.6, consisting of a row corresponding to each element of a set of intervals of values of d. A common technique for limiting this table to a single row is *prescaling*, whereby x and d are multiplied by the same factor (thus preserving the quotient) in order to confine d to a small neighborhood of 1. In the present case, this allows the same four comparison constants m_k, $-1 \leq k \leq 2$, to be used for all values of d. This technique has been employed in an Arm radix-4 floating-point divider to ensure that the divisor lies within the range

$$\frac{63}{64} \leq d \leq \frac{9}{8}, \tag{10.9}$$

as described below. Let A and B be the numerical values of the divider's operands, corresponding to the dividend and divisor, respectively. Figure 10.1 defines a multiplier M as a function of the leading 3 fractional bits of the significand of B, i.e., $\lfloor 8(sig(B) - 1) \rfloor$. The prescaled dividend and divisor are computed as

$$x = M \cdot sig(A)$$

and

$$d = M \cdot sig(B).$$

$\lfloor 8(sig(B)-1) \rfloor$	M
000	$1 = \frac{1}{4} + \frac{1}{4} + \frac{1}{2}$
001	$\frac{7}{8} = \frac{1}{8} + \frac{1}{4} + \frac{1}{2}$
010	$\frac{13}{16} = \frac{1}{4} + \frac{1}{16} + \frac{1}{2}$
011	$\frac{3}{4} = \frac{1}{4} + 0 + \frac{1}{2}$
100	$\frac{11}{16} = \frac{1}{8} + \frac{1}{16} + \frac{1}{2}$
101	$\frac{5}{8} = \frac{1}{8} + 0 + \frac{1}{2}$
110	$\frac{9}{16} = 0 + \frac{1}{16} + \frac{1}{2}$
111	$\frac{9}{16} = 0 + \frac{1}{16} + \frac{1}{2}$

Fig. 10.1 Prescaling multiplier M

Note that each of the eight possible values of M may be expressed as a sum of at most 3 powers of 2. This means that the multiplications require at most three partial products each and were found to incur no significant timing penalty.

The bounds on d given by (10.9) may be established directly for each index. For example, if $\lfloor 8(sig(B) - 1)\rfloor = 3$, then $3 \leq 8(sig(B) - 1) < 4$ or

$$\frac{11}{8} \leq sig(B) < \frac{3}{2}.$$

In this case, $d = M sig(B) = \frac{3}{4} sig(B)$ and

$$\frac{63}{64} < \frac{33}{32} = \frac{3}{4} \cdot \frac{11}{8} \leq d < \frac{3}{4} \cdot \frac{3}{2} = \frac{9}{8}.$$

The other seven cases are similar.

m_2	m_1	m_0	m_{-1}	m_{-2}
$\frac{13}{8}$	$\frac{4}{8}$	$-\frac{3}{8}$	$-\frac{12}{8}$	$-\infty$

Fig. 10.2 Comparison constants m_k

Now according to Lemma 10.6, we seek a value of t for which there exist m_k such that $2^t m_k \in \mathbb{Z}$ and

$$L_k(d) + 2^{-t} \leq m_k \leq U_{k-1}(d)$$

for all d satisfying (10.9). Let

$$\overline{L}_k = \max\left(L_k\left(\frac{63}{64}\right), L_k\left(\frac{9}{8}\right)\right)$$

and

$$\underline{U}_k = \min\left(U_k\left(\frac{63}{64}\right), U_k\left(\frac{9}{8}\right)\right).$$

Since L_k and U_k are linear functions of d, it follows from (10.9) that

$$L_k(d) \leq \overline{L}_k$$

and

$$U_k(d) \geq \underline{U}_k.$$

Our requirement, therefore, may be reduced to

$$\overline{L}_k + 2^{-t} \leq m_k \leq \underline{U}_{k-1}, \tag{10.10}$$

or

$$2^t \overline{L}_k + 1 \leq 2^t m_k \leq 2^t \underline{U}_{k-1},$$

where $2^t m_k \in \mathbb{Z}$. Thus, we must select t to be large enough that there exists an integer in the interval $[2^t \overline{L}_k + 1, 2^t \underline{U}_{k-1}]$, or equivalently,

$$\lceil 2^t \overline{L}_k + 1 \rceil \leq \lfloor 2^t \underline{U}_{k-1} \rfloor,$$

for $-1 \leq k \leq 2$. It is easily verified that $\overline{L}_k < \underline{U}_{k-1}$ for each k, which implies the satisfiability of this constraint, and that the smallest solution is $t = 3$. It is also easily verified that (10.10) is satisfied for this value of t by the constants m_k displayed in Fig. 10.2. Thus, we have the following instantiation of Lemma 10.6:

Lemma 10.7 *Let $r = 4$ and $a = 2$ and let m_k be as listed in Fig. 10.2 for $-2 \leq k \leq 2$. Assume that*

$$\frac{63}{64} \leq d \leq \frac{9}{8}.$$

Let $j \geq 0$ and assume that the following conditions hold:

(a) $|R_j| \leq \frac{2}{3}d$;
(b) $A_j \in \mathbb{Q}$ satisfies $8A_j \in \mathbb{Z}$ and $|A_j - 4R_j| < \frac{1}{8}$;
(c) q_{j+1} is the greatest $k \in \{-2, \ldots, 2\}$ such that $m_k \leq A_j$.

Then $|R_{j+1}| \leq \frac{2}{3}d$.

10.3   Minimally Redundant Radix-8 Division

In the successor of the processor in which the radix-4 divider of the preceding section was implemented, the radix was increased to 8, thereby providing an extra quotient bit per iteration at the expense of increased complexity. This divider is also minimally redundant, with $a = 4$ and $\rho = \frac{4}{7}$. The present section provides the basis for its design, which is described in detail in Chap. 18.

Effective prescaling in this case would require stricter bounds on the divisor and consequently a multiplier with seven partial products (as opposed to three in the radix-4 case), which was found to be infeasible with respect to timing. Instead, we merely assume that

$$\frac{1}{2} \leq d < 1. \tag{10.11}$$

Consequently, the comparison constants are dependent on d. Specifically, they are based on a partition of the range of d into 64 subintervals. Let

$$i = \left\lfloor 128 \left(d - \frac{1}{2} \right) \right\rfloor,$$

so that $0 \le i < 64$ and

$$\frac{1}{2} + \frac{i}{128} \le d < \frac{1}{2} + \frac{i+1}{128}. \tag{10.12}$$

The comparison constants are determined by i; that is, we define a set of constants $m_k(i)$ for each of these intervals.

For $-4 \le k \le 4$, let

$$\overline{L}_k(i) = \max \left(L_k \left(\frac{1}{2} + \frac{i}{128} \right), L_k \left(\frac{1}{2} + \frac{i+1}{128} \right) \right)$$

and

$$\underline{U}_k(i) = \min \left(U_k \left(\frac{1}{2} + \frac{i}{128} \right), U_k \left(\frac{1}{2} + \frac{i+1}{128} \right) \right).$$

Once again, $L_k(d)$ and $U_k(d)$ are linear functions of d, and hence

$$L_k(d) \le \overline{L}_k(i) \text{ and } U_k(d) \ge \underline{U}_k(i). \tag{10.13}$$

Thus, we seek $t \in \mathbb{N}$ and constants $m_k(i)$, where $-3 \le k \le 4$ and $0 \le i < 64$, such that $2^t m_k(i) \in \mathbb{Z}$ and

$$\overline{L}_k(i) + 2^{-t} \le m_k(i) \le \underline{U}_{k-1}(i). \tag{10.14}$$

By the same reasoning as used in the radix-4 case, a solution exists if and only if

$$\lceil 2^t \overline{L}_k(i) + 1 \rceil \le \lfloor 2^t \underline{U}_{k-1}(i) \rfloor$$

for all i. By direct computation, the minimal value for which this condition holds is $t = 6$.

According to Lemma 10.5, each constant satisfies

$$|m_k(i)| < r\rho d < 8 \cdot \frac{4}{7} < 5$$

and may therefore be represented with 10 bits, 4 integer and 6 fractional. It follows that we can construct a table of size $64 \times 8 \times 10 = 5120$ bits, from which the appropriate eight 10-bit constants may be extracted according to the leading 6 bits of d.

But closer inspection reveals that the constants may be chosen to be independent of the least significant bit of i with very few exceptions, an observation that may be exploited effectively to reduce the table size by half. We shall define $m_k(i)$ according to Fig. 10.3. Note that for a given value of i, the constants $m_k(i)$ are derived by a table access based on the 5-bit value $\lfloor i/2 \rfloor$ and require possible adjustment according to the 6th bit only in the four cases $i < 4$.

Since these constants are readily shown to satisfy (10.14) for $t = 6$, we have the following consequence of Lemma 10.6:

Lemma 10.8 *Let $r = 8$ and $a = 4$, and let $m_k(i)$ be as listed in Fig. 10.3 for $-4 \leq k \leq 4$, where*

$$i = \left\lfloor 128 \left(d - \frac{1}{2} \right) \right\rfloor.$$

Let $j \geq 0$ and assume that the following conditions hold:

(a) $|R_j| \leq \frac{4}{7}d$;
(b) $A_j \in \mathbb{Q}$ satisfies $64A_j \in \mathbb{Z}$ and $|A_j - 8R_j| < \frac{1}{64}$;
(c) q_{j+1} is the greatest $k \in \{-4, \ldots, 4\}$ such that $m_k(i) \leq A_j$.

Then $|R_{j+1}| \leq \frac{4}{7}d$.

10.4 SRT Square Root

Let r be a fixed power of 2. Given a positive rational number x in the range $\frac{1}{4} \leq x < 1$, our objective is to construct a sequence of *root digits* $q_j \in \mathbb{Z}$ and a corresponding sequence of *partial roots*

$$Q_j = 1 + \sum_{i=1}^{j} r^{-i} q_i, \tag{10.15}$$

which converge to $\sqrt{x} \in [\frac{1}{2}, 1)$. Note that for all $j \in \mathbb{N}$, $r^j Q_j \in \mathbb{Z}$.

The digits are selected from a set $\{-a, \ldots, a\}$, where a is again chosen so that the redundancy factor $\rho = a/(r-1)$ satisfies (10.5). We define the *partial remainders*,

$$R_j = r^j (x - Q_j^2), \tag{10.16}$$

which are computed by the following recurrence relation:

$\lfloor i/2 \rfloor$	m_4	m_3	m_2	m_1	m_0	m_{-1}	m_{-2}	m_{-3}	m_{-4}
0	115/64*	82/64	50/64	16/64	$-16/64$	$-48/64$	$-81/64$	-112/64*	$-\infty$
1	118/64*	84/64	50/64	16/64	$-16/64$	$-50/64$	$-83/64$	-116/64*	$-\infty$
2	121/64	86/64	52/64	16/64	$-16/64$	$-52/64$	$-86/64$	-120/64	$-\infty$
3	125/64	90/64	54/64	18/64	$-18/64$	$-54/64$	$-88/64$	-124/64	$-\infty$
4	128/64	92/64	54/64	18/64	$-18/64$	$-54/64$	$-90/64$	-127/64	$-\infty$
5	132/64	94/64	56/64	18/64	$-18/64$	$-56/64$	$-94/64$	-131/64	$-\infty$
6	135/64	96/64	58/64	18/64	$-18/64$	$-58/64$	$-96/64$	-134/64	$-\infty$
7	139/64	100/64	60/64	20/64	$-20/64$	$-60/64$	-98/64	-138/64	$-\infty$
8	142/64	102/64	60/64	20/64	$-20/64$	$-60/64$	-100/64	-141/64	$-\infty$
9	146/64	104/64	62/64	20/64	$-20/64$	$-62/64$	-104/64	-144/64	$-\infty$
10	150/64	106/64	64/64	20/64	$-20/64$	$-64/64$	-106/64	-148/64	$-\infty$
11	152/64	108/64	64/64	20/64	$-20/64$	$-64/64$	-108/64	-152/64	$-\infty$
12	156/64	112/64	66/64	22/64	$-22/64$	$-66/64$	-112/64	-156/64	$-\infty$
13	160/64	114/64	68/64	22/64	$-22/64$	$-68/64$	-114/64	-158/64	$-\infty$
14	164/64	116/64	70/64	24/64	$-24/64$	$-70/64$	-116/64	-162/64	$-\infty$
15	166/64	118/64	70/64	24/64	$-24/64$	$-70/64$	-118/64	-166/64	$-\infty$
16	170/64	120/64	72/64	24/64	$-24/64$	$-72/64$	-120/64	-170/64	$-\infty$
17	173/64	124/64	73/64	24/64	$-24/64$	$-72/64$	-124/64	-172/64	$-\infty$
18	176/64	126/64	76/64	24/64	$-24/64$	$-76/64$	-124/64	-176/64	$-\infty$
19	180/64	128/64	76/64	24/64	$-24/64$	$-76/64$	-128/64	-180/64	$-\infty$
20	184/64	132/64	78/64	24/64	$-24/64$	$-78/64$	-132/64	-184/64	$-\infty$
21	188/64	134/64	80/64	28/64	$-28/64$	$-80/64$	-134/64	-188/64	$-\infty$
22	190/64	136/64	82/64	28/64	$-28/64$	$-82/64$	-136/64	-190/64	$-\infty$
23	194/64	138/64	82/64	28/64	$-28/64$	$-82/64$	-138/64	-194/64	$-\infty$
24	198/64	140/64	84/64	28/64	$-28/64$	$-84/64$	-140/64	-198/64	$-\infty$
25	200/64	142/64	84/64	28/64	$-28/64$	$-84/64$	-142/64	-200/64	$-\infty$
26	204/64	146/64	86/64	28/64	$-28/64$	$-86/64$	-146/64	-204/64	$-\infty$
27	208/64	148/64	88/64	28/64	$-28/64$	$-88/64$	-148/64	-208/64	$-\infty$
28	212/64	152/64	90/64	28/64	$-28/64$	$-90/64$	-152/64	-212/64	$-\infty$
29	214/64	152/64	90/64	28/64	$-28/64$	$-90/64$	-152/64	-214/64	$-\infty$
30	218/64	154/64	94/64	28/64	$-28/64$	$-94/64$	-154/64	-218/64	$-\infty$
31	222/64	158/64	94/64	32/64	$-32/64$	$-94/64$	-158/64	-222/64	$-\infty$

* Exceptions: $m_4(0) = \frac{113}{64}$, $m_{-3}(1) = -\frac{114}{64}$, $m_4(2) = \frac{117}{64}$, $m_{-3}(3) = -\frac{117}{64}$

Fig. 10.3 Comparison constants $m_k(i)$, where $i = \left\lfloor 128 \left(d - \frac{1}{2} \right) \right\rfloor$

Lemma 10.9 $R_0 = x - 1$, *and for* $j \geq 0$,

$$R_{j+1} = rR_j - q_{j+1}(2Q_j + r^{-(j+1)}q_{j+1}).$$

Proof The claim is trivial for $j = 0$, and for $j > 0$,

$$
\begin{aligned}
R_{j+1} &= r^{j+1}(x - Q_{j+1}^2) \\
&= r^{j+1}\left(x - (Q_j + r^{-(j+1)}q_{j+1})^2\right) \\
&= r^{j+1}\left(x - (Q_j^2 + 2Q_j r^{-(j+1)}q_{j+1} + r^{-2(j+1)}q_{j+1}^2)\right) \\
&= r^{j+1}(x - Q_j^2) - r^{j+1}(2Q_j r^{-(j+1)}q_{j+1} + r^{-2(j+1)}q_{j+1}^2) \\
&= rR_j - q_{j+1}(2Q_j + r^{-(j+1)}q_{j+1}).
\end{aligned}
$$ $\square$

For $j \geq 0$, let

$$\underline{B}(j) = -2\rho Q_j + \rho^2 r^{-j}$$

and

$$\overline{B}(j) = 2\rho Q_j + \rho^2 r^{-j}.$$

The root digits will be selected with the goal of preserving the invariant

$$\underline{B}(j) \leq R_j \leq \overline{B}(j). \tag{10.17}$$

To motivate this choice of bounds, note that if

$$\sqrt{x} = \lim_{j \to \infty} Q_j = \lim_{j \to \infty}\left(1 + \sum_{i=1}^{j} r^{-i}q_i\right) = 1 + \sum_{i=1}^{\infty} r^{-i}q_i,$$

then

$$|\sqrt{x} - Q_j| = \left|\sum_{i=j+1}^{\infty} r^{-i}q_i\right| \leq a\sum_{i=j+1}^{\infty} r^{-i} = \frac{ar^{-j}}{r-1} = \rho r^{-j},$$

or

$$Q_j - \rho r^{-j} \leq \sqrt{x} \leq Q_j + \rho r^{-j}.$$

Since

$$Q_j = 1 + \sum_{i=1}^{j} r^{-i} q_i \geq 1 - \sum_{i=1}^{j} r^{-i} a = 1 - \frac{a(1 - r^{-j})}{r - 1} = 1 - \rho + \rho r^{-j} \geq \rho r^{-j},$$

this may be expressed in rational terms as

$$(Q_j - \rho r^{-j})^2 \leq x \leq (Q_j + \rho r^{-j})^2.$$

Thus, (10.17) is equivalent to convergence:

Lemma 10.10 *For $j \geq 0$,*

$$(Q_j - \rho r^{-j})^2 \leq x \leq (Q_j + \rho r^{-j})^2 \Leftrightarrow \underline{B}(j) \leq R_j \leq \overline{B}(j).$$

Proof

$$(Q_j - \rho r^{-j})^2 \leq x \leq (Q_j + \rho r^{-j})^2$$
$$\Leftrightarrow Q_j^2 - 2Q_j \rho r^{-j} + \rho^2 r^{-2j} \leq x \leq Q_j^2 + 2Q_j \rho r^{-j} + \rho^2 r^{-2j}$$
$$\Leftrightarrow -2\rho Q_j + \rho^2 r^{-j} \leq r^j (x - Q_j^2) \leq 2\rho Q_j + \rho^2 r^{-j}$$
$$\Leftrightarrow \underline{B}(j) \leq R_j \leq \overline{B}(j). \qquad \square$$

Note that the invariant holds trivially for $j = 0$:

Lemma 10.11 $\underline{B}(0) \leq R_0 \leq \overline{B}(0).$

Proof Since $R_0 = x - 1$ and $\frac{1}{4} \leq x < 1$, we have $-\frac{3}{4} \leq R_0 < 0$,

$$\underline{B}(0) = -2\rho Q_0 + \rho^2 = -2\rho + \rho^2 = (\rho - 1)^2 - 1 \leq \left(\frac{1}{2} - 1\right)^2 - 1 = -\frac{3}{4} \leq R_0,$$

and

$$\overline{B}(0) = 2\rho Q_0 + \rho^2 = 2\rho + \rho^2 \geq 2 \cdot \frac{1}{2} + \frac{1}{2}^2 = \frac{5}{4} > R_0. \qquad \square$$

For $j \geq 0$ and $k \in \{-a, \ldots, a\}$, the *selection interval* $[L_k(j), U_k(j)]$, defined by

$$U_k(j) = 2(k + \rho)Q_j + (k + \rho)^2 r^{-(j+1)} \qquad (10.18)$$

and

$$L_k(j) = 2(k - \rho)Q_j + (k - \rho)^2 r^{-(j+1)}, \qquad (10.19)$$

is so named because if the shifted partial remainder rR_j lies in this interval, then the invariant (10.17) may be ensured by choosing $q_{j+1} = k$:

Lemma 10.12 *Let $j \geq 0$ and $-a \leq k \leq a$. If $L_k(j) \leq rR_j \leq U_k(j)$ and $q_{j+1} = k$, then*

$$\underline{B}(j+1) \leq R_{j+1} \leq \overline{B}(j+1).$$

Proof By hypothesis and Lemma 10.9,

$$\begin{aligned}
R_{j+1} &= rR_j - 2Q_j k - r^{-(j+1)}k^2 \\
&\leq U_k(j) - 2Q_j k - r^{-(j+1)}k^2 \\
&= 2(k+\rho)Q_j + (k+\rho)^2 r^{-(j+1)} - 2Q_j k - r^{-(j+1)}k^2 \\
&= 2\rho Q_j + (2k\rho + \rho^2)r^{-(j+1)} \\
&= 2\rho(Q_{j+1} - r^{-(j+1)}k) + (2k\rho + \rho^2)r^{-(j+1)} \\
&= 2\rho Q_{j+1} + \rho^2 r^{-(j+1)} \\
&= \overline{B}(j+1),
\end{aligned}$$

and similarly,

$$\begin{aligned}
R_{j+1} &\geq L_k(j) - 2Q_j k - r^{-(j+1)}k^2 \\
&= 2(k-\rho)Q_j + (k-\rho)^2 r^{-(j+1)} - 2Q_j k - r^{-(j+1)}k^2 \\
&= -2\rho Q_j + (-2k\rho + \rho^2)r^{-(j+1)} \\
&= -2\rho(Q_{j+1} - r^{-(j+1)}k) + (-2k\rho + \rho^2)r^{-(j+1)} \\
&= -2\rho Q_{j+1} + \rho^2 r^{-(j+1)} \\
&= \underline{B}(j+1). \qquad \square
\end{aligned}$$

Thus, the existence of a root digit q_{j+1} that preserves (10.17) is guaranteed if the selection intervals cover the entire range of rR_j, i.e.,

$$\left[r\underline{B}(j), r\overline{B}(j) \right] \subseteq \bigcup_{k=-a}^{a} [L_k(j), U_k(j)].$$

Unlike the corresponding intervals for division, the overlapping of successive selection intervals for the square root depends on the choice of parameters, but the following important property holds in the general case.

Lemma 10.13 *For $j \geq 0$, $U_a(j) = r\overline{B}(j)$ and $L_{-a}(j) = r\underline{B}(j)$.*

Proof As noted in the proof of Lemma 10.5, $a + \rho = r\rho$. Thus,

$$U_a(j) = 2(a + \rho)Q_j + (a + \rho)^2 r^{-(j+1)} = 2r\rho Q_J + r\rho^2 r^{-j} = r\overline{B}(j)$$

and

$$L_{-a}(j) = 2(-a - \rho)Q_j + (-a - \rho)^2 r^{-(j+1)} = -2r\rho Q_J + r\rho^2 r^{-j} = r\underline{B}(j).\ \square$$

Since the selection intervals depend on the iteration j, and in particular on Q_j, so do the comparison constants that govern digit selection, which we shall denote here as $m_k(j)$. In practice, however, they vary only for the smallest values of j, until the partial root stabilizes. We have the following analog of Lemma 10.6:

Lemma 10.14 *Let* $j \in \mathbb{N}$, $t \in \mathbb{N}$, *and* $A_j \in \mathbb{Q}$. *Let* $m_{-a}(j) = -\infty$, *and for* $-a < k \le a$, *let* $m_k(j) \in \mathbb{Q}$ *such that*

$$A_j < m_k(j) \Rightarrow rR_j < m_k(j) \tag{10.20}$$

and

$$A_j \ge m_k(j) \Rightarrow rR_j > m_k(j) - 2^{-t}. \tag{10.21}$$

Let $q = q_{j+1}$. *Assume that* q *is the greatest* $k \in \{-a, \dots, a\}$ *such that* $m_k(j) \le A_j$ *and that*

$$q \ne a \Rightarrow m_{q+1}(j) \le U_q(j) \tag{10.22}$$

and

$$q \ne -a \Rightarrow L_q(j) + 2^{-t} \le m_q(j). \tag{10.23}$$

If $\underline{B}(j) \le R_j \le \overline{B}(j)$, *then* $\underline{B}(j + 1) \le R_{j+1} \le \overline{B}(j + 1)$.

Proof We shall show that $L_q(j) \le rR_j \le U_q(j)$ and invoke Lemma 10.12. Since Lemma 10.13 ensures that $L_{-a}(j) \le rR_j \le U_a(j)$, the required bounds are reduced to

$$q \ne a \Rightarrow rR_j \le U_q(j)$$

and

$$q \ne -a \Rightarrow rR_j \ge L_q(j).$$

But according to hypothesis, if $q \neq a$, then $A_j < m_{q+1}(j)$, which implies

$$r R_j < m_{q+1}(j) \leq U_q(j),$$

and if $q \neq -a$, then $A_j \geq m_q(j)$, which implies

$$r R_j > m_q(j) - 2^{-t} \geq (L_q(j) + 2^{-t}) - 2^{-t} = L_q(j). \qquad \square$$

We note that the hypothesis of Lemma 10.14 is weaker than that of Lemma 10.6 in two respects. First, instead of the general assumption that

$$L_k(j) + 2^{-t} \leq m_k(j) \leq U_{k-1}(j) \tag{10.24}$$

for all k, we have the special cases (10.22) and (10.23). The only reason for this is that it is often the case, as illustrated in the following sections, that positive values of k cannot occur as the first digit q_1, and therefore the corresponding cases of (10.24) are irrelevant for $j = 0$ and may not be satisfied.

Second, instead of the explicit assumptions that $2^t A_j \in \mathbb{Z}$, $2^t m_k(j) \in \mathbb{Z}$, and

$$|A_j - r R_j| < 2^{-t}, \tag{10.25}$$

we have the weaker conditions (10.20) and (10.21). (It is clear that (10.21) is a consequence of (10.25), and when $2^t A_j$ and $2^t m_k(j)$ are integers, so is (10.20).) This is intended to account for cases of interest in which these conditions hold but (10.25) does not (see Lemma 10.15).

10.5   Minimally Redundant Radix-4 Square Root

The results of this section are the basis of the square root component of the divider of Sect. 10.2. Once again, we have $r = 4$, $a = 2$, $\rho = \frac{2}{3}$. The problem of parameter optimization for the square root is not as straightforward as for division, but it has been determined experimentally that for this case, the required accuracy of the approximation A_j is given by $t = 5$, with only 3 fractional bits needed for the comparison constants.

Since prescaling of the radicand is not feasible, the constants must depend on the partial root Q_j. Given $j \geq 0$, let $j' = \min(j, 2)$, and let i be the integer defined by $Q_{j'} = \frac{1}{2} + \frac{i}{16}$. We shall show that each partial root satisfies the invariant $\frac{1}{2} \leq Q_j \leq 1$ so that $0 \leq i \leq 8$. The constants, which are displayed in Fig. 10.4, depend only on i and j and are denoted as $m_k(i, j)$. In fact, as noted in the figure,

i	m_2	m_1	m_0	m_{-1}	m_{-2}
0	$\frac{12}{8}$	$\frac{4}{8}$	$-\frac{4}{8}$	$-\frac{12}{8}*$	$-\infty$
1	$\frac{13}{8}*$	$\frac{4}{8}$	$-\frac{4}{8}$	$-\frac{13}{8}$	$-\infty$
2	$\frac{15}{8}$	$\frac{4}{8}$	$-\frac{4}{8}$	$-\frac{15}{8}$	$-\infty$
3	$\frac{16}{8}$	$\frac{6}{8}$	$-\frac{6}{8}$	$-\frac{16}{8}$	$-\infty$
4	$\frac{18}{8}$	$\frac{6}{8}$	$-\frac{6}{8}$	$-\frac{18}{8}$	$-\infty$
5	$\frac{20}{8}$	$\frac{8}{8}$	$-\frac{6}{8}$	$-\frac{20}{8}$	$-\infty$
6	$\frac{20}{8}$	$\frac{8}{8}$	$-\frac{8}{8}$	$-\frac{20}{8}$	$-\infty$
7	$\frac{22}{8}$	$\frac{8}{8}$	$-\frac{8}{8}$	$-\frac{22}{8}$	$-\infty$
8	$\frac{24}{8}$	$\frac{8}{8}$	$-\frac{8}{8}$	$-\frac{24}{8}*$	$-\infty$

* Exceptions: $m_{-1}(0,1) = -\frac{11}{8}$, $m_2(1,2) = \frac{15}{8}$, $m_{-1}(8,0) = -\frac{20}{8}$

Fig. 10.4 Comparison constants $m_k(i,j)$

$m_k(i, j)$ is determined solely by i with the three exceptions $m_{-1}(0, j)$, $m_2(1, j)$, and $m_{-1}(8, j)$. Also note that for fixed i and j,

$$m_{-2}(i, j) < m_{-1}(i, j) < m_0(i, j) < m_1(i, j) < m_2(i, j).$$

The following result is the inductive step in a proof of the required bounds on the partial root and remainder. Note that the induction is based on a slightly different scheme from that used for division: each of Lemmas 10.7 and 10.8 derives the conclusion that a set of properties hold for an index $j + 1$ from the assumption that they hold for j, whereas the proof below requires the stronger inductive hypothesis that the desired properties hold for all indices between 0 and j.

Lemma 10.15 *Let $r = 4$ and $a = 2$ and let $m_k(i, j)$ be as specified in Fig. 10.4. Given $j \geq 0$, suppose that the following conditions hold for all ℓ, $0 \leq \ell \leq j$, with $i = 64(Q_{\min(\ell, 2)} - \frac{1}{2})$:*

(a) $\frac{1}{2} \leq Q_\ell \leq 1$;
(b) $\underline{B}(\ell) \leq R_\ell \leq \overline{B}(\ell)$;
(c) For all $k \in \{-1, \dots, 2\}$, $A_\ell \in \mathbb{Q}$ satisfies

$$A_\ell < m_k(i, \ell) \Rightarrow 4R_\ell < m_k(i, \ell)$$

and

$$A_\ell \geq m_k(i, \ell) \Rightarrow 4R_\ell > m_k(i, \ell) - \frac{1}{32};$$

(d) $q_{\ell+1}$ is the greatest $k \in \{-2, \ldots, 2\}$ such that $m_k(i, \ell) \leq A_\ell$.
Then (a) and (b) also hold for $\ell = j + 1$.

Proof

(a) If $Q_j = \frac{1}{2}$, then

$$R_j = 4^j \left(x - \frac{1}{4} \right) \geq 0.$$

It follows that $q_{j+1} \geq 0$, for otherwise $A_j < m_0(i, j)$, which implies

$$4R_j < m_0(i, j) < 0.$$

Thus,

$$Q_{j+1} = Q_j + 4^{-(j+1)} q_{j+1} \geq Q_j = \frac{1}{2}.$$

On the other hand, suppose $Q_j > \frac{1}{2}$. If $j = 0$, then

$$Q_{j+1} = Q_1 = Q_0 + 4^{-1} q_1 \geq 1 + 4^{-1}(-2) = \frac{1}{2},$$

and if $j > 0$, then since $4^j Q_j \in \mathbb{Z}$, $Q_j \geq \frac{1}{2} + 4^{-j}$ and

$$Q_{j+1} = Q_j + 4^{-(j+1)} q_{j+1} \geq \frac{1}{2} + 4^{-j} + 4^{-(j+1)}(-2) > \frac{1}{2}.$$

The proof of the upper bound is similar.

(b) To prove that $\underline{B}(j + 1) \leq R_{j+1} \leq \overline{B}(j + 1)$, we shall invoke Lemma 10.14. Let
$k = q_{j+1}$. We need only show that

$$k \neq 2 \Rightarrow m_{k+1}(i, j) \leq U_k(j) \tag{10.26}$$

and

$$k \neq -2 \Rightarrow L_k(j) + \frac{1}{32} \leq m_k(i, j). \tag{10.27}$$

First suppose $j \leq 2$. If $j = 0$, then $i = 8$, and since $m_k(8, 0) \leq A_0$, (c) implies

$$m_k(8, 0) < 4R_0 + \frac{1}{32} = 4(x - 1) + \frac{1}{32} < \frac{1}{32},$$

which in turn implies $k = q_1 \in \{-2, -1, 0\}$. Thus, if $j = 1$, then $i \in \{0, 4, 8\}$. Subject to these constraints, (10.26) and (10.27) may be verified by direct computation for all values of i, j, and k, since $U_k(j)$ and $L_k(j)$ are determined by j, k, and

$$Q_j = \frac{1}{2} + \frac{i}{16}.$$

Thus, we may assume $j \geq 3$. Since

$$|Q_j - Q_2| = \left| \sum_{\ell=3}^{j} 4^{-\ell} q_\ell \right| < 2 \sum_{\ell=3}^{\infty} 4^{-\ell} = \frac{2}{3} 4^{-2} = \frac{1}{24},$$

we have the following bounds on Q_j:

$$\frac{1}{2} + \frac{i}{16} - \frac{1}{24} = \frac{11}{24} + \frac{i}{16} < Q_j < \frac{1}{2} + \frac{i}{16} + \frac{1}{24} = \frac{13}{24} + \frac{i}{16}. \qquad (10.28)$$

This turns out to be sufficient for all cases except $i = 1$, but in this case, a better upper bound is possible. We have $Q_2 = \frac{9}{16}$, which implies $q_1 = -2$, $q_2 = 1$, and $Q_1 = \frac{1}{2}$. Since $q_2 < 2$, we must have

$$4R_1 < m_2(0, 1) = \frac{3}{2}$$

and

$$R_1 = 4(x - Q_1^2) = 4\left(x - \frac{1}{4}\right) < \frac{3}{8},$$

which implies $x < \frac{11}{32}$. It follows that $q_3 < 2$, for otherwise

$$4R_2 \geq m_2(1, 2) - \frac{1}{32} = \frac{15}{8} - \frac{1}{32} = \frac{59}{32} > \frac{7}{4},$$

$$R_2 = 4^2(x - Q_2^2) = 16\left(x - \frac{81}{256}\right) > \frac{7}{16},$$

and therefore

$$x > \frac{7}{256} + \frac{81}{256} = \frac{11}{32},$$

a contradiction. Thus, $q_3 \leq 1$ and

$$Q_j < \frac{9}{16} + 4^{-3}q_3 + 2\sum_{\ell=4}^{\infty} 4^{-\ell} \leq \frac{9}{16} + 4^{-3} + \frac{2}{3} \cdot 4^{-3} = \frac{113}{192}.$$

Combining this with (10.28), along with the assumption that $\frac{1}{2} \leq Q_j \leq 1$, we have $Q_{\min}(i) \leq Q_j \leq Q_{\max}(i)$, where

$$Q_{\min}(i) = \max\left(\frac{1}{2}, \frac{11}{24} + \frac{i}{16}\right)$$

and

$$Q_{\max}(i) = \begin{cases} \frac{113}{192} & \text{if } i = 1 \\ \min\left(1, \frac{13}{24} + \frac{i}{16}\right) & \text{if } i \neq 1. \end{cases}$$

Applying (10.18) and (10.19) with $r = 4$ and $\rho = \frac{2}{3}$, we conclude that for $k > 0$,

$$U_{k-1}(j) = 2Q_j\left(k - 1 + \frac{2}{3}\right) + 4^{-(j+1)}\left(k - 1 + \frac{2}{3}\right)^2 \geq 2Q_{\min}(i)\left(k - \frac{1}{3}\right)$$

and

$$L_k(j) = 2Q_j\left(k - \frac{2}{3}\right) + 4^{-(j+1)}\left(k - \frac{2}{3}\right)^2 \leq 2Q_{\max}(i)\left(k - \frac{2}{3}\right) + \frac{1}{256}\left(k - \frac{2}{3}\right)^2,$$

and similarly, for $k \leq 0$,

$$U_{k-1}(j) \geq 2Q_{\max}(i)\left(k - \frac{1}{3}\right)$$

and

$$L_k(j) \leq 2Q_{\min}(i)\left(k - \frac{2}{3}\right) + \frac{1}{256}\left(k - \frac{2}{3}\right)^2.$$

The required inequalities (10.26) and (10.27) follow from these inequalities in all cases.   □

10.6   Minimally Redundant Radix-8 Square Root

Our final example addresses the minimally redundant radix-8 case: $r = 8$, $a = 4$, and $\rho = \frac{4}{7}$. The results of this section are the basis of the design of Chap. 20. It has been determined experimentally that the required accuracy of the approximation A_j of $8R_j$ for this case is given by $t = 7$, with 6 fractional bits needed for the comparison constants. Since the initial remainder is stored in nonredundant form, we shall assume that the approximation is precise for $j = 0$.

m_4	m_3	m_2	m_1	m_0	m_{-1}	m_{-2}	m_{-3}	m_{-4}
0	0	0	0	$-64/64$	-176/64	-272/64	-352/64	$-\infty$

Fig. 10.5   Comparison constants $m_k(32, 0)$

Given $j \geq 0$, let $j' = \min(j, 2)$, and let i be the integer defined by $Q_{j'} = \frac{1}{2} + \frac{i}{64}$. The constants $m_k(j)$, which are displayed in Figs. 10.5, 10.6, and 10.7, are denoted as $m_k(i, j)$ but depend only on i and j'. Thus, $m_k(i, j) = m_k(i, 2)$ for $j > 2$.

Clearly, for $j = 0$, $i = 32$. Since $A_0 = 8R_0 = 8(x - 1) < 0 = m_1(0)$, $q_1 \leq 0$. It follows that $m_k(32, 0)$ is irrelevant for $k > 1$ and that for $j = 1$, $i \in \{0, 8, 16, 24, 32\}$. We shall show that each partial root satisfies the invariant $\frac{1}{2} \leq Q_j \leq 1$, and therefore, for $j \geq 2$, $0 \leq i \leq 32$.

Lemma 10.16 *Let $r = 8$ and $a = 4$, and let $m_k(i, j)$ be as specified in Figs. 10.5, 10.6, and 10.7. Given $j \geq 0$, suppose that the following conditions hold for all ℓ, $0 \leq \ell \leq j$, with $i = 64(Q_{\min(\ell,2)} - \frac{1}{2})$:*

(a) $\frac{1}{2} \leq Q_\ell \leq 1$;
(b) $\underline{B}(\ell) \leq R_\ell \leq \overline{B}(\ell)$;
(c) If $\ell = 0$, then $A_\ell = 8R_0$, and if $\ell > 0$, then for all $k \in \{-3, \dots, 4\}$,

$$A_\ell < m_k(i, \ell) \Rightarrow 8R_\ell < m_k(i, \ell)$$

and

$$A_\ell \geq m_k(i, \ell) \Rightarrow 8R_\ell > m_k(i, \ell) - \frac{1}{128};$$

(d) $q_{\ell+1}$ is the greatest $k \in \{-4, \dots, 4\}$ such that $m_k(i, \ell) \leq A_\ell$.

Then (a) and (b) hold for $\ell = j + 1$.

Proof

(a) If $Q_j = \frac{1}{2}$, then

$$R_j = 8^j \left(x - \frac{1}{4} \right) \geq 0.$$

It follows that $q_{j+1} \geq 0$, for otherwise $A_j < m_0(i, j)$ and

$$8R_j < A_j + \frac{1}{128} < m_0(i, j) + \frac{1}{128} < 0.$$

i	m_4	m_3	m_2	m_1	m_0	m_{-1}	m_{-2}	m_{-3}	m_{-4}
0	237/64	167/64	97/64	32/64	−32/64	-92/64	-152/64	-212/64	$-\infty$
8	292/64	207/64	122/64	42/64	−42/64	-122/64	-192/64	-267/64	$-\infty$
16	352/64	242/64	142/64	47/64	−47/64	-142/64	-232/64	-322/64	$-\infty$
24	407/64	282/64	172/64	62/64	−62/64	-172/64	-277/64	-377/64	$-\infty$
32	462/64	327/64	192/64	62/64	−62/64	-192/64	-317/64	-442/64	$-\infty$

Fig. 10.6 Comparison constants $m_k(i, 1)$

Thus,

$$Q_{j+1} = Q_j + 8^{-(j+1)} q_{j+1} \geq Q_j = \frac{1}{2}.$$

On the other hand, suppose $Q_j > \frac{1}{2}$. If $j = 0$, then

$$Q_{j+1} = Q_1 = Q_0 + 8^{-1} q_1 \geq 1 + 8^{-1}(-4) = \frac{1}{2},$$

and if $j > 0$, then since $8^j Q_j \in \mathbb{Z}$, $Q_j \geq \frac{1}{2} + 8^{-j}$ and

$$Q_{j+1} = Q_j + 8^{-(j+1)} q_{j+1} \geq \frac{1}{2} + 8^{-j} + 8^{-(j+1)}(-4) > \frac{1}{2}.$$

The proof of the upper bound is similar.

(b) To prove that $\underline{B}(j+1) \leq R_{j+1} \leq \overline{B}(j+1)$, we shall invoke Lemma 10.14. Let $k = q_{j+1}$. We need only show that

$$k \neq 4 \Rightarrow m_{k+1}(i, j) \leq U_k(j) \tag{10.29}$$

and

$$k \neq -4 \Rightarrow L_k(j) + \frac{1}{128} \leq m_k(i, j). \tag{10.30}$$

First suppose $j \leq 2$. As noted above, if $j = 0$, then $i = 32$ and $k \leq 0$, and if $j = 1$, then $i \in \{0, 4, 8, 16, 24, 32\}$. Subject to these constraints, (10.29) and (10.30) may be verified by direct computation for all possible values of i, j, and k, since $U_k(j)$ and $L_k(j)$ are determined by j, k, and

$$Q_j = \frac{1}{2} + \frac{i}{64}.$$

Thus, we may assume $j \geq 3$. Since

$$|Q_j - Q_2| = \left| \sum_{\ell=3}^{j} 8^{-\ell} q_\ell \right| < 4 \sum_{\ell=3}^{\infty} 8^{-\ell} = \frac{4}{7} \cdot 8^{-2} = \frac{1}{112},$$

we have the following bounds on Q_j:

$$\frac{1}{2} + \frac{i}{64} - \frac{1}{112} = \frac{55}{112} + \frac{i}{64} < Q_j < \frac{1}{2} + \frac{i}{64} + \frac{1}{112} = \frac{57}{112} + \frac{i}{16}. \tag{10.31}$$

Thus, $Q_{\min}(i) \leq Q_j \leq Q_{\max}(i)$, where

$$Q_{\min}(i) = \max \left(\frac{1}{2}, \frac{55}{112} + \frac{i}{64} \right)$$

and

$$Q_{\max}(i) = \min \left(1, \frac{57}{112} + \frac{i}{64} \right).$$

Applying (10.18) and (10.19) with $r = 8$ and $\rho = \frac{4}{7}$, we conclude that for $k > 0$,

$$U_{k-1}(j) = 2Q_j \left(k - 1 + \frac{4}{7} \right) + 8^{-(j+1)} \left(k - 1 + \frac{4}{7} \right)^2 \geq 2Q_{\min}(i) \left(k - \frac{3}{7} \right)$$

and

$$L_k(j) = 2Q_j \left(k - \frac{4}{7} \right) + 8^{-(j+1)} \left(k - \frac{4}{7} \right)^2 \leq 2Q_{\max}(i) \left(k - \frac{4}{7} \right) + 8^{-4} \left(k - \frac{4}{7} \right)^2,$$

i	m_4	m_3	m_2	m_1	m_0	m_{-1}	m_{-2}	m_{-3}	m_{-4}
0	226/64	161/64	97/64	32/64	−32/64	−97/64	-161/64	-225/64	$-\infty$
1	231/64	165/64	99/64	33/64	−33/64	−99/64	-165/64	-231/64	$-\infty$
2	238/64	170/64	102/64	34/64	−34/64	-102/64	-170/64	-238/64	$-\infty$
3	245/64	175/64	105/64	35/64	−35/64	-105/64	-175/64	-245/64	$-\infty$
4	252/64	180/64	108/64	36/64	−36/64	-108/64	-180/64	-252/64	$-\infty$
5	259/64	185/64	111/64	37/64	−37/64	-111/64	-185/64	-259/64	$-\infty$
6	266/64	190/64	114/64	38/64	−38/64	-114/64	-190/64	-266/64	$-\infty$
7	273/64	195/64	117/64	39/64	−39/64	-117/64	-195/64	-273/64	$-\infty$
8	280/64	200/64	120/64	40/64	−40/64	-120/64	-200/64	-280/64	$-\infty$
9	287/64	205/64	123/64	41/64	−41/64	-123/64	-205/64	-287/64	$-\infty$
10	294/64	210/64	126/64	42/64	−42/64	-126/64	-210/64	-294/64	$-\infty$
11	301/64	215/64	129/64	43/64	−43/64	-129/64	-215/64	-301/64	$-\infty$
12	308/64	220/64	132/64	44/64	−44/64	-132/64	-220/64	-308/64	$-\infty$
13	315/64	225/64	135/64	45/64	−45/64	-135/64	-225/64	-315/64	$-\infty$
14	322/64	230/64	138/64	46/64	−46/64	-138/64	-230/64	-322/64	$-\infty$
15	329/64	235/64	141/64	47/64	−47/64	-141/64	-235/64	-329/64	$-\infty$
16	336/64	240/64	144/64	48/64	−48/64	-144/64	-240/64	-336/64	$-\infty$
17	343/64	245/64	147/64	49/64	−49/64	-147/64	-245/64	-343/64	$-\infty$
18	350/64	250/64	150/64	50/64	−50/64	-150/64	-250/64	-350/64	$-\infty$
19	357/64	255/64	153/64	51/64	−51/64	-153/64	-255/64	-357/64	$-\infty$
20	364/64	260/64	156/64	52/64	−52/64	-156/64	-260/64	-364/64	$-\infty$
21	371/64	265/64	159/64	53/64	−53/64	-159/64	-265/64	-371/64	$-\infty$
22	378/64	270/64	162/64	54/64	−54/64	-162/64	-270/64	-378/64	$-\infty$
23	385/64	275/64	165/64	55/64	−55/64	-165/64	-275/64	-385/64	$-\infty$
24	392/64	280/64	168/64	56/64	−56/64	-168/64	-280/64	-392/64	$-\infty$
25	398/64	285/64	171/64	57/64	−57/64	-171/64	-285/64	-399/64	$-\infty$
26	406/64	290/64	174/64	58/64	−58/64	-174/64	-290/64	-406/64	$-\infty$
27	413/64	295/64	177/64	59/64	−59/64	-177/64	-295/64	-413/64	$-\infty$
28	420/64	300/64	180/64	60/64	−60/64	-180/64	-300/64	-420/64	$-\infty$
29	427/64	305/64	183/64	61/64	−61/64	-183/64	-305/64	-427/64	$-\infty$
30	434/64	310/64	186/64	62/64	−62/64	-186/64	-310/64	-434/64	$-\infty$
31	441/64	315/64	189/64	63/64	−63/64	-189/64	-315/64	-441/64	$-\infty$
32	447/64	319/64	191/64	64/64	−63/64	-191/64	-318/64	-446/64	$-\infty$

Fig. 10.7 Comparison constants $m_k(i, j)$ for $j \geq 2$

and similarly, for $k \leq 0$,

$$U_{k-1}(j) \geq 2Q_{\max}(i)\left(k - \frac{3}{7}\right)$$

and

$$L_k(j) \leq 2Q_{\min}(i)\left(k - \frac{4}{7}\right) + 8^{-4}\left(k - \frac{4}{7}\right)^2.$$

The required inequalities (10.29) and (10.30) follow in all cases with the exception of the lower bound on $m_k(i, j)$ in the case $i = 1$, $k = -3$. In this case, however, a better upper bound on $L_k(j)$ is possible: since $Q_2 = \frac{1}{2} + \frac{1}{64} = \frac{33}{64}$ and

$$Q_j = Q_2 + \sum_{\ell=3}^{j} 8^{-j} q_j \geq \frac{33}{64} - 4\sum_{\ell=3}^{j} 8^{-j} = \frac{33}{64} - \frac{4}{7}(8^{-2} - 8^{-j}),$$

$$L_{-3}(j) = 2Q_j\left(-3 - \frac{4}{7}\right) + 8^{-(j+1)}\left(-3 - \frac{4}{7}\right)^2$$

$$\leq 2\left(\frac{33}{64} - \frac{4}{7}(8^{-2} - 8^{-j})\right)\left(-\frac{25}{7}\right) + 8^{-(j+1)}\left(-\frac{25}{7}\right)^2$$

$$= -\frac{5675}{1568} - \frac{975}{393} \cdot 8^{-j}$$

$$< -\frac{5675}{1568},$$

which is sufficient to ensure that

$$L_{-3}(j) + \frac{1}{128} < -\frac{231}{64} = m_{-3}(i, j). \qquad \square$$

Chapter 11
Division Based on Fused Multiply-Add

Multiplicative division algorithms are typically based on a sequence of approximations of the reciprocal of the divisor, derived by the Newton–Raphson method. Given a differentiable function f and a sufficiently accurate initial approximation y_0 of a root of the equation $f(y) = 0$, the Newton–Raphson recurrence formula

$$y_{k+1} = y_k - \frac{f(y_k)}{f'(y_k)}$$

computes a convergent sequence of approximations. For the case

$$f(y) = \frac{1}{y} - b,$$

the root of which is the reciprocal of b, this yields

$$y_{k+1} = y_k + \frac{\frac{1}{y_k} - b}{\frac{1}{y_k^2}} = y_k(2 - by_k). \tag{11.1}$$

Since the relative error

$$\frac{\frac{1}{b} - y_k}{\frac{1}{b}} = 1 - by_k$$

satisfies

$$1 - by_{k+1} = 1 - by_k(2 - by_k) = (1 - by_k)^2,$$

© The Author(s), under exclusive license to Springer Nature Switzerland AG 2022
D. M. Russinoff, *Formal Verification of Floating-Point Hardware Design*,
https://doi.org/10.1007/978-3-030-87181-9_11

the convergence is quadratic, i.e., the number of bits of accuracy of the approximation doubles with each iteration. This means that convergence is achieved in fewer iterations than required by the digit-recurrence approach, but the complexity of each iteration is greater. On the other hand, the hardware requirement may be minimized by utilizing existing multiplication hardware.

The subject of this chapter is a multiplicative technique for floating-point division that was developed for IBM RISC processors in the late 1980s and remains in widespread use today. The initial approximation of the reciprocal of the divisor is derived from tables in read-only memory. The sequence of Newton–Raphson refinements of this approximation is interleaved with a sequence of refinements of the quotient. Central to this process is a hardware *fused multiplication-addition* (FMA) operation, which is assumed to be implemented in support of the standard FMA machine instructions. The significance of this operation is that it has the effect of performing two arithmetic operations with a single rounding and is therefore both more efficient and more accurate than two separate instructions.

Implementations of this method are generally slower than those based on the SRT algorithms of Chap. 10 but have the advantage of lower hardware requirements. In fact, the computations are typically performed by either microcode or software.

We shall present proofs of correctness of two representative algorithms based on this approach, which operate on single- and double-precision floating-point numbers, respectively. The inputs to each are a pair of operands, a and b, and a rounding mode, $\mathcal{R}$. The operands are p-exact real numbers in the interval $[1, 2)$, where $p = 24$ or 53. $\mathcal{R}$ may be any of the IEEE rounding modes, although since it is applied here only to positive arguments, there is no distinction between *RTZ* and *RDN*. The returned value, as specified by IEEE 754 (see Fig. 3.1), is the rounded quotient $\mathcal{R}(\frac{a}{b}, p)$.

The algorithms are based on two primitive functions, which are assumed to be implemented in hardware:

1. A function *rcp24*, which computes a 24-exact approximation of the reciprocal of a 24-exact number in the interval $[1, 2)$ with relative error bounded by 2^{-23}. The definition of *rcp24*, which is based on table reference and interpolation, is presented in Sect. 11.1.
2. An atomic FMA operation, which computes the rounded value $\mathcal{R}(xy + z, p)$ for p-exact operands x, y, and z and any IEEE rounding mode $\mathcal{R}$. No restriction is imposed on the exponents of the operands.

The relevant theory, which is largely based on the work of Markstein [19] and Harrison [7], is developed in Sects. 11.2 and 11.3. The algorithms are presented in Sect. 11.4 along with proofs of their correctness.

11.1 Reciprocal Approximation

We shall define a function *rcp24* that computes an approximation of the reciprocal of a 24-exact number b, $1 \leq b < 2$, by the method of *minimax quadratic interpolation* [26]. The computation is based on a partition of the interval $[1, 2)$ into 2^k subintervals, $I_i = [1 + 2^{-k}i, 1 + 2^{-k}(i+1))$, where $0 \leq i < 2^k$, and a quadratic function defined on each subinterval. Thus, if $b \in I_i$ and $x = b - (1 + 2^{-k}i)$ is its offset within that subinterval, then the approximation is computed as

$$\frac{1}{b} \approx C_0 + C_1 x + C_2 x^2$$

		$i[6{:}4]$							
		0	1	2	3	4	5	6	7
	0	7FFFFFD	71C71C6	6666666	5D1745E	5555556	4EC4EC4	4924924	4444444
	1	7F01FC0	70FE3BE	65C393E	5C90A1E	54E4254	4E6470A	48D159D	43FBC04
	2	7E07E08	70381C0	6522C3E	5C0B814	54741FA	4E04E06	487EDE0	43B3D5A
	3	7D11968	6F74AE0	6483ED0	5B87DDC	5405402	4DA637C	482D1C0	436C82C
	4	7C1F079	6EB3E42	63E7062	5B05B06	539782A	4D4873E	47DC120	4325C54
	5	7B301EA	6DF5B0C	634C064	5A84F32	532AE22	4CEB917	478BBCE	42DF9BA
	6	7A44C6C	6D3A06C	62B2E44	5A05A04	52BF5A8	4C8F8D2	473C1AA	429A044
$i[3{:}0]$	7	795CEB0	6C80D90	621B97E	5987B1A	5254E7A	4C3463E	46ED290	4254FCE
	8	7878786	6BCA1AE	6186186	590B214	51EB852	4BDA130	469EE58	4210842
	9	77975B9	6B15C08	60F25DE	588FEA0	51832F2	4B80970	46514DE	41CC984
	A	76B981E	6A63BD8	6060606	5816058	511BE18	4B27ED2	4604602	4189376
	B	75DED92	69B406A	5FD0180	579D6EC	50B5989	4AD012C	45B81A2	41465FE
	C	7507504	6906908	5F417D0	572620A	5050506	4A79049	456C798	4104104
	D	7432D64	685B4FC	5EB4882	56B015C	4FEC050	4A22C04	45217C4	40C2470
	E	73615A2	67B23A6	5E2931E	563B48C	4F88B2E	49CD430	44D7204	4081020
	F	7292CC0	670B450	5D9F736	55C7B4C	4F26566	497889E	448D639	4040404

Fig. 11.1 $2^{27} C_0(i), 0 \leq i < 2^7$

and rounded to 24 bits, where the coefficients $C_j = C_j(i)$ are read from tables in read-only memory indexed by i, which have been designed to minimize the maximum error over each subinterval. Naturally, the value of k and the precisions of the coefficients are chosen to be as small as possible while providing the desired accuracy of the approximation.

For our present purpose, it has been determined that $k = 7$ and coefficients C_0, C_1, and C_2 of bit-widths 27, 17, and 12, respectively, are sufficient. The resulting tables, which are displayed in Figs. 11.1, 11.2, and 11.3, occupy $2^7(27 + 17 + 12)/8 = 896$ bytes of ROM. Since the reciprocal function is positive, decreasing, and convex, $C_0 > 0$, $C_1 < 0$, and $C_2 > 0$. Furthermore, each coefficient satisfies

$|C_j| < 1$, i.e., all bits are fractional. Note that the table entries are represented with the radix points and the sign of C_1 is omitted. That is, the displayed values are $2^{27}C_0$, $-2^{17}C_1$, and $2^{12}C_2$, all in hexadecimal notation. The approximation is computed as follows:

Definition 11.1 Given a 24-exact number b, $1 \leq b < 2$, let $i = \lfloor 2^7(b-1) \rfloor$ and $x = b - (1 + 2^{-7}i)$. Then

$$rcp24(b) = RNE(C_0(i) + C_1(i)x + C_2(i)x^2, 24),$$

where $C_0(i)$, $C_1(i)$, and $C_2(i)$ are defined as shown in Figs. 11.1, 11.2, and 11.3.

For details pertaining to the construction of such tables and the hardware implementation of the computation of Definition 11.1, the reader is referred to [26]. For our purpose, the following properties of *rcp24* are readily verified by straightforward exhaustive computation, without appealing to the derivation of the tables or the underlying theory:

i[3:0] \ i[6:4]	0	1	2	3	4	5	6	7
0	1FFFB	19488	147AC	10ECE	0E38E	0C1E4	0A72E	091A2
1	1F814	18EF8	1439E	10BC0	0E134	0C00A	0A5B3	0906E
2	1F05A	18986	13FA2	108C0	0DEE2	0BE38	0A43C	08F3C
3	1E8CE	18430	13BB8	105CE	0DC9A	0BC6A	0A2CA	08E10
4	1E16C	17EF6	137E2	102E8	0DA5C	0BAA4	0A15E	08CE6
5	1DA36	179D6	1341E	1000C	0D826	0B8E5	09FF6	08BC0
6	1D32A	174D2	1306A	0FD3E	0D5F8	0B72C	09E92	08A9E
7	1CC44	16FE6	12CC8	0FA7C	0D3D4	0B578	09D34	08980
8	1C586	16B12	12936	0F7C4	0D1B6	0B3CC	09BDA	08864
9	1BEEE	16658	125B4	0F51A	0CFA2	0B224	09A84	0874C
A	1B87A	161B4	12242	0F278	0CD94	0B082	09932	08638
B	1B228	15D28	11EE0	0EFE2	0CB8F	0AEE8	097E6	08526
C	1ABFA	158B2	11B8C	0ED56	0C992	0AD50	0969D	08418
D	1A5EE	15450	11848	0EAD6	0C79C	0ABC0	09558	0830E
E	1A002	15006	11510	0E85E	0C5AC	0AA36	09418	08206
F	19A36	14BCE	111E8	0E5F0	0C3C4	0A8B0	092DB	08102

Fig. 11.2 $-2^{17}C_1(i), 0 \leq i < 2^7$

$$i[6:4]$$

		0	1	2	3	4	5	6	7
	0	FCF	B1C	81C	618	4B8	3B4	2F4	268
	1	F74	AE0	7F8	5FC	4A4	3A4	2EE	264
	2	F18	AA8	7D4	5E4	48C	39C	2E0	258
	3	EC4	A74	7A8	5CC	47C	388	2D8	254
	4	E6A	A40	788	5B8	46C	37C	2D0	24C
	5	E18	A08	768	598	45C	372	2C8	244
	6	DCC	9DC	744	584	448	368	2BC	23C
$i[3:0]$	7	D7C	9A8	724	570	43C	358	2B4	23C
	8	D32	974	704	554	428	350	2AC	230
	9	CEB	948	6E4	544	41C	340	2A4	228
	A	CA4	918	6C4	52C	408	334	298	224
	B	C5C	8F0	6A8	518	3FA	330	294	21C
	C	C18	8C4	688	500	3EC	31E	28A	214
	D	BD8	894	670	4F0	3E0	314	280	210
	E	B98	870	64C	4D8	3CC	310	27C	20C
	F	B5C	844	634	4C4	3C0	304	272	208

Fig. 11.3 $2^{12}C_2(i), 0 \le i < 2^7$

Lemma 11.1 *If b is 24-exact, $1 \le b < 2$, and $y_0 = rcp24(b)$, then $\frac{1}{2} < y_0 \le 1$ and*

$$|1 - by_0| < 2^{-23}.$$

11.2 Quotient Refinement

The first step of each algorithm is the computation of an initial approximation y_0 of $\frac{1}{b}$ as an application of the function *rcp24*. An initial approximation of the quotient $\frac{a}{b}$ is then computed as

$$q_0 = RNE(ay_0, p).$$

The accuracy of q_0 may be derived from that of y_0:

Lemma 11.2 *Let $y > 0$, $a > 0$, $b > 0$, and $p > 1$. Assume that $|1 - by| \le \epsilon$, and let $q = RNE(ay, p)$. Then*

$$\left| 1 - \frac{b}{a}q \right| \le \epsilon + 2^{-p}(1 + \epsilon).$$

Proof Since

$$\left|1 - \frac{b}{a}ay\right| = |1 - by| \le \epsilon$$

and

$$|ay - q| \le 2^{-p}a|y| \le 2^{-p}\frac{a}{b}(1 + \epsilon),$$

$$\left|1 - \frac{b}{a}q\right| \le \left|1 - \frac{b}{a}ay\right| + \frac{b}{a}|ay - q| \le \epsilon + 2^{-p}(1 + \epsilon). \qquad \square$$

Each of the initial values y_0 and q_0 undergoes a series of refinements, culminating in the final rounded quotient q. Each refinement q_{k+1} of the quotient is computed from the preceding approximation q_k and the current reciprocal approximation y as follows:

$$r_k = RNE(a - bq_k, p)$$

$$q_{k+1} = RNE(q_k + r_k y, p)$$

In the final step, the input rounding mode $\mathcal{R}$ is used instead of *RNE*:

$$r_k = RNE(a - bq_k, p)$$

$$q = \mathcal{R}(q_k + r_k y, p)$$

Our main lemma, due to Markstein [19], ensures that the final quotient is correctly rounded under certain assumptions pertaining to the accuracy of the reciprocal and quotient approximations from which it is derived.

Lemma 11.3 *Let a, b, y, and q be p-exact, where $p > 1$, $1 \le a < 2$, and $1 \le b < 2$. Assume that the following inequalities hold:*

(i) $\left|\frac{a}{b} - q\right| < 2^{e+1-p}$, *where* $e = \begin{cases} 0 \ if \ a > b \\ -1 \ if \ a \le b; \end{cases}$

(ii) $|1 - by| < 2^{-p}$.

Let $r = a - bq$. Then r is p-exact, and for any IEEE rounding mode $\mathcal{R}$,

$$\mathcal{R}(q + ry, p) = \mathcal{R}\left(\frac{a}{b}, p\right).$$

Proof We may assume $r \ne 0$, for otherwise $\frac{a}{b} = q = q + ry$ and the claim holds trivially. We may also assume $a \ne b$, for otherwise,

$$|1 - q| = \left|\frac{a}{b} - q\right| < 2^{e+1-p} = 2^{-p}$$

implies $q = 1 = \frac{a}{b}$ and $r = 0$. It follows from the bounds on a and b that $e = expo(\frac{a}{b})$. We shall show that $e = expo(q)$ as well.

If $a > b$, then

$$\frac{a}{b} \geq \frac{b + 2^{1-p}}{b} = 1 + \frac{2^{1-p}}{b} > 1 + 2^{-p}$$

and

$$q > \frac{a}{b} - 2^{e+1-p} = \frac{a}{b} - 2^{1-p} > 1 + 2^{-p} - 2^{1-p} = 1 - 2^{-p},$$

which implies $q \geq 1$. On the other hand,

$$q < \frac{a}{b} + 2^{1-p} \leq 2 - 2^{1-p} + 2^{1-p} = 2,$$

and hence, $expo(q) = 0 = e$.

Similarly, if $a < b$, then

$$\frac{a}{b} \geq \frac{1}{2 - 2^{1-p}} = \frac{(1 - 2^{-p}) + 2^{-p}}{2 - 2^{1-p}} = \frac{1}{2} + \frac{2^{-p-1}}{1 - 2^{-p}} > \frac{1}{2} + 2^{-p-1}$$

and

$$q > \frac{a}{b} - 2^{e+1-p} = \frac{a}{b} - 2^{-p} > \frac{1}{2} + 2^{-p-1} - 2^{-p} = \frac{1}{2} - 2^{-p-1},$$

which implies $q \geq \frac{1}{2}$. On the other hand,

$$\frac{a}{b} \leq \frac{a}{a + 2^{1-p}} = 1 - \frac{2^{1-p}}{a + 2^{1-p}} \leq 1 - \frac{2^{1-p}}{2} = 1 - 2^{-p},$$

$$q < \frac{a}{b} + 2^{-p} < 1 - 2^{-p} + 2^{-p} = 1,$$

and $expo(q) = -1 = e$.

To establish p-exactness of r, since

$$|r| = b \left| \frac{a}{b} - q \right| < 2 \cdot 2^{e+1-p} = 2^{e+2-p},$$

either $r = 0$ or $expo(r) \leq e + 1 - p$, and it suffices to show that

$$2^{p-1-(e+1-p)}r = 2^{2p-2-e}r = 2^{2p-2-e}a - 2^{2p-2-e}bq \in \mathbb{Z}.$$

Since a is p-exact, $expo(a) = 0$, and $2p - 2 - e \geq 2p - 2 \geq p - 1$, $2^{2p-2-e}a \in \mathbb{Z}$; since b and q are p-exact, $expo(b) = 0$, and $expo(q) \geq e$,

$$2^{2p-2-e}bq = (2^{p-1}b)(2^{p-1-e}q) \in \mathbb{Z}.$$

For the proof of the second claim, we shall focus on the case $r > 0$; the case $r < 0$ is similar. Let $q' = q + 2^{e+1-p}$. The quotient $\frac{a}{b}$ lies in the interval (q, q'), and its rounded value is either q or q'. For the directed rounding modes (RUP, RDN, and RTZ), we need only show that $q + ry$ also belongs to this interval, i.e., $ry < 2^{1+e-p}$. Since $\frac{a}{b} = q + \frac{r}{b}$, this condition may be expressed as

$$\frac{r}{b} < 2^{e+1-p} \Rightarrow ry < 2^{e+1-p}, \tag{11.2}$$

or

$$2^{p-1-e}r < b \Rightarrow 2^{p-1-e}r < \frac{1}{y}.$$

Since (ii) implies

$$\frac{1}{y} = \frac{b}{by} > \frac{b}{1 + 2^{-p}},$$

this will follow from

$$2^{p-1-e}r < b \Rightarrow 2^{p-1-e}r \leq \frac{b}{1 + 2^{-p}}.$$

Since $2^{p-1-e}r$ and b are both p-exact and $expo(b) = 0$, we have

$$2^{p-1-e}r < b \Rightarrow 2^{p-1-e}r \leq b - 2^{1-p},$$

and it will suffice to show that

$$b - 2^{1-p} \leq \frac{b}{1 + 2^{-p}},$$

but this reduces to $b \leq 2 + 2^{1-p}$, and we have assumed that $b < 2$.

For the remaining case, $\mathcal{R} = RNE$, the proof may be completed by showing that $\frac{a}{b}$ and $q + ry$ lie on the same side of the midpoint $m = q + 2^{e-p}$ of the interval (q, q'). Note that $\frac{a}{b} = m$ is impossible, for if this were true, then since $a = \frac{a}{b} \cdot b$ is p-exact, Lemma 4.14 would imply that $\frac{a}{b} = m$ is also p-exact, but this is not the case. Thus, we must show that

$$\frac{r}{b} < 2^{e-p} \Rightarrow ry < 2^{e-p} \tag{11.3}$$

and

$$\frac{r}{b} > 2^{e-p} \Rightarrow ry > 2^{e-p} \tag{11.4}$$

The proof of (11.3) is the same as that of (11.2), and we may similarly show that (11.4) is a consequence of

$$b + 2^{1-p} \geq \frac{b}{1 - 2^{-p}}.$$

But this is equivalent to $b \leq 2 - 2^{1-p}$, which follows from the assumptions that b is p-exact and $b < 2$.   □

IEEE compliance of the algorithms of interest will be proved as applications of Lemma 11.3 by establishing the two hypotheses (i) and (ii) for appropriate values of y and q.

The next lemma consists of two results. The first specifies the accuracy of an intermediate quotient approximation;[1] the second addresses the final approximation, supplying the first inequality (i) required by Lemma 11.3.

Lemma 11.4 *Let* $1 \leq a < 2$, $1 \leq b < 2$, *and* $p > 0$. *Assume that* $|1 - by| \leq \epsilon$ *and* $|1 - \frac{b}{a}q_0| \leq \delta$. *Let* $r = RNE(a - bq_0, p)$ *and* $q = RNE(q_0 + ry, p)$. *Then*

$$\left| 1 - \frac{b}{a}q \right| \leq 2^{-p} + (1 + 2^{-p})\delta\epsilon + 2^{-p}\delta(1 + \epsilon) + 2^{-2p}\delta(1 + \epsilon).$$

If $\delta\epsilon + 2^{-p}\delta(1 + \epsilon) < 2^{-p-1}$, *then*

$$\left| q - \frac{a}{b} \right| < 2^{e+1-p},$$

where $e = \begin{cases} 0 \ \text{if } a > b \\ -1 \ \text{if } a \leq b. \end{cases}$

Proof Let $u = 1 - by$, $v = 1 - \frac{b}{a}q_0$, $r' = av = a - bq_0$, and $q' = q_0 + ry$. Then

$$q_0 + r'y = \frac{a}{b}(1 - v) + \frac{av}{b}(1 - u) = \frac{a}{b}(1 - uv)$$

and

$$q' = q_0 + r'y + (r - r')y = \frac{a}{b}(1 - uv) + (r - r')y,$$

---

[1] The first result is not used in the analysis of the algorithms presented in Sect. 11.4, each of which involves only two quotient approximations.

where

$$|(r - r')y| \le 2^{-P}|r'y| \le 2^{-P} \cdot a\delta \cdot \frac{1}{b}(1 + \epsilon) = \frac{a}{b} \cdot 2^{-P}\delta(1 + \epsilon).$$

Thus,

$$q' \le \frac{a}{b}(1 + \delta\epsilon) + \frac{a}{b} \cdot 2^{-P}\delta(1 + \epsilon)$$

and

$$\left| q' - \frac{a}{b} \right| = \left| (r - r')y - \frac{a}{b}uv \right| \le \frac{a}{b}|uv| + |(r - r')y| \le \frac{a}{b}(\delta\epsilon + 2^{-P}\delta(1 + \epsilon)).$$

For the proof of the first claim, we have

$$\left| q - \frac{a}{b} \right| \le |q - q'| + \left| q' - \frac{a}{b} \right|$$

$$\le 2^{-P}q' + \left| q' - \frac{a}{b} \right|$$

$$\le \frac{a}{b}(2^{-P}(1 + \delta\epsilon) + 2^{-2P}\delta(1 + \epsilon) + \delta\epsilon + 2^{-P}\delta(1 + \epsilon))$$

$$= \frac{a}{b}(2^{-P} + (1 + 2^{-P})\delta\epsilon + \cdot 2^{-P}\delta(1 + \epsilon) + \cdot 2^{-2P}\delta(1 + \epsilon)).$$

For the proof of the second claim, since $\frac{a}{b} \le 2^{e+1}$, we have

$$\left| q' - \frac{a}{b} \right| \le 2^{e+1}(\delta\epsilon + 2^{-P}\delta(1 + \epsilon)) < 2^{e+1}2^{-P-1} = 2^{e-P}$$

and

$$\left| q - \frac{a}{b} \right| \le |q - q'| + \left| q' - \frac{a}{b} \right| < 2^{expo(q')-P} + 2^{e-P}.$$

If $expo(q') \le e$, then the claim follows trivially. Thus, we may assume that $expo(q') > e$. But then

$$2^{e+1} \le q' = \frac{a}{b} + \left(q' - \frac{a}{b} \right) < 2^{e+1} + 2^{e-P}$$

implies $q = 2^{e+1}$. It follows that $\frac{a}{b} \le q \le q'$ and

$$\left| q - \frac{a}{b} \right| =\le \left| q' - \frac{a}{b} \right| < 2^{e-P}. \qquad \square$$

11.3 Reciprocal Refinement

A refinement y_{k+1} of a given reciprocal approximation y_k may be derived according to (11.1) in two steps:

$$e_k = RNE(1 - by_k, p)$$

$$y_{k+1} = RNE(y_k + e_k y_k, p)$$

Alternatively, as illustrated by the double-precision algorithm of Sect. 11.4, an approximation may be computed from the preceding two approximations as follows. This results in lower accuracy but allows the two steps to be executed in parallel:

$$e_{k+1} = RNE(1 - by_{k+1}, p)$$

$$y_{k+2} = RNE(y_k + e_k y_{k+1}, p)$$

The following lemma may be applied to either of these computations.

Lemma 11.5 *Assume that* $|1-by_1| \leq \epsilon_1$ *and* $|1-by_2| \leq \epsilon_2$. *Let* $e_1 = RNE(1-by_1, p)$ *and* $y_3 = RNE(y_3', p)$, *where* $y_3' = y_1 + e_1 y_2$ *and* $p > 0$. *Let*

$$\epsilon_3' = \epsilon_1(\epsilon_2 + 2^{-p}(1 + \epsilon_2)).$$

and

$$\epsilon_3 = \epsilon_3' + 2^{-p}(1 + \epsilon_3').$$

Then (a) $|1 - by_3'| \leq \epsilon_3'$ *and (b)* $|1 - by_3| \leq \epsilon_3$.

Proof Let $u_1 = 1 - by_1$ and $u_2 = 1 - by_2$. Then $|u_1| < \epsilon_1$, $|u_2| < \epsilon_2$, and

$$e_1 = (1 - by_1)(1 + v) = e_1(1 + v),$$

where $|v| \leq 2^{-p}$. Thus

$$\begin{aligned}
|1 - by_3'| &= |1 - b(e_1 y_2 + y_1)| \\
&= |(1 - by_1) - e_1(by_2)| \\
&= |u_1 - u_1(1 + v)(1 - u_2)| \\
&= |u_1(u_2 + v(u_2 - 1))| \\
&\leq \epsilon_3'
\end{aligned}$$

and

$$|1 - by_3| \leq |1 - by_3'| + b|y_3 - y_3'| \leq \epsilon_3' + 2^{-p}|by_3'| \leq \epsilon_3' + 2^{-p}(1 + \epsilon_3') = \epsilon_3. \quad \square$$

The inequality (b) of Lemma 11.5 provides a significantly reduced error bound for a refined reciprocal approximation y_3 as long as the bounds ϵ_1 and ϵ_2 for the earlier approximations y_1 and y_2 are large in comparison to 2^{-p}. To establish the bound 2^{-p} for the final approximation as required by Lemma 11.3, we shall use the inequality (a), pertaining to the corresponding unrounded value, in conjunction with the following additional lemma, which is a variation by Harrison [7] of another result of Markstein [19]. In practice, the application of this lemma involves explicitly checking a small number of excluded cases.

Lemma 11.6 *Let b be p-exact. Assume that $y = RNE(y', p)$ and $|1 - by'| \leq \epsilon' < 2^{-p-1}$. Let $d = \lceil 2^{2p}\epsilon' \rceil$. Then $|1 - by| < 2^{-p}$, with the possible exceptions $b = 2 - 2^{1-p}k$, $k = 1, \ldots, d$.*

Proof If $b = 1$, then $|1 - y'| < 2^{-p-1}$ implies $y = 1$ and $|1 - by| = 0$. Thus we may assume $b > 1$, and therefore $b \geq 1 + 2^{1-p}$. Consequently,

$$y' \leq \frac{1}{b}(1 + 2^{-p-1}) < \frac{1 + 2^{-p-1}}{1 + 2^{1-p}} < 1,$$

and it follows that $|y - y'| \leq 2^{expo(y')-p} \leq 2^{-p-1}$. Since

$$|1 - by'| \leq \epsilon' = 2^{-2p}2^{2p}\epsilon' \leq 2^{-2p}d,$$

and apart from the allowed exceptions, $b < 2 - 2^{1-p}d$, we have

$$|1 - by| \leq |1 - by'| + b|y - y'| < 2^{-2p}d + (2 - 2^{1-p}d)2^{-p-1} = 2^{-p}. \quad \square$$

11.4 Examples

The single-precision algorithm is given by the following sequence of operations. The spacing of the steps is intended to denote groups of operations that may be executed in parallel. Note that the rounding mode *RNE* is used for all intermediate steps and the input mode $\mathcal{R}$ is used for the final step.

Definition 11.2 $q = divsp(a, b, \mathcal{R})$ is the result of the following sequence of computations:

$$y_0 = rcp24(b)$$

$$q_0 = RNE(ay_0, 24)$$

$$e_0 = RNE(1 - by_0, 24)$$

$$y_1 = RNE(y_0 + e_0 y_0, 24)$$

$$r_0 = RNE(a - bq_0, 24)$$

$$q_1 = RNE(q_0 + r_0 y_1, 24)$$

$$r_1 = RNE(a - bq_1, 24)$$

$$q = \mathcal{R}(q_1 + r_1 y_1, 24)$$

The following lemma, which has been verified by exhaustive computation, will be used in conjunction with Lemma 11.6 in the proof of Theorem 11.1.

Lemma 11.7 *For $k = 1, \ldots, 7$, let $b = 2 - 2^{-23}k$. If y_1 is computed as in Definition 11.2, then $|1 - by_3| < 2^{-24}$.*

Theorem 11.1 *Let a and b be 24-exact with $1 \leq a < 2$ and $1 \leq b < 2$. If $\mathcal{R}$ is an IEEE rounding mode and $q = divsp(a, b, \mathcal{R})$, then*

$$q = \mathcal{R}\left(\frac{a}{b}, 24\right).$$

Proof According to Lemma 11.3, we need only establish the two inequalities (ii) $|1 - by_3| < 2^{-24}$ and (i) $|\frac{a}{b} - q_1| < 2^{e-23}$, where e is defined as in the lemma.
   Let

$$\epsilon_0 = 2^{-23},$$

$$\epsilon_1' = \epsilon_0(\epsilon_0 + 2^{-24}(1 + \epsilon_0)),$$

$$\epsilon_1 = \epsilon_1' + 2^{-24}(1 + \epsilon_1'),$$

$$y_1' = y_0 + e_0 y_0,$$

and $\quad d = \lceil 2^{48}\epsilon_1' \rceil.$

By Lemma 11.1, $|1 - by_0| \leq \epsilon_0$. By Lemma 11.5 (under the substitutions of y_0 for both y_1 and y_2, ϵ_0 for both ϵ_1 and ϵ_2, and y_1' for y_3'), $|1 - by_1'| \leq \epsilon_1'$ and $|1 - by_1| \leq \epsilon_1$. It is easily verified by direct computation that $\epsilon_1' < 2^{-25}$ and $d = 6$. The required inequality (ii) then follows from Lemmas 11.6 and 11.7.
   Let $\delta_0 = \epsilon_0 + 2^{-24}(1 + \epsilon_0)$. By Lemma 11.2,

$$\left|1 - \frac{b}{a}q_0\right| \leq \delta_0.$$

Since $\delta_0\epsilon_1 + 2^{-24}\delta_0(1 + \epsilon_1) < 2^{-25}$ (by direct computation), we may apply Lemma 11.4 (substituting y_1, q_0, ϵ_1, and δ_0 for y, q_2, ϵ, and δ, respectively) to conclude that (i) holds as well.   □

The double-precision algorithm follows:

Definition 11.3 $q = divdp(a, b, \mathcal{R})$ is the result of the following sequence of computations:

$$y_0 = rcp24(RTZ(b, 24))$$

$$q_0 = RNE(ay_0, 53)$$

$$e_0 = RNE(1 - by_0, 53)$$

$$r_0 = RNE(a - bq_0, 53)$$
$$y_1 = RNE(y_0 + e_0y_0, 53)$$

$$e_1 = RNE(1 - by_1, 53)$$
$$y_2 = RNE(y_0 + e_0y_1, 53)$$
$$q_1 = RNE(q_0 + r_0y_1, 53)$$

$$y_3 = RNE(y_1 + e_1y_2, 53)$$
$$r_1 = RNE(a - bq_1, 53)$$

$$q = \mathcal{R}(q_1 + r_1y_3, 53)$$

Note that in this case, the initial approximation is based on a truncation of the denominator, which increases its relative error[2]:

Lemma 11.8 *If b is 53-exact, $1 \le b < 2$, and $y_0 = rcp24(RTZ(b, 24))$, then*

$$|1 - by_0| \le 2^{-22}.$$

Proof Let $b_0 = RTZ(b, 24)$. Then $b_0 \le b < b_0 + 2^{-23}$ and by Lemma 11.8,

$$|1 - by_0| \le |1 - b_0y_0| + |y_0(b - b_0)| \le 2^{-23} + 2^{-23} = 2^{-22}.   □$$

The relative error of the final reciprocal approximation must be computed explicitly in 1027 cases:

---

[2] By exhaustive computation, we could establish an error bound slightly less than $1.5 \cdot 2^{-23}$, which would reduce the number of special cases to be checked from 1027 to 573, but the bound 2^{-22} is sufficient for the proof of Theorem 11.2.

Lemma 11.9 *For $k = 1, \ldots, 1027$, let $b = 2 - 2^{-52}k$. If y_3 is computed as in Definition 11.3, then $|1 - by_3| < 2^{-53}$.*

Theorem 11.2 *Let a and b be 53-exact with $1 \le a < 2$ and $1 \le b < 2$. If $\mathcal{R}$ is an IEEE rounding mode and $q = divdp(a, b, \mathcal{R})$, then*

$$q = \mathcal{R}\left(\frac{a}{b}, 53\right).$$

Proof Applying Lemma 11.3 once again, we need only establish the two inequalities (ii) $|1 - by_3| < 2^{-53}$ and (i) $|\frac{a}{b} - q_1| < 2^{e-52}$.

Let

$$\epsilon_0 = 2^{-22},$$

$$\epsilon_1' = \epsilon_0(\epsilon_0 + 2^{-53}(1 + \epsilon_0)),$$

$$\epsilon_1 = \epsilon_1' + 2^{-53}(1 + \epsilon_1'),$$

$$\epsilon_2' = \epsilon_0(\epsilon_1 + 2^{-53}(1 + \epsilon_1)),$$

$$\epsilon_2 = \epsilon_2' + 2^{-53}(1 + \epsilon_2'),$$

$$\epsilon_3' = \epsilon_1(\epsilon_2 + 2^{-53}(1 + \epsilon_2)),$$

$$y_3' = y_1 + e_1 y_2,$$

$$\text{and} \quad d = \lceil 2^{106} \epsilon_3' \rceil.$$

Let $b_0 = RTZ(b, 24)$. Then $b_0 \le b < b_0 + 2^{-23}$ and by Lemma 11.8,

$$|1 - by_0| \le |1 - b_0 y_0| + |y_0(b - b_0)| \le 2^{-23} + 2^{-23} = \epsilon_0.$$

By Lemma 11.8, $|1 - by_0| \le \epsilon_0$. By repeated applications of Lemma 11.5, $|1 - by_1| \le \epsilon_1$, $|1 - by_2| \le \epsilon_2$, and $|1 - by_3'| \le \epsilon_3'$. It is easily verified by direct computation that $\epsilon_3' < 2^{-54}$ and $d = 1027$ and (ii) follows from Lemmas 11.6 and 11.9.

Let $\delta_0 = \epsilon_0 + 2^{-53}(1 + \epsilon_0)$. Direct computation yields $\delta_0 \epsilon_1 + 2^{-53}\delta_0(1 + \epsilon_1) < 2^{-54}$ and (i) again follows from Lemma 11.4. $\quad\square$

Part IV
Comparative Architectures: SSE, x87, and Arm

While the principle of correct rounding defines the value of an arithmetic operation under normal conditions, there are a variety of exceptional cases that require special consideration, including invalid and denormal operands, overflow, underflow, and inexact results. Since the advent of floating-point hardware, there has been general agreement on the desirability of an industry standard for exception handling in order to ensure consistent results across all computing platforms. This was the objective of the original IEEE floating-point specification, Standard 754-1985[8], which was developed in parallel with Intel's x87 instruction set, the first "IEEE-compliant" architecture. In the 1990s, a number of competing floating-point architectures emerged. Among these are the *Streaming SIMD (single instruction, multiple data) Extensions*, or *SSE* instructions, which were added by Intel to support multimedia and graphics applications. Of equal importance is the Arm family of reduced instruction set computing architectures, which has dominated the mobile device market.

Unfortunately, these newer architectures did not strictly adhere to IEEE-prescribed behavior. For example, in the event of trapped overflow or underflow, 754-1985 dictates the return of a result generated from the rounded value by a specified shift into the normal range, intended to allow the trap handler to perform scaled arithmetic. This feature of the x87 instructions was not replicated in later architectures, which typically do not return any value in this case.

In 2008, IEEE issued an updated version of Standard 754 [9]. Although it is claimed that "numerical results and exceptions are uniquely determined" by the new standard, an apparent conflicting objective is to accommodate the diverse architectural behaviors that arose in the interim. Consequently, it exhibits a number of ambiguities and deficiencies pertaining, for example, to the detection and handling of underflow, the response to a denormal operand, the order of precedence of the pre-computation conditions, the precedence of operands when more than one is a NaN, and the interaction of exceptions reported by the component operations of a SIMD instruction. Each of these issues has been resolved independently by the architectures mentioned earlier with inconsistent results. Consequently, no single standard can possibly serve the needs of a designer or verifier of an implementation of a particular architecture, for which strict backward compatibility is a necessity.

The ACL2 RTL library addresses this problem for the three floating-point architectures of interest—SSE, x87, and Arm—by providing formal executable specifications of the primary elementary arithmetic operations: addition, multiplication, division, square root extraction, and fused multiplication-addition (FMA). These specifications are presented informally but in complete detail in the following chapters. We begin with the SSE instructions (Chap. 12), which have the most commonality with the other two. In our presentation of the x87 and Arm specifications (Chaps. 13 and 14), we emphasize their points of departure from SSE behavior.

Chapter 12
SSE Floating-Point Instructions

The SSE floating-point instructions were introduced by Intel in 1998 and have continually expanded ever since. They operate on single-precision or double-precision data (Definition 5.3) residing in the 128-bit *XMM* registers or the 256-bit *YMM* registers. Some SSE instructions are *packed*, i.e., they partition their operands into several floating-point encodings to be processed in parallel; others are *scalar*, performing a single operation, usually on data residing in the low-order bits of their register arguments. The specifications presented in this chapter apply to both scalar and packed instructions that perform the operations of addition, multiplication, division, square root extraction, and FMA.

A single dedicated 16-bit register, the *MXCSR*, controls and records the response to exceptional conditions that arise during the execution of SSE floating-point operations and controls the rounding of floating-point values.

12.1 SSE Control and Status Register

The MXCSR bits are named as displayed in Fig. 12.1.

- The least significant 6 bits are the *exception flags*, corresponding to the pre-computation exceptions, invalid operand (IE), denormal operand (DE), and division by zero (ZE); and the post-computation exceptions, overflow (OE), underflow (UE), and inexact result (PE).

15	14 13	12	11	10	9	8	7	6	5	4	3	2	1	0
F T Z	RC	P M	U M	O M	Z M	D M	I M	D A Z	P E	U E	O E	Z E	D E	I E

Fig. 12.1 MXCSR: SSE floating-point control and status register

© The Author(s), under exclusive license to Springer Nature Switzerland AG 2022   241
D. M. Russinoff, *Formal Verification of Floating-Point Hardware Design*,
https://doi.org/10.1007/978-3-030-87181-9_12

Table 12.1 x86 rounding control

Encoding	Rounding mode
00	*RNE*
01	*RDN*
10	*RUP*
11	*RTZ*

- Bit 6 is the *denormal-as-zero* bit, which, if set, coerces all denormal inputs to ±0.
- Bits 12:7 are the *exception masks* corresponding to the flags, which determine whether an exceptional condition results in the return of a default value or the generation of an exception.
- Bits 14:13 form the *rounding control* field, which encodes a rounding mode as displayed in Table 12.1.
- Bit 15 is the *force-to-zero* bit, which, if set, coerces any denormal output to ±0.

12.2   Overview of SSE Floating-Point Exceptions

If an exceptional floating-point condition is detected during the execution of an SSE instruction, then depending on the condition, one of the exception flags MXCSR[5:0] may be set. If a flag is set and the exception is *unmasked*, i.e., the corresponding mask bit in MXCSR[12:7] is 0, then execution of the instruction is terminated, no value is written to the destination, and control is passed to a trap handling routine. If either no flag is set or the exception is *masked*, then depending on the exceptional condition, either the instruction proceeds normally or a default value is determined, allowing execution of the program to proceed. For a packed instruction, if any of the component operations results in an unmasked exception, then no result is written for any operation. Otherwise, a result is written for each operation.

Instruction execution consists of three phases: pre-computation, computation, and post-computation. The exceptional conditions are partitioned into two classes, which are detected during the first and third of these phases. The following procedure is executed by both packed and scalar SSE floating-point instructions. Note that in the case of a packed instruction, the procedure is complicated by the requirement of parallel execution:

- Pre-computation (Sect. 12.3): The operands of each operation are examined in parallel for a set of conditions, some of which result in the setting of a flag, IE, DE, or ZE. If a flag is set and the corresponding mask is clear, then execution is terminated for all operations, and no value is written to the destination. Otherwise, for each operation, either a QNaN is selected as a default value (but not yet written) or the computation proceeds.

- Computation (Sect. 12.4): Unless an unmasked exception is detected during the pre-computation phase, for each operation that has not terminated with a default value, a computation is performed. If the value is infinite or 0, then a result is determined (but not yet written). Otherwise, execution proceeds.
- Post-computation (Sect. 12.5): For each remaining operation, the computed value is rounded and the result is examined for a set of conditions, which may result in the setting of one or two of the flags OE, UE, and PE. If a flag is set and the corresponding mask bit is 1, then a result is determined. If a flag is set by any operation and the corresponding mask bit is 0, then an exception is generated, and no value is written to the destination for any operation. Otherwise, the result that has been determined for each operation is written.

These three phases and the pre- and post-computation SSE exceptions are discussed in detail in the following sections.

12.3 Pre-computation Exceptions

The first step in the execution of any SSE floating-point instruction, before any exception checking is performed, is to examine the DAZ bit of MXCSR. If this bit is set, then any denormal operand is replaced by a zero of the same sign.

The conditions that may cause an exception flag to be set, or the operation to be terminated with a QNaN value, or both, prior to an SSE computation are as follows:

- SNaN operand: IE is set and the operation is terminated. If the first NaN operand is a QNaN, then the value is that operand; if the first NaN operand is an SNaN, then the value is that operand converted to a QNaN. For this purpose, in the case of an FMA $a \cdot b + c$, the operands are ordered as a, b, c.
- QNaN operand and no SNaN operand: No flag is set, but the operation is terminated. The value is the first NaN operand.
- Undefined operation: IE is set, the operation is terminated, and the value is the real indefinite QNaN (Definition 5.23). The operands for which this condition holds depend on the operation:

  - Addition: Two infinities with opposite signs;
  - Subtraction: Two infinities with the same sign;
  - Multiplication: Any infinity and any zero;
  - Division: Any two infinities or any two zeroes;
  - Square root extraction: Any operand with negative sign, excluding negative zero;
  - Multiply-accumulate: A product of an infinity and a zero (with no restriction on the other operand) or a product of an infinity and any non-NaN added to an infinity with sign opposite to that of the product.

- A division operation with any zero as divisor and any finite numerical dividend: ZE is set, but the operation proceeds (resulting in an infinity) unless an unmasked exception occurs.
- Any denormal operand (with $DAZ = 0$) and none of the above conditions: DE is set, but the numerical computation proceeds unless an unmasked exception occurs.

Note that these conditions are prioritized in the order listed: if any condition holds for a given operation, then any other of lower priority is ignored for that operation. For a packed instruction, all operands are examined in parallel for pre-computation exception conditions. Consequently, it is possible for different flags to be set for different operations.

If any exception flag is set during this process and the corresponding mask bit is clear, then all operations are terminated before any computation is performed, no result is written to the destination, and an exception is generated.

If an operation of a packed instruction is terminated with a default value, the value is not written to the destination until execution of the instruction is completed, since no value is written in the event of an unmasked post-computation exception.

The setting of status flags and the default values are summarized in Table 12.2.

Table 12.2 SSE pre-computation exceptions

Exception or termination condition	Flag set	QNaN result (masked case)
SNaN operand	IE	QNaNized operand
QNaN operand	None	Operand
Undefined operation	IE	Indefinite QNaN
Zero exception	ZE	None
Denormal operand	DE	None

12.4 Computation

Unless terminated in response to a pre-computation exceptional condition, each operation of an SSE arithmetic instruction computes an unrounded value, which is then processed according to the contents of the MCXSR and the floating-point format of the instruction as described below. In the case of a packed instruction, all operations for which a default QNaN value has not been determined in the pre-computation stage are similarly processed in parallel.

For each of these operations, a value is computed. If this value is infinite or 0, then no flags are set and the sign of the result is determined by the signs of the operands and the rounding mode $\mathcal{R} = $ MXCSR [14:13]:

- Infinity: The result is an infinity with sign determined according to the operation:

  - Addition: The sign of the infinite operand or operands.

- Subtraction: The sign of the minuend if it is infinite, and otherwise the inverse of the sign of the subtrahend.
- Multiplication or division: The product (xor) of the signs of the operands.
- Square root: The sign of the operand, which must be positive.
- Multiply-accumulate: The sign of the addend if it is infinite, and otherwise the product (xor) of the signs of the factors.

• Zero: The result is a zero with sign determined according to the operation:

- Addition: The sign of operands if they agree; if not, then negative if $\mathcal{R} = RDN$, and otherwise positive.
- Subtraction: The sign of the minuend if it is the inverse of that of the subtrahend; if not, then negative if $\mathcal{R} = RDN$, and otherwise positive.
- Multiplication or division: The product (xor) of the signs of the operands.
- Square root: The sign of the operand.
- Multiply-accumulate: The product of the signs of the factors if it agrees with that of the addend; if not, then negative if $\mathcal{R} = RDN$, and otherwise positive.

Otherwise, execution proceeds to the next phase with the unrounded computed value, which is finite and nonzero.

12.5 Post-computation Exceptions

The procedure described in this section is applied in the same way to all operations under consideration. For each operation that reaches this phase, the precise mathematical result (which, of course, need not be computed explicitly by an implementation) is a finite nonzero value u, which is rounded according to the rounding mode $\mathcal{R}$ and the precision p of the data format F, producing a value $r = \mathcal{R}(u, p)$. This value is subjected to the following case analysis, which may result in the setting of one or more exception flags. If any operation produces an unmasked exception, no result is written for any operation. Otherwise, a final result is written for each operation.

• Overflow (r is above the normal range of the target format, i.e., $|r| > lpn(F)$): in all cases, OE is set.

- Masked Overflow (OM $= 1$):
    PE is set. The final result, which is valid only if PM $= 1$, depends on $\mathcal{R}$ and the sign of r.
    If (a) $\mathcal{R} = RNE$, (b) $\mathcal{R} = RUP$ and $r > 0$, or (c) $\mathcal{R} = RDN$ and $r < 0$, then the final result is an infinity with the sign of r.
    Otherwise, the result is the encoding of the maximum normal value for the target format, $\pm lpn(F)$, with the sign of r.
- Unmasked Overflow (OM $= 0$):
    No final result is returned. If $r \neq u$, then PE is set.

- Underflow (r is below the normal range, i.e., $0 < |r| < spn(F)$):
  - Masked Underflow (UM $= 1$):
    If FTZ $= 1$, then UE and PE are set. The final result, which is valid only if PM $= 1$, is a zero with the sign of r.
    If FTZ $= 0$, then u is rounded again to produce $d = drnd(u, \mathcal{R}, F)$, which may be a denormal value, 0, or the smallest normal, $\pm spn(F)$. If $d \neq u$, then both UE and PE are set; otherwise, neither flag is modified. The final result, which is valid unless PE is set and PM $= 1$, is the encoding of d, with the sign of u if $d = 0$.
  - Unmasked Underflow (UM $= 0$):
    UE is set. No final result is returned. If $r \neq u$, then PE is set.

- Normal Case (r is within the normal range, i.e., $spn(F) \leq |r| \leq lpn(F)$):
  If $r \neq u$, then PE is set. The final result, which is valid unless PE is set and PM $= 1$, is the normal encoding of r.

Thus, PE indicates either that the rounded result is an inexact approximation of an intermediate result or that some other value has been written to the destination. In the case of masked underflow, it may not be obvious that the behavior described above is consistent with the definition of the auxiliary ACL2 function *sse-round*, which sets PE if either $r \neq u$ or $d \neq u$. However, Lemma 6.107 guarantees that if the first inequality holds, then so does the second.

Also note there is one case in which underflow occurs and UE is not modified: UM $= 1$ and the unrounded value is returned as a denormal.

Chapter 13
x87 Instructions

The x87 instruction set was Intel's first floating-point architecture, introduced in 1981 with the 8087 hardware coprocessor. The architecture provides an array of eight 80-bit data registers, which is managed as a circular stack. Each numerical operand is located either in an x87 data register or in memory and is interpreted according to one of the data formats defined by Definition 5.3. Data register contents are always interpreted according to the double extended precision format (EP). Memory operands may be encoded in the single (SP), double (DP), or double extended precision format. Numerical results are written only to the data registers in the EP format. The architecture also provides distinct 16-bit control and status registers, the FCW and the FSW, corresponding to the single SSE register MXCSR.

The x87 instructions have largely been replaced by the newer SSE architecture but remain important for applications that require high-precision computations. Our analysis of these instructions benefits from two simplifications relative to the SSE architecture: all instructions are scalar and there is no FMA instruction. Thus, every instruction to be considered here performs a single operation on one or two operands.

13.1 x87 Control Word

The x87 control word, FCW, allows software to manage the precision and rounding of floating-point operations and to control the response to exceptional conditions that may arise during their execution.

The control word bits are named as shown in Fig. 13.1. The least significant 6 bits are the *exception masks*, which represent the same classes of exceptional conditions that are encoded in the MXCSR: invalid operand (IM), denormal operand (DM), division by zero (ZM), overflow (OM), underflow (UM), and inexact result (PM).

© The Author(s), under exclusive license to Springer Nature Switzerland AG 2022   247
D. M. Russinoff, *Formal Verification of Floating-Point Hardware Design*,
https://doi.org/10.1007/978-3-030-87181-9_13

The 2-bit RC field FCW[11:10] encodes a rounding mode according to the same scheme used for the SSE instructions (Table 12.1).

15	14	13	12	11	10	9	8	7	6	5	4	3	2	1	0
0	0	0	Y	RC		PC		0	1	PM	UM	OM	ZM	DM	IM

Fig. 13.1 x87 control word

Table 13.1 x87 precision control

Encoding	Precision
00	24
01	
14	53
11	64

The control word also includes a 2-bit PC field FCW[9:8], which controls the precision of rounded results. These results are written only to x87 data registers in the EP format, but they are rounded to 24, 53, or 64 bits of precision as determined Table 13.1.

The remaining 6 bits of FCW are unused. The 5 bits FCW[15:13] and FCW[7:6] are reserved: any attempt to alter them is ignored. Bit 6 is always set; the other four are always clear. FCW[12], labeled Y and known as the *infinity bit*, may be read or written by software, but its value has no predefined meaning. This bit was used in interpreting floating-point infinities in pre-386 processors but is now obsolete.

Note that the x87 control word contains neither a *denormal-as-zero* (DAZ) bit nor a *force-to-zero* (FTZ) bit.

13.2 x87 Status Word

The x87 status word, FSW[15:0], is used by hardware to record exceptional and other conditions that arise during the execution of x87 instructions. It also contains a pointer to the top of the x87 data register stack. The status word bits, as shown in Fig. 13.2, are as follows:

- Exception flags: The least significant 7 bits, FSW[6:0], are the exception flags. The 6 bits FSW[5:0] correspond to the mask bits FCW[5:0] and record exceptional conditions of the six types listed in Sect. 13.1. Bit 6, labeled SF, is set by hardware to indicate a *stack fault*, which may occur during access of the

15	14	13	12	11	10	9	8	7	6	5	4	3	2	1	0
B	C3		TOP		C2	C1	C0	ES	SF	PE	UE	OE	ZE	DE	IE

Fig. 13.2 x87 status word

instruction operands (stack underflow) or destination register (stack overflow) prior to execution of the instruction. There is no mask bit in FCW corresponding to SF. Whenever hardware sets SF, it also sets IE.

- Exception summary: Bit 7 is the *exception summary* bit ES, indicating an unmasked exception. In fact, this bit is redundant: ES = 1 if and only if at least one of the exception flags FSW[5:0] is set with the corresponding mask bit in FCW[5:0] clear. This invariant is strictly maintained by hardware.
- Condition codes: Bits 8, 9, 10, and 14 are the *condition codes*, C0, C1, C2, and C3, which are modified by x87 instructions. The elementary arithmetic instructions modify only C1, which is cleared by default and set in the event that an instruction returns a result that is larger in absolute value than its precise mathematical value, i.e., the value has been rounded away from 0 or replaced by an infinity.
- Stack top: The 3 bits FSW[13:11] encode a pointer to the top of the x87 circular data stack. When a value is loaded from memory, it is pushed onto the stack, i.e., the stack top in decremented (modulo 8), and the value is written to the new top of the stack; when the value at the top of stack is stored to memory, it is popped, i.e., the stack top is incremented. The x87 tag word, a related register that maintains tags indicating which stack entries are currently valid, is adjusted accordingly.
- Busy bit: Bit 15 is included only for backward compatibility with the 8087, in which it indicated that the floating-point coprocessor was busy. It is architecturally defined to have the same value as ES.

13.3   Overview of x87 Exceptions

An essential difference between x87 and SSE exceptions is that when an unmasked exceptional condition is detected during execution of an x87 instruction, the exception summary (ES) bit and busy bit (B) are set along with the indicated exception flag, but the exception itself is postponed until the next x87 instruction is encountered. That is, whenever an x87 instruction is initiated, other than the control instructions, the ES bit is examined, and if set, execution is aborted and an exception is generated.

In the case of an unmasked pre-computation exception, execution is terminated, and, as in the SSE case, no value is written to the destination. For an unmasked

post-computation exception, however, in a departure from SSE behavior, execution
proceeds and a value is written.

Another distinctive feature of the x87 architecture is the *stack fault* (SF) flag.
The stack top and tag word are examined upon initiation of any instruction that
requires access to the stack. Stack overflow occurs when an instruction attempts a
push to a valid stack entry; underflow is the result of a read access of an invalid
entry. When either of these conditions is detected, SF is set along with IE. If IM
$= 0$, then ES and B are set as well and execution is terminated; otherwise, the real
indefinite QNaN is written to the destination. Thus, a stack fault may be viewed as
a pre-computation exceptional condition with priority over all others. However, we
shall view the stack access as well as the initial check of ES as extraneous to the
execution of an instruction and exclude these features from our formal model.

Thus, we have the following three phases of execution:

- Pre-computation: The operands are examined for a set of conditions, some of
  which result in the setting of a flag, IE, DE, or ZE. If a flag is set and the
  corresponding mask bit, IM, DM, or ZM, is clear, then ES and B are set,
  execution is terminated, and no value is written to the destination. If no unmasked
  exception is detected, then either a QNaN is written to the destination or the
  computation proceeds.
- Computation: Unless execution terminates during the pre-computation phase, a
  computation is performed. If the value is infinite or 0, then a result is written to
  the destination. Otherwise, execution proceeds.
- Post-computation: The computed result is rounded and examined for a set of
  conditions, which may result in the setting of one or two of the flags OE, UE,
  and PE. If a flag is set and the corresponding mask bit, OM, UM, or PM, is clear,
  then ES and B are set. In any case, a value is written to the destination.

13.4   Pre-computation Exceptions

In addition to those already noted, there are three differences between x87 and SSE
pre-computation exceptions:

- Unsupported operand: This class of encodings does not exist in the SSE formats.
  This condition has priority over all other pre-computation exceptions (with the
  exception of the stack fault). If an unsupported operand (see Definition 5.6) is
  detected, then IE is set. If IM $= 0$, then ES and B are set as well and execution is
  terminated; otherwise the real indefinite QNaN is written to the destination.
- NaN operand: For an SNaN or a QNaN input, the priorities and the setting of IE
  are the same as for SSE instructions, but there are differences in the computation
  of the default value. First, any SP or DP NaN operand is converted to the EP
  format by inserting the integer bit and appending the appropriate number of 0s.

In the case of two NaN operands, the order of the operands is irrelevant. If the operands are distinct NaNs, then the one with the greater significand field is selected, but if the significand fields coincide, then the zero sign field is selected.
- Pseudo-denormal operands (Definition 5.13): This is another class of encodings that does not exist in the SSE formats. A pseudo-denormal operand triggers a denormal exception in the same way as a denormal operand (and in the event of a masked denormal exception, its value is computed by the same formula).

13.5 Post-computation Exceptions

In the absence of an unmasked exception, the computational and post-computational behavior of the x87 instructions is the same as that of the corresponding SSE instructions with FTZ $= 0$, except that (a) results are rounded to the precision specified in the FCW and encoded in the double extended precision format, and (b) if a result is rounded away from 0 or replaced by an infinity, then the condition code C1 is set, and otherwise C1 is cleared.

Here we describe the response of an x87 instruction to an unmasked post-computation exceptional condition. We assume that the computation produces a finite nonzero value u, which is rounded according to the rounding mode $\mathcal{R}$ and the precision p indicated by the FCW, producing a value $r = \mathcal{R}(u, p)$. If r lies outside the normal range, then the objective is to return a normal encoding of a shifted version r' of r from which r can be recovered and which may be used in computations performed by the trap handler. The shift is intended to produce a result near the center of the normal range in order to minimize the chance of further overflow or underflow.

- Unmasked overflow ($|r| > lpn(EP)$ and OM $= 0$): OE, ES, and B are set.

  Let $r' = 2^{-3 \cdot 2^{13}} r$. If r' is still above the normal range, then the final result is an infinity with the sign of r; PE and C1 are set.

  Otherwise, the final result is the normal encoding of r'. In this case, PE is set only if $r \neq u$. If $|r| > |u|$, then C1 is set; otherwise, C1 is cleared.
- Unmasked underflow ($0 < |r| < spn(EP)$ and UM $= 0$): UE, ES, and B are set.

  Let $r' = 2^{3 \cdot 2^{13}} r$. If r' is still below the normal range, then the final result is a zero with the sign of r; PE is set and C1 is cleared.

  Otherwise, the final result is the normal encoding of r'. In this case, PE is set only if $r \neq u$. If $|r| > |u|$, then C1 is set; otherwise, C1 is cleared.
- Unmasked precision exception: The only effect of PM is that if PE is set and PM $= 0$, then ES and B are set.

Chapter 14
Arm Floating-Point Instructions

The first Arm *floating-point accelerator*, which appeared in 1993, resembled the x87 coprocessor in its use of an 80-bit EP register file. This was succeeded by the *vector floating-point* (VFP) architecture, which included 64-bit registers implementing the single-, double-, and half-precision data formats. The *NEON Advanced SIMD* extension later added 128-bit instructions for media and signal processing applications.

The current version of the Arm architecture as of this writing, which has evolved from the VFP and NEON instruction sets, is designated *Armv8*. The behavior of the elementary arithmetic instructions of this architecture is similar to that of the SSE instructions of Chap. 12, the primary difference being in the handling of floating-point underflow: as described in Sect. 6.7, the Arm instructions base the detection of underflow on the unrounded result of an operation rather than the rounded result.

Armv8 supports two distinct *execution states*, AArch32 and AArch64, which are based on 32-bit and 64-bit register files, respectively. From the perspective of floating-point design, the main difference between the two is that the AArch32 state uses a single 32-bit *Floating-Point Status and Control Register* (FPSCR), while AArch64 includes separate expanded 64-bit registers for status (FPSR) and control (FPCR). The FPCR in particular includes several extra bits that provide extended functionality, most notably an "Intel compatibility mode" of underflow detection, in which the Arm method of detection is overridden by SSE conventions. In this chapter, we shall focus on the simpler AArch32 state.

Among the bits of the FPSCR are six *cumulative exception flags* and the corresponding *trap enable bits*, analogous to the exception flags and masks of the SSE MXCSR. An implementation may or may not support the generation of exceptions, i.e., the transfer of control by hardware to an exception handler in the event of an exceptional condition. For a processor that supports this feature, when an exceptional condition is detected, the response depends on the relevant trap enable bit: if this bit is set, control is transferred to the handler with all registers restored to their states prior to initiation of the instruction, with the possible exception of

D. M. Russinoff, *Formal Verification of Floating-Point Hardware Design*,
https://doi.org/10.1007/978-3-030-87181-9_14

31	25 24 23 22	15	12 11 10 9 8 7	4 3 2 1 0
	DN FZ RC	IDE	IXE UFE OFE DZE IOE IDC	IXC UFC OFC DZC IOC

Fig. 14.1 FPSCR: Arm floating-point status and control register

the FPSCR; otherwise, the appropriate flag is set and execution of the instruction proceeds with a default behavior. If this feature is not supported, then the trap enable bits are effectively read as zero.

Although the architecture provides the option of trap handling, the behavior of this process is underspecified in various ways and left to the implementation, especially with regard to the interaction between competing exceptions and the restoration of the exception flags. On the other hand, no processors designed by Arm, including those represented in later chapters, implement trap handling. It will serve our purpose, therefore, to base our discussion on the assumption that trapping is not supported. Thus, we assume that an Arm floating-point operation always executes to completion and returns a data result.

Consequently, there is no interaction between the component operations of a SIMD instruction of the sort described in Sect. 12.2. We may confine our attention, therefore, to the behavior of scalar instructions. The specifications of the instructions presented in this section are formalized by a set of three ACL2 functions, which belong to the library books/rtl in the ACL2 repository [13]:

- *arm-binary-spec(op, a, b, FPSCR, F)*
- *arm-sqrt-spec(a, FPSCR, F)*
- *arm-fma-spec(a, b, c, FPSCR, F)*

Each of these functions takes one or more operands (single-, double-, or half-precision floating-point encodings), the initial contents of the FPSCR register, and the data format F of the instruction, SP, DP, or HP. The function *arm-binary-spec*, which applies to the four binary arithmetic operations, takes an additional argument, *op*, representing the operation (*ADD, SUB, MUL,* or *DIV*). Each function returns two values: the data result and the final value of the FPSCR.

In Chaps. 16–22, we shall present a set of floating-point modules that have been implemented in Arm processors. The above functions are the basis of the statements of correctness of these modules.

Table 14.1 Arm rounding control

Encoding	Rounding mode
00	*RTZ*
01	*RDN*
10	*RUP*
11	*RNE*

14.1   Floating-Point Status and Control Register

The FPSCR bits that are relevant to the instructions of interest are named as displayed in Fig. 14.1.

- Bits 4:0 and 7 are the *cumulative exception flags*, indicating invalid operand (IOC), division by zero (DZC), overflow (OFC), underflow (UFC), inexact result (IXC), and denormal operand (IDE),
- Bits 12:8 and 15 are the *trap enables* corresponding to the flags. In a processor that implements trap handling, these bits determine whether, in the event of exceptional condition, the flag is set by hardware or control is passed to a trap handler. Since we are not considering such implementations, we assume that these bits are effectively held at 0.
- Bits 23:22 form the *rounding control* field (RC), which encodes a rounding mode as displayed in Table 14.1. Note the difference between this encoding and that of Table 12.1.
- Bit 24 is the *force-to-zero* bit (FZ), which, if set, coerces both denormal inputs and (except in the half-precision case; see Sect. 14.3) denormal results to ± 0. Thus, this bit plays the roles of both the DAZ and FTZ bits of the SSE MXCSR (Sect. 12.1).
- Bit 25 is the *default NaN* bit (DN). If asserted, any NaN result of an instruction is replaced by the real indefinite QNaN (Definition 5.23).

14.2   Pre-computation Exceptions

The floating-point pre-computation behavior of the Arm architecture is formalized by three functions:

- *arm-binary-pre-comp*($op, a, b, FPSCR, F$)
- *arm-sqrt-pre-comp*($a, FPSCR, F$)
- *arm-fma-pre-comp*($a, b, c, FPSCR, F$)

Each of these returns an optional data value and an updated FPSCR. If a data value is returned, then execution is terminated; otherwise the computation proceeds.

We have noted that in an Arm-designed processor, a pre-computation exception never prevents the return of a value. The other departures from SSE pre-computation exception handling are in the setting of the denormal flag, the returned value in the case of a NaN operand, and the precedence of an undefined operation over a QNaN operand. Note, however, that the only case in which an undefined operation and a NaN operand can simultaneously occur is an FMA operation with a product of an infinity and a zero with a NaN addend.

The conditions that may cause an exception flag to be set, or the operation to be terminated with a QNaN value, or both, prior to an Arm floating-point computation are as follows:

- Denormal operand: If FZ $= 1$, then the operand is forced to ± 0 and, unless the format is *HP*, IDC is asserted; otherwise, neither of these actions is taken. (Note the contrast with the detection of SSE denormal exceptions.) The detection of an undefined operation or division by zero (see below) is based on the result of this step. Note that a denormal exception is not suppressed by another exceptional condition; it is possible for two flags to be set.
- SNaN operand: IOC is set. If DN $= 1$, then the real indefinite QNaN is returned (Definition 5.23). Otherwise, the first SNaN operand is converted to a QNaN and returned. For this purpose, in the case of an FMA $a + b \cdot c$, the operands are ordered as a, b, c.
- Undefined operation and no SNaN operand: The conditions are as specified in Sect. 12.3. IOC is set and the real indefinite QNaN is returned.
- QNaN operand and no SNaN operand or undefined operation: If DN $= 1$, then the real indefinite QNaN is returned. Otherwise, the first NaN operand is returned (with FMA operands ordered as in the SNaN case).
- A division operation with any zero as divisor and any finite numerical dividend: IDZ is set, but the operation proceeds (resulting in an infinity).

14.3 Post-computation Exceptions

If a final result is not produced during the pre-computation phase, then control is passed to one of the following, which computes an unrounded value:

- *arm-binary-comp(op, a, b, FPSCR, F)*
- *arm-sqrt-comp(a, FPSCR, F)*
- *arm-fma-comp(a, b, c, FPSCR, F)*

If the computed value is infinite or 0, then execution is terminated. No flags are set and the sign of the result is determined by the signs of the operands and the rounding mode $\mathcal{R} = \text{FPSCR}[23 : 22]$ as described in Sect. 12.4.

Otherwise, the precise mathematical result of the operation is a finite nonzero value u. This value is passed to the common function

$$arm\text{-}post\text{-}comp(u, FPSCR, F),$$

which performs the rounding and detects exceptions as described below. Note that in addition to the absence of exception masks, there are several departures from SSE behavior in the detection and handling of underflow.

Unless u is a denormal, it is rounded according to the rounding mode $\mathcal{R}$ and the precision p of the data format f, producing a value $r = \mathcal{R}(u, p)$. The returned value and the setting of exception flags are determined by the following case analysis. In all cases, the setting of a flag is understood to be contingent on the value of the corresponding trap enable bit, except in a certain case of underflow as noted below.

- Overflow (r is above the normal range of the target format, i.e., $|r| > lpn(F)$):

  In all cases, OFC and IXC are set. The result depends on $\mathcal{R}$ and the sign of r.

  If (a) $\mathcal{R} = RNE$, (b) $\mathcal{R} = RUP$ and $r > 0$, or (c) $\mathcal{R} = RDN$ and $r < 0$, then the final result is an infinity with the sign of r.

  Otherwise, the result is the encoding of the maximum normal value for the target format, $\pm lpn(F)$, with the sign of r.

- Underflow, which is detected before rounding (u is below the normal range, i.e., $0 < |u| < spn(F)$):

  If $FZ = 1$ and F is SP or DP, then UFC is set. (IXC is not set.) UFE is ignored in this case, and FZ is ignored if F is HP. The final result is a zero with the sign of u.

  If $FZ = 0$ or F is HP, then u is rounded to produce $d = drnd(u, \mathcal{R}, F)$, which may be a denormal value, 0, or the smallest normal, $\pm spn(F)$. If $d \neq u$, then both UFC and IXC are set; otherwise, neither flag is modified. The final result is the encoding of d, with the sign of u if $d = 0$.

- Normal case (u and r are both within the normal range):

  If $r \neq u$, then IXC is set. The final result is the normal encoding of r.

Part V
Formal Verification of RTL Designs

The practical significance of the foregoing development is its utility in bringing mathematical analysis and interactive theorem proving to bear on the formal verification of arithmetic RTL designs, thereby addressing the inherent limitations of more conventional automatic methods. As a vehicle for combining theorem proving with sequential logic equivalence checking, we have introduced the *RAC* (*Restricted Algorithmic C*) modeling language. RAC is essentially a limited subset of C augmented by the C++ register class templates of Algorithmic C [20], an ANSI standard library intended for system and hardware design. A Verilog RTL design may be represented in this language in a form that is more compact, abstract, and amenable to formal analysis. RAC is designed to support a functional programming style, which we have exploited by implementing a translator to the logical language of the ACL2 theorem prover [13]. This provides a path to mechanical verification of the correctness of a model with respect to a high-level formal specification of correctness of the sort presented in Part IV.

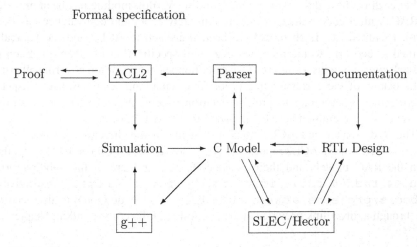

Fig. V.1 Verification methodology

Our verification methodology is diagrammed in Fig. V.1. Ideally, the RAC model is produced in the collaboration between a floating-point architect and a verification engineer, in advance of the RTL design. This allows algorithmic bugs to be detected earlier in the design process than is possible with traditional verification methods, which ordinarily cannot begin until stable RTL is available. A special-purpose RAC parser produces the ACL2 translation as well as a more readable pseudocode version, which serves as documentation. The model may be tested through simulation in both C++ and ACL2, and its architectural correctness formally verified by the ACL2 prover. The RTL may be developed in parallel with the correctness proof. Once the RTL and the proof are both in place, the verification process is completed by sequential logic equivalence checking of the model against the RTL with a commercial tool, such as Hector [36] or SLEC [21]. This methodology has been used by architects, designers, and verification engineers at Intel and Arm in the specification and formal verification of arithmetic algorithms and their implementations.

In Chap. 15, we summarize the features of RAC and its translation to ACL2. The remaining chapters illustrate the methodology with applications to a variety of floating-point modules of state-of-the-art Arm processors. Chapters 16–20 address the essential elementary operations of the FPU of a typical "big core" processor, designed to maximize execution speed. Chapters 16 and 17 describe a double-precision multiplier and adder and their combination in the implementation of a fused multiply-add operation. Chapters 18–20 describe the divider, which performs multi-precision floating-point division and square root extraction and 64-bit integer division. The remaining two chapters are based on the FPUs of smaller cores and illustrate the different design decisions that are driven by an emphasis on power consumption and silicon area over execution speed: Chap. 21 describes a low-area radix-2 divider and the special problems encountered in its design; the subject of Chap. 22 is a multi-purpose fused multiply-add unit of a graphics processor.

For each of these designs, the corresponding Verilog module has been modeled in RAC, with a code reduction of approximately 85%, and equivalence has been verified with SLEC. Each model has been translated to ACL2 and mechanically verified to conform to the relevant behavioral specification of Chap. 14. Each of these chapters contains a comprehensive mathematical exposition covering all arithmetic details of the corresponding proof. The remaining relatively trivial aspects of correctness, pertaining to pre-computation exceptions and non-computational special cases, are omitted here but included in the ACL2 proofs.

The RAC parser and ACL2 translator reside in the directory `books/projects/rac/` of the ACL2 repository. The proof scripts for Chaps. 16–22, together with the RAC models and their pseudocode versions, are in the subdirectories `fmul64`, `fadd64`, `fdiv8`, `idiv8`, `fsqrt4`, `fdiv2`, and `fma32`, respectively, of `books/projects/arm/second/`. Each pseudocode model is also accessible through a direct link provided at the beginning of the corresponding chapter.

Chapter 15
The RAC Modeling Language

The design of RAC—its features as well as the restrictions that we impose on it—is driven by the following goals:

- Documentation: C++ is a natural candidate in view of its versatility and widespread use in system modeling. For our purpose as a specification language, we require a subset that is simple enough to allow a clear and easily understood semantic definition but sufficiently expressive for detailed encoding of complex arithmetic algorithms.
- RTL modeling and equivalence checking: This is the motivation for incorporating the Algorithmic C data types, which model integer and fixed-point registers of arbitrary width and provide the basic bit manipulation features of Verilog, thereby closing the gap between an algorithm and its RTL implementation and easing the burden of equivalence checking. All language features are supported by both Hector [36] and SLEC [21].
- Formal analysis: The objectives of mathematical analysis and translation to ACL2 dictate an applicative programming paradigm, which we promote by eliminating side effects, replacing the pointers and reference parameters of C++ with other suitable extensions.

The construction of a RAC model is generally a compromise between two opposing objectives. On one hand, a higher-level model is more susceptible to mathematical analysis and allows a simpler correctness proof. On the other hand, successful equivalence checking of a complex design generally requires a significant amount of proof decomposition, using techniques that depend on structural similarities between the model and the design. As a rule of thumb, the model should be as abstract as possible while performing the same essential computations as the design.

RAC is supported by a special-purpose parser, written in C++ and based on Flex and Bison [16], which performs the following functions:

- Following a check to ensure that a model conforms to the prescribed restrictions, a pseudocode version is generated for the purpose of documentation. This is

D. M. Russinoff, *Formal Verification of Floating-Point Hardware Design*,
https://doi.org/10.1007/978-3-030-87181-9_15

intended to be more readable than the executable RAC code, especially the arcane syntax of C++ methods. In particular, expressions pertaining to the register classes are replaced with a bit vector notation that is more familiar to Verilog programmers.

- An S-expression representation of the model is generated. This is a first step toward translation to ACL2, which is completed by a translation program written in ACL2 itself.

The parser and ACL2 translator reside in the subdirectories `src` and `lisp`, respectively, of the ACL2 directory `books/projects/rac/`.

The present chapter summarizes the features of the language, assuming a basic understanding of C. For a more thorough treatment of the register classes of Algorithmic C, the reader is referred to [20]. The description of the translator (Sect. 15.6) presupposes familiarity with ACL2 but is not a prerequisite for subsequent chapters.

15.1   Language Overview

Program Structure

A RAC program consists of a sequence of elements of the following three varieties. These may appear in any order, except that an element may not refer to another that follows it in the sequence.

- Type declarations, constructed with the standard C keywords `typedef`, `struct`, and `enum`, each of which associates a type with an identifier.
- Global constant declarations, each of which associates a new constant with an identifier, a type, and a value. Note that global variables are not permitted.
- Function definitions, one of which has special status as the top-level function: its arguments are the inputs of the model, and its return value comprises the outputs. All other functions are called, directly or indirectly, by the top-level function. Recursion (including mutual recursion) is disallowed.

Data Types

All program data are of the following types. Note that pointer types are not included.

- Three basic numerical types: boolean values (`bool`), unsigned integers (`uint`), and signed integers (`int`).
- The standard C composite types: arrays, structures (`struct`), and enumeration (`enum`) types. Note that since pointers are disallowed, C arrays may not occur as function parameters
- Two class templates of the C++ Standard Template Library [10], both of which are intended to compensate for the absence of pointers and reference parameters: the `array` template, which allows arrays to be passed by value, and the `tuple` template, which provides the effect of multiple-valued functions.

- Integer and fixed-point register types: support for the register class templates of Algorithmic C [20] is provided.

Statements
The body of a function is composed of statements of the following forms:

- Local variable and constant declarations: Note that a program constant may be either global (declared at the top level) or local (declared within the body of a function), while all variables are local.
- Assignments: In another departure from standard C, *assignments*, which are classified as statements, are distinct from *expressions*, which may occur only within statements of the forms listed here.
- The standard C control statements corresponding to the keywords if, if...else, for, switch, and return (under the limitations specified in Sect. 15.5).
- Statement blocks: arbitrary sequences of statements delineated by "{" and "}".
- Assertions, indicated by the keyword assert. An assertion has no semantic import but may signal a run-time error in either C++ or ACL2.
- Type declarations: As in standard C, data types may be declared globally or locally.

Functions
A function definition consists of the following components:

- A return type, which may be a simple type or an array or tuple class. Since functions are free of side effects, every function returns a value; the keyword void may not appear in lieu of a return type.
- An identifier, the *name* of the function.
- A list of formal parameters, each of which is represented simply by a type and an identifier, with no qualifiers. In particular, reference parameters are disallowed.
- A statement block, the *body* of the function.

15.2 Parameter Passing

The stipulation that function parameters are passed only by value dictates that native C arrays may be used only as global constants and in instances where an array is used only locally by a function. The effect of passing arrays as value parameters is achieved by means of the standard C++ array class template.

As an illustration of its use, suppose that we define a function as follows:

```
array<int, 8> Sum8 (array<int, 8> a, array<int, 8> b) {
    for (uint i=0; i<8; i++) {
        a[i] += b[i];
    }
    return a;
}
```

If a and b are variables of type array<int, 8>, then the result of the assignment

```
b = Sum8 (a, b);
```

which does not affect the value of a, is that each entry of the array b is incremented by the corresponding entry of a.

Aside from the restriction on parameter passing, there is no semantic difference between ordinary C arrays and instances of an array template class. The pseudocode printer, therefore, simply converts array objects to C arrays and prints the above definition as follows:

```
int [8]  Sum8 (int a [8], int b [8]) {
  for (uint i=0; i<8; i++) {
    a [i] += b [i];
  }
  return a;
}
```

The effect of multiple-valued functions is achieved through the tuple class template. While the same effect could be achieved by means of an ordinary struct return type, this feature provides a convenient means of simultaneously assigning the components of a returned value to local variables of the caller.

For example, the following function performs integer division and returns a quotient and remainder as a tuple:

```
tuple<uint, uint> Divide (m uint, n uint) {
  assert (n != 0);
  uint quot = 0, rem = m;
  while (rem >= n) {
    quot++;
    rem -= n;
  }
  return tuple<uint, uint>(quot, rem);
}
```

A call to this function has the following syntax:

```
uint q, r;
tie (q, r) = Divide (23, 5);
```

The pseudocode printer provides a slightly simpler syntax, printing the above definition as

```
<uint, uint> Divide (m uint, n uint) {
  assert (n != 0);
  uint quot = 0, rem = m;
  while (rem >= n) {
    quot++;
    rem -= n;
  }
  return <quot, rem>;
}
```

and the invocation as

```
uint q, r;
<q, r> = Divide(23, 5);
```

Note that our use of the `tuple` template is intended only for the purpose of parameter passing and is not recognized by the parser in any other context.

15.3 Registers

RAC includes the signed and unsigned integer and fixed-point register class templates of Algorithmic C, which are fully documented on the Mentor Graphics Web site [20]. The unsigned integer registers are the simplest of these and are generally sufficient for modeling any RTL design. The other classes, while not strictly necessary, are sometimes useful in documenting the intended meaning of a register, and their more complicated semantics may be convenient in performing arithmetic computations. In the RAC models of Chaps. 16–22, we use signed and unsigned integer registers but avoid the fixed-point classes.

The register class templates may be instantiated as follows:

- `ac_int<n, false>` and `ac_int<n, true>`: unsigned and signed integer types of width n, where n may be any positive integer;
- `ac_fixed<n, m, false>` and `ac_fixed<n, m, true>`: unsigned and signed fixed-point register types of width n with m integer bits, where n may be any positive integer and m is any integer.

By convention, we use the names ui$n$, si$n$, uf$n$im, and sf$n$im for the register types `ac_int<n, false>`, `ac_int<n, true>`, `ac_fixed<n, m, false>`, and `ac_fixed<n, m, true>`, respectively. These types are expected to be declared as needed at the beginning of a program, as in the following examples:

```
typedef ac_int<96, false> ui96;
typedef ac_int<48, true> si48;
typedef ac_fixed<64, 4, false> uf64i4;
typedef ac_fixed<32, 16, false> uf32i16;
```

A register is associated with two values: a *raw value*, which is a bit vector of the same width as the register, and an integer or rational *interpreted value*. The latter is used when a register is evaluated as an argument of an arithmetic operation or assigned to another variable; the former is used in all other contexts. The interpreted value is derived from the raw value according to the register's type as follows:

- ui$n$ (unsigned integer): The interpreted value is the same as the raw value, an integer in the interval $[0, 2^n)$.
- si$n$ (signed integer): A signed version of ui$n$, with the leading bit interpreted as the sign. Thus, the represented range is the interval of integers $[-2^{n-1}, 2^{n-1})$.

- $\text{uf}nim$ (unsigned fixed-point): Interpreted with an implicit binary point following the most significant m bits. The represented values are rational numbers of the form $2^{m-n}k$, where k is an integer and $0 \leq k < 2^n$.
- $\text{sf}nim$ (signed fixed-point): A signed version of $\text{uf}nim$, representing rational numbers of the form $2^{m-n}k$, where $-2^{n-1} \leq k < 2^{n-1}$.

Thus, the interpreted value of a register of any of these types is related to its raw value r according to Definitions 2.5, 2.6, and 2.8, which are collected here for reference:

- $ui(r) = r$;
- $si(r, n) = \begin{cases} r & \text{if } r < 2^{n-1} \\ r - 2^n & \text{if } r \geq 2^{n-1}. \end{cases}$;
- $uf(r, n, m) = 2^{m-n}ui(r) = 2^{m-n}r$;
- $sf(r, n, m) = 2^{m-n}si(r, n) = \begin{cases} 2^{m-n}r & \text{if } r < 2^{n-1} \\ 2^{m-n}r - 2^m & \text{if } r \geq 2^{n-1}, \end{cases}$

As discussed in Sect. 2.5, while the width n of a register must be positive, there is no restriction on the number m of integer bits of a fixed-point register. If $m > n$, then the interpreted value is an integer with $m - n$ trailing zeroes; if $m < 0$, then the interpreted value is a fraction with $-m$ leading zeroes.

The Algorithmic C register types include a *bit select* operator and *slice read* and *write* methods. Bit selection is independent of the register type (signed vs. unsigned, integer vs. fixed-point) and uses the familiar syntax. Thus, a bit x[i] of a register x may be extracted or assigned in the natural way. The syntax of bit slices is less straightforward. The slice read method slc requires a constant template parameter indicating the width of the slice and a variable argument that represents the base index. Thus,

```
x.slc<4>(n)
```

determines the slice of x of width 4 based at index n, which is rendered in pseudocode as

```
x[n+3:n].
```

The value of this method is an integer register (independent of whether the register is integer or fixed-point) with the same width parameter as the method and the same sign parameter as the operand. Thus, the numerical value of a slice is a signed or unsigned integer according to the type of the register from which it is extracted.

A bit slice is written by a method of two arguments, the base index and an integer register containing the value to be written. The width of the slice is not specified explicitly but is inferred from the type of the second argument. For example, the assignment

```
x.set_slc(n, ui4(6))
```

replaces the slice of x of width 4 based at index n with the value 6 (binary 0110) and is represented in pseudocode as

```
x[n+3:n] = 6.
```

15.4 Arithmetic

The following numerical operators are inherited from native C:

- The unary arithmetic operators + and - and the binary arithmetic operators +, -, *, and /.
- The shift operators, << and >>.
- The modulus operator, %.
- The bit-wise logical unary complement operator ~ and binary operators &, |, and ^.
- The binary boolean-valued arithmetic relational operators <, >, <=, >=, ==, and !=.
- The boolean unary operator ! and binary operators && and ||.
- The ternary conditional operator ?.

These operators are extended to integer and fixed-point registers under the restrictions specified in [20]. In addition to these restrictions, we disallow the application of the division operator (/) to fixed-point arguments. When a register occurs as an argument of an arithmetic or relational operator, its interpreted value is used; when it occurs as an argument of a logical operator, its raw value is used.

Assignment statements may use the basic assignment operator = or any of the assignment operators corresponding to the binary arithmetic and logical operators listed above: +=, *=, &=, etc. An application of either of the unary arithmetic operators ++ and -- of C (which are not recognized as operators, i.e., may not occur within an expression) is also admitted as an assignment statement. When a numerical value is assigned to a register, it is truncated by discarding fractional and integer bits as required to fit into the register format, as specified in [20].

Arithmetic operations on registers are unbounded and performed with absolute precision. This is consistent with the semantics of ACL2, which is based on unbounded rational arithmetic. However, special care must be taken with arithmetic performed on native C integer data, the inherent limitations of which are not addressed by the ACL2 translator. It is the responsibility of the programmer to avoid arithmetic overflow and match the simpler semantics of ACL2, i.e., to ensure that the results of all such computations and the values of all assignments lie within the bounds of the relevant data formats.

On the other hand, precision may be lost through assignment. When the result of a computation is assigned to an integer or fixed-point register, the least significant fractional bits and most significant integer bits are discarded as necessary for the result to fit into the target format.

15.5   Control Restrictions

The syntax of control statements is constrained in order to minimize the difficulty
of translating from an imperative to a functional programming paradigm. Thus, the
only control statements supported are those listed in Sect. 15.1. In particular, `while`
and `do...while` are not included. Moreover, a number of restrictions are imposed
on the supported constructs.

As described in Sect. 15.6, the ACL2 translator converts a `for` loop to an
auxiliary recursive function. In order for this function to be admitted into the ACL2
logic, the prover must be able to establish that execution of the function always
terminates. To guarantee that this proof succeeds, we require a `for` loop to have the
form

```
for (init; test; update) { ... }
```

under the following restrictions:

- *init* is either a variable assignment

  ```
  var = val
  ```

  or a declaration

  ```
  type var = val
  ```

  where *var* is the *loop variable*, of type `uint` or `int`, and *val* is a numerical
  expression.
- *test* is either a comparison between the loop variable and a numerical expression
  of the form *var op limit*, where *op* is `<`, `<=`, `>`, or `>=`, or a conjunction of the form
  test$_1$ && test$_2$, where *test$_1$* is such a comparison.
- *update* is an assignment to the loop variable *var*. The combination of *test* and
  update must guarantee termination of the loop. (The translator derives an ACL2
  `:measure` declaration from *test*, which is used to establish the admissibility of
  the generated recursive function.)

Neither `break` nor `continue` may occur in a `for` loop. In some cases, the loop
test may be used to achieve the functionality of `break`. For example, instead of

```
for (uint i=0; i<N; i++) {
  if (expr) break;
  ...
}
```

we may write

```
for (uint i=0; i<N && !expr; i++) {
  ...
}
```

This feature may also be used in cases where the equivalence checker is unable to
establish an absolute upper bound on the number of iterations executed by the loop,

which is required for its "unrolling". Thus, while ACL2 has no trouble establishing termination of either of the above loops, SLEC may require something like the following, which may be used when N is known never to exceed 128:

```
for (uint i=0; i<N && i<128; i++) { ... }
```

The ACL2 translator converts a `switch` statement to the Lisp `case` macro. We require that each case of a switch have one of two forms:

> `case` *label*: *stmt*$_1$ ... *stmt*$_k$
> `default`: *stmt*$_1$ ... *stmt*$_k$

where

(1) the second form may occur only as the final case of the statement,
(2) with the possible exception of the final case, *stmt*$_k$ is `break`, and
(3) otherwise, `break` does not occur in the statement.

Further restrictions are imposed on the placement of `return` statements. We require every function body to be a statement block that recursively satisfies the following conditions:

(1) The statement block consists of a non-empty sequence of statements;
(2) None of these statements except the final one contains a `return` statement;
(3) The final statement of the block is either a `return` statement or an `if...else` statement of which each branch is either a `return` statement or a statement block satisfying all three of these conditions.

15.6   Translation to ACL2

Translation of a C++ model to ACL2 is performed in two steps: (1) the parser generates a representation of the model as a set of S-expressions, and (2) an ACL2 program converts this representation to an ACL2 program.

Most of the recognized C primitives correspond naturally to built-in ACL2 functions. The rest are implemented by a set of functions defined in an RTL library book that is included in any book generated by the translator. These include the following:

- `AG` and `AS` extract and set entries of arrays, which are implemented in ACL2 as association lists;
- `BITN`, `BITS`, `SETBITN`, and `SETBITS` access and set bits and slices of bit vectors;
- `LOG=`, `LOG<>`, `LOG<`, `LOG>`, `LOG<=`, and `LOG>=` are boolean comparators corresponding to the C operators `==`, `!=`, `<`, etc., based on the values 1 and 0 instead of the Lisp symbols `T` and `NIL`;
- `LOGIOR1`, `LOGAND1`, and `LOGNOT1` are boolean functions similarly corresponding to the C operators `||`, `&&`, and `!`;

- IF1 is a macro with the semantics of IF, except that it compares its first argument to 0 instead of NIL.

The S-expression generated by the parser for a function definition has the form

$$(\text{FUNCDEF } name \ (arg_1 \dots arg_k) \ body)$$

where *name* is the name of the function, $arg_1, \dots, arg_k$ are its formal parameters, and *body* is an S-expression derived from its body, which is assumed to be a statement block. The parser generates an S-expression for each statement as follows:

- Statement block: (BLOCK $stmt_1 \dots stmt_k$).
- Simple assignment: (ASSIGN *var term*).
- Multiple-value assignment: (MV-ASSIGN $(var_1 \dots var_k)$ *term*), where *term* corresponds to a call to a multiple-valued function.
- Variable or constant declaration: (DECLARE *var term*) or (ARRAY *var term*), where *term* is optional.
- Conditional branch: (IF *term left right*), where *left* is a block and *right* is either a block or NIL.
- Return statement : (RETURN *term*).
- For loop: (FOR (*init test update*) *body*), where *init* is a declaration or an assignment, *test* is a term, *update* is an assignment, and *body* is a statement block.
- Switch statement: (SWITCH *test* $(lab_1 \ . \ stmts_1) \dots (lab_k \ . \ stmts_k)$)), where lab_i is either an integer or a list of integers and $stmts_i$ is a list of statements.
- Assertion: (ASSERT *fn term*), where *fn* is the name of the function in which the assertion occurs and *term* is a term that is expected to have a nonzero value.

Note that variable types are not explicitly preserved in the translation. Instead, they are used by the parser to inform the translation of terms. Consider, for example, the statement block

```
{ sf8i2 x = -145;
  ui8 y = 100, z = 3;
  z = y[4:2] * x; }
```

In the evaluation of the expression on the right side of the final assignment, the type of x dictates that its value is interpreted as a signed rational with 6 fractional bits, and according to the type of y and z, their assigned values must be truncated to 8 integer bits. Thus, the above code produces the following S-expression:

```
(BLOCK (DECLARE X (BITS (* -145 (EXPT 2 6)) 7 0))
       (LIST (DECLARE Y (BITS 100 7 0))
             (DECLARE Z (BITS 3 7 0)))
       (ASSIGN Z
               (BITS (FL (* (BITS Y 4 2)
                            (/ (SI 8 X) (EXPT 2 6))))
                     7 0)))
```

The translation is completed by an ACL2 program that operates on the output of the parser. The overall strategy of this program is to convert the body of a function to a nest of LET, LET∗, and MV-LET terms. For each statement in the body, the translator generates the following:

- *ins*: a list of the variables whose values (prior to execution of the statement) are read by the statement;
- *outs*: a list of the variables (nonlocal to the statement) that are written by the statement;
- *term*: an expression of which (a) the unbound variables are *ins* and (b) the value is a multiple value consisting of the updated values of the variables of *outs* or a single value if *outs* is a singleton.

Each statement except the last corresponds to a level of the nest in which the variables of *outs* are bound to the value of *term*, except that as an optimization to improve readability, adjacent LETs are combined into a single LET or LET∗ whenever possible. The *term* of the final statement of the body becomes the body of the nest.

As a trivial (and nonsensical) example, the C++ code that generates the pseudocode function

```
uint foo(uint x, uint y, uint z) {
  uint u = y + z, v = u * x;
  <x, y, z> = bar(u, v);
  y = x > y ? 2 * u : v;
  if (x >= 0) {
    u = 2*u;
  }
  else {
    v = 3 * u;
  }
  if (x < y) {
    return u;
  }
  else {
  return y + v;
  }
}
```

also generates the corresponding ACL2 function

```
(DEFUN FOO (X Y Z)
  (LET* ((U (+ Y Z)) (V (* U X)))
    (MV-LET (X Y Z) (BAR U V)
      (LET ((Y (IF1 (LOG> X Y) (* 2 U) V)))
        (MV-LET (V U)
          (IF1 (LOG>= X 0)
               (MV V (* 2 U))
               (MV (* 3 U) U))
          (IF1 (LOG< X Y) U (+ Y V)))))))
```

Note that an if statement generates a call to the macro IF1. Similarly, a switch statement generates a CASE term. Assertions, which do not affect any program variables, are handled specially. An expression (ASSERT *fn term*) generated by the parser results in a binding of the dummy variable ASSERT to the value

(IN-FUNCTION fn term), where IN-FUNCTION is a macro that throws an error if the value of *term* is 0, with a message indicating the function in which the error occurred.

In addition to the top-level ACL2 function corresponding to a C++ function, a separate recursive function is generated for each for loop. Its returned values are those of the nonlocal variables that are assigned within the loop. Its arguments include these variables, along with any variables that are required in the execution of the loop, as well as any variables that occur in the loop initialization or test. The construction of this function is similar to that of the top-level function, but the final statement of the loop body is not treated specially. Instead, the body of the nest of bindings is a recursive call in which the loop variable is replaced by its updated value. The resulting term becomes the left branch of an IF expression, of which the right branch is simply the returned variable (if there is only one) or a multiple value consisting of the returned variables (if there are more than one). The test of the IF is the test of the loop.

For example, the function

```
uint baz(uint x, uint y, uint z) {
  uint u = y + z, v = u * x;
  for (uint i=0; i<u && u < v; i+=2) {
    v--;
    for (int j=5; j>=-3; j--) {
      assert(v > 0);
      u = x + 3 * u;
    }
  }
  return u + v;
}
```

generates three ACL2 functions:

```
(DEFUN BAZ-LOOP-0 (J V X U)
  (DECLARE (XARGS :MEASURE (NFIX (- J (1- -3)))))
  (IF (AND (INTEGERP J) (>= J -3))
      (LET ((ASSERT (IN-FUNCTION BAZ (> V 0)))
            (U (+ X (* 3 U))))
           (BAZ-LOOP-0 (- J 1) V X U))
    U))

(DEFUN BAZ-LOOP-1 (I X V U)
  (DECLARE (XARGS :MEASURE (NFIX (- U I))))
  (IF (AND (INTEGERP I) (INTEGERP U) (INTEGERP V)
           (AND (< I U) (< U V)))
      (LET* ((V (- V 1)) (U (BAZ-LOOP-0 5 V X U)))
            (BAZ-LOOP-1 (+ I 2) X V U))
    (MV V U)))

(DEFUN BAZ (X Y Z)
  (LET* ((U (+ Y Z)) (V (* U X)))
        (MV-LET (V U)
                (BAZ-LOOP-1 0 X V U)
                (+ U V))))
```

Chapter 16
Double-Precision Multiplication and Scaling

The first illustration of our verification methodology is a proof of correctness of a double-precision floating-point multiplier that supports the binary FMUL and FSCALE instructions and the ternary FMA. In a typical implementation of the last, addition is combined with multiplication in a single pipeline by inserting the addend into the multiplier's compression tree as an additional term. The resulting FMA latency may be somewhat greater than that of a simple multiplication but is less than two successive operations. One drawback of this integrated approach is that the computation cannot be initiated until all three operands are available. Another is that in order to conserve area, hardware is typically shared with the pure addition and multiplication operations, each of which is implemented as a degenerate case of FMA, resulting in increased latencies.

The Arm floating-point design team has pursued the alternative scheme of two distinct successive operations. Of course, the primary benefit of the FMA instruction is that it performs these operations with a single rounding. Thus, the multiplier must produce the full unrounded product of the significands, which is passed to the adder. The resulting FMA latency is not optimal, but in practice this operation does not occur in isolation [18]. More often, the result of one FMA is passed as an operand to another in the computation of a sum of products $x_1 y_1 + \ldots + x_n y_n$, as in a dot product. In this situation, our scheme allows the products to be computed independently and overlapped with the sums, resulting in a lower overall latency.

In this chapter and the next, we state and prove specifications of correctness of a multiplier and an adder pertaining to pure double-precision multiplication and addition, respectively, as well as the FMA support provided by each. In the DP case,

Supplementary Information The online version contains supplementary material available at (https://doi.org/10.1007/978-3-030-87181-9_16).

the significand product is a 106-bit vector. For the FMUL and FSCALE instructions, this product is rounded to 53 bits. The reason that all bits are required in the FMA case is that if cancellation occurs in the computation of a difference, a normalizing left shift may be required, moving some or all of the low bits of the product into the result significand. In the case of a denormal product, however, no normalizing left shift is performed, and these low bits need not be accurate. As seen in Sect. 16.3, the multiplier design exploits this observation.

The multiplier is represented by the module `fmul64`, a RAC program that was derived from a Verilog RTL design. The pseudocode for this function is available online at

https://go.sn.pub/fmul64.

Functional equivalence between the model and the RTL has been established with SLEC [21]. The model is significantly more compact and readable than the RTL, consisting of 20 kb of code as compared to 140 kb. This reduction is in part a reflection of the more concise syntax of RAC but was primarily achieved through the elimination of timing and other optimizations. On the other hand, the essential computations performed by the RTL are replicated in the model in order to facilitate the equivalence check. As usual, this involved some experimentation. For example, the multiplier uses a compression tree to reduce a sum of 29 partial products to a pair of vectors, which are then added. In an initial version of the model, this was replaced by a simple sum, but the resulting burden on the tool proved unmanageable. In the final version, the entire tree, consisting of 27 3:2 compressors, is replicated by the auxiliary function `compress`.

In a similar experiment, the function `CLZ53`, a logarithmic-time leading zero counter based on Lemma 8.9, was replaced by a simpler and more transparent linear-time version. In this case, the tool was able to manage the complexity of the equivalence check, but the overall execution time suffered markedly, increasing from 2 min to 22 min, and the more faithful version was ultimately used.

Sections 16.1–16.3 discuss the parameters of `fmul64` and the computation of the unrounded product. Here we limit our analysis to the case of two nonzero numerical operands, as the remaining cases, involving a NaN, an infinity, or a zero, are handled trivially by the auxiliary function `specialCase`. We apply these results to the FMA case in Sect. 16.4 and to FMUL and FSCALE in Sect. 16.5. The results of Sect. 16.4 will be combined with those of Chap. 17 to establish the correctness of the FMA operation.

Notation In our analysis of RAC models in this and subsequent chapters, we adopt the convention of using italics to denote the mathematical function represented by a function of the model as well as the numerical value of a program variable or expression. If x is a variable of a signed integer type, then x, occurring either in isolation or as an argument of an arithmetic operation, is understood to represent the interpreted value.

16.1 Parameters

The inputs of `fmul64` are as follows:

- `ui64 opa`: The double-precision encoding of the first operand.
- `ui64 opb`: For FMUL and FMA, the DP encoding of the second operand; for FSCALE, this operand is required to be the DP encoding of 1.
- `si64 scale`: The scale factor as a signed integer, valid only for FSCALE.
- `bool fz, dn`: The FZ and DN fields of the FPSCR (Sect. 14.1).
- `ui2 rmode`: The RC field of the FPSCR, a 2-bit encoding of an IEEE rounding mode (Table 14.1), which we shall denote as $\mathcal{R}$.
- `bool fma`: An indication that the operation is FMA.
- `bool fscale`: An indication that the operation is FSCALE.

Thus, the FMUL case is the default, indicated by $fma = fscale = 0$. Our assumption that opa represents a nonzero numerical value means that $opa[62:52] < 2^{11} - 1$, $opa[62:0] \neq 0$, and if $fz = 1$ then $opa[62:52] \neq 0$; the same restrictions apply to opb.

The following results are returned:

- `ui117 D`: The data result. In the FMUL and FSCALE cases, the double-precision result is $D[63:0]$, with $D[116:64] = 0$; in the FMA case, $D[116]$ is the sign, $D[115:105]$ is the biased exponent, and $D[104:0]$ is the mantissa.
- `ui8 flags`: The exception flags, $FPSCR[7:0]$. We use the mnemonics defined in Fig. 14.1 to refer to the bits of this vector.
- `bool piz`: An indication of a product of an infinity and a zero, producing the default NaN. Valid only for FMA.
- `bool inz`: An indication that the result is an infinity, a NaN, or a zero, returned in $D[116:53]$ (with $D[52:0] = 0$). Valid only for FMA.
- `bool expOvfl`: An indication that the exponent of the product is too large to be represented in the 11-bit format. Valid only for FMA when $inz = 0$.

Among the local variables are the sign, exponent, and mantissa fields of the operands:

$signa = opa[63]$, $signb = opb[63]$;
$expa = opa[62:52]$, $expb = opb[62:52]$;
$mana = opa[51:0]$, $manb = opb[51:0]$.

We make the following additional definitions pertaining to the operands: the shifted significands

$$
s_A = \begin{cases} 2^{52} + mana & \text{if } expa > 0 \\ mana & \text{if } expa = 0 \end{cases}
$$

$$
s_B = \begin{cases} 2^{52} + manb & \text{if } expb > 0 \\ manb & \text{if } expb = 0; \end{cases}
$$

the unbiased exponents

$$e_A = \begin{cases} expa - (2^{10} - 1) \text{ if } expa > 0 \\ 1 - (2^{10} - 1) \text{ if } expa = 0 \end{cases}$$

$$e_B = \begin{cases} scale \text{ if } fscale = 1 \\ expb - (2^{10} - 1) \text{ if } fscale = 0 \text{ and } expb > 0 \\ 1 - (2^{10} - 1) \text{ if } fscale = 0 \text{ and } expb = 0; \end{cases}$$

and the numerical values

$$A = (-1)^{signa} 2^{e_A - 52} s_A$$

$$B = (-1)^{signb} 2^{e_B - 52} s_B.$$

It follows trivially from the above definitions, the input assumptions, and the definition of *decode* that

$$A = decode(opa, DP)$$

and

$$B = \begin{cases} decode(opb, DP) \text{ if } fscale = 0 \\ 2^{scale} \text{ if } fscale = 1. \end{cases}$$

Our objective is to compute the product AB.

16.2   Booth Multiplier

The 53×53 integer multiplier, as represented by the function *computeProduct*, operates on *mana* and *manb* and takes as additional parameters boolean indications of the conditions $expa = 0$ and $expb = 0$. It returns the product of s_A and s_B.

Lemma 16.1 $prod = s_A \cdot s_B$.

Proof The proof is based on a straightforward application of Corollary 9.6 under the substitutions $n = 53$, $m = 27$, $x = mana$, and $y = manb$, which yields

$$pp[0] + \sum_{i=1}^{26} 2^{2(i-1)} pp[i] \bmod 2^{106} = mana \cdot manb.$$

In order to account for the leading integer bits of s_A and s_B, *computeProduct* inserts two additional terms, $ia[100:0]$ and $ib[101:0]$, defined by

$$ia[100:0] = \begin{cases} 2^{49} manb & \text{if } expa \neq 0 \\ 0 & \text{if } expa = 0, \end{cases}$$

$$ib[100:0] = \begin{cases} 2^{49} mana & \text{if } expb \neq 0 \\ 0 & \text{if } expb = 0, \end{cases}$$

and

$$ib[101] = \begin{cases} 0 & \text{if } expa = expb = 0 \\ 1 & \text{otherwise,} \end{cases}$$

each shifted left by 3 bits. The resulting sum is readily seen to satisfy

$$pp[0] + \sum_{i=1}^{26} 2^{2(i-1)} pp[i] + 2^{52}(ia + ib) \bmod 2^{106} = s_A \cdot s_B.$$

The function *compress* reduces the 29-term sum on the left to a sum of two terms, *ppa* and *ppb*, by means of 27 3:2 compressors. The equivalence of these two sums may be established by 27 simple (although tedious) applications of Lemma 8.7. Thus

$$prod = (ppa + ppb) \bmod 2^{106}$$

$$= \left(pp[0] + \sum_{i=1}^{26} 2^{2(i-1)} pp[i] + 2^3(ia + ib) \right) \bmod 2^{106}$$

$$= s_A \cdot s_B. \qquad \square$$

Corollary 16.2 $|AB| = 2^{e_A + e_B - 104} prod.$

Proof This follows from the definitions of A and B and Lemma 16.1. $\qquad \square$

16.3 Unrounded Product

In order to accommodate the range of exponents that occur in the execution of FSCALE, the design represents exponents internally in a 14-bit signed integer format with a bias of $2^{10} - 1$. The variables *hugePosScale* and *hugeNegScale* indicate extreme values of e_B in the FSCALE case that ensure that the product is outside of the normal range:

$$hugePosScale = 1 \Leftrightarrow e_B \geq 2^{12}$$

and

$$hugeNegScale = 1 \Leftrightarrow e_B < -2^{12}.$$

When neither of these variables is asserted, *expBiased* is the internal representation of $e_A + e_B$:

Lemma 16.3 *If hugePosScale = hugeNegScale = 0, then*

$$-2^{12} < expBiased = e_A + e_B + (2^{10} - 1) < 2^{12} + 2^{11}.$$

Proof Since $-2^{12} \leq e_B < 2^{12}$ and $1 \leq e_A + (2^{10} - 1) \leq 2^{11} - 2$,

$$e_A + e_B + (2^{10} - 1) < 2^{11} - 2 + 2^{12} < 2^{12} + 2^{11}$$

and

$$e_A + e_B + (2^{10} - 1) > 1 - 2^{12} > -2^{12}.$$

It is clear that

$$expbUnbiased = e_B,$$

$$expaBiased = e_A + (2^{10} - 1),$$

and by Lemma 2.32,

$$expBiased = e_A + e_B + (2^{10} - 1). \qquad \square$$

Corollary 16.4 *If hugePosScale = hugeNegScale = 0, then*

$$|AB| = 2^{expBiased - (2^{10} - 1) - 104} prod.$$

The case *hugePosScale* $= 1$ is handled specially. In the remaining case, depending on the sign of the biased sum $e_A + e_B + (2^{10} - 1)$, one of the functions *rightShft* and *leftShft* is called to perform a shift of the product. When the biased sum is positive and one of the operands is denormal (they cannot both be denormal in this case), a normalizing left shift is performed. In this case, the function *CLZ53* is used to estimate the leading zero count of the product. This function is a straightforward implementation of Definition 8.9, and Lemma 8.29 yields the following:

Lemma 16.5 *If s is a nonzero 53-bit vector, then CLZ53(s) = 52 − expo(s).*

When *expBiased* ≤ 0, a right shift of at least 1 bit is performed. It follows from Corollary 16.4 that the product is normal iff *expBiased* $= 0$ and *prod*[105] $= 1$. As noted at the beginning of this chapter, the lower bits of the product are not required in the denormal case. More precisely, only the most significant 53 fractional bits (which include a guard bit) are relevant, except insofar as whether the lower bits are all 0. Consequently, if the right shift exceeds 53, its exact value is immaterial. This allows the design to simplify the shifter by limiting the shift to 63 bits.

The values returned by each of the functions *leftShft* and *rightShft* include the mantissa of the shifted product, *frac105*; a boolean indication of imprecision, *stkShft*; the resulting biased exponent, *expShft*; and a boolean *expInc*, which indicates an overflow condition requiring that *expShft* be incremented. These values are characterized by Lemma 16.6. The two functions also return 3 bits of rounding information relevant to the FMUL and FSCALE instructions, which are discussed in Sect. 16.5.

We define

$$e_P = expShft + expInc - (2^{10} - 1). \tag{16.1}$$

This is the unbiased exponent of the product. Clearly, $e_P + (2^{10} - 1) \geq 0$. It is a consequence of the following lemma that $e_P + (2^{10} - 1) = 0$ iff the product is subnormal.

Lemma 16.6 *Assume hugePosScale* $= 0$.

(a) *If* $e_P + (2^{10} - 1) > 0$, *then stkShft* $= 0$ *and*

$$|AB| = 2^{e_P}(1 + 2^{-105}frac105);$$

(b) *If* $e_P + (2^{10} - 1) = 0$, *then*

$$2^{-52-(2^{10}-1)}frac105[104:52] \leq |AB| < 2^{-52-(2^{10}-1)}(frac105[104:52] + 1)$$

and

$$|AB| = 2^{-52-(2^{10}-1)}frac105[104:52] \Leftrightarrow frac105[51:0] = stkShft = 0.$$

Proof The proof is a case analysis based on the sign of the biased exponent sum:

Case 1: $e_A + e_B + (2^{10} - 1) \leq 0$.

In this case, the function *rightShft* is called. It always returns *expShft* $= 0$ and therefore $e_P + (2^{10} - 1) = expInc$. Thus, either (a) or (b) may apply depending on the value of *expInc*.

Note that *expDeficit* is the amount that must be added to *expBiased* to produce a result of 1. If either *hugeNegScale* $= 1$ or *expDeficit* ≥ 64, then *shift* ≥ 62, and otherwise, by Lemma 16.3,

$$expBiased = e_A + e_B + (2^{10} - 1) \leq 0$$

and

$$shift = expDeficit = 1 - expBiased > 0.$$

As noted in the preceding discussion, if $expBiased = 0$ (i.e., $shift = 1$) and $prod[105] = 1$ (Case 1.2.1), then the result is normal and all bits of the product must be preserved. In order to accommodate this case, prior to the shift, $prod$ is padded with a zero on the right, producing the vector $prod0[106 : 0] = 2prod[105 : 0]$. The result of the shift is

$$frac105 = prodShft[104 : 0]$$

$$= prod0[106 : shift][104 : 0]$$

$$= \begin{cases} prod0[106 : shift] = prod[105 : shift - 1] & \text{if } shift > 1 \\ prod0[105 : shift] = prod[104 : 0] & \text{if } shift = 1. \end{cases}$$

When $shift > 1$, the result may be inexact, as indicated by the bit $stkShft$, which is computed in advance of the shift and satisfies

$$stkShft = 0 \Leftrightarrow prod[shift-2 : 0] = 0 \Leftrightarrow prod0[shift-1 : 0] = 0.$$

Case 1.1: $hugeNegScale = 1$ or $expDeficit > 54$.
    In this case, $shift > 54$, which implies $expInc = 0$, and we must prove the claims of (b). Since

$$frac105 = prod0[106 : shift] < 2^{106-shift+1} \leq 2^{52},$$

$$frac105[104 : 52] = 0$$

and

$$frac105[51 : 0] = frac105 = prod0[106 : shift].$$

Since $AB \neq 0$, $prod \neq 0$, and it follows that $frac105[51 : 0]$ and $stkShft$ are not both 0. Thus, we need only show that $|AB| < 2^{-52-(2^{10}-1)}$. If $hugeNegScale = 1$, then

$$|AB| < lpn(DP)2^{-2^{12}} < 2^{-52-(2^{10}-1)}.$$

Otherwise, $expDeficit \geq 55$, and by Corollary 16.2, since $prod < 2^{106}$, it suffices to show that $e_A + e_B \leq -54 - (2^{10} - 1)$. But

$$e_A + e_B = expBiased - (2^{10} - 1) = 1 - expDeficit - (2^{10} - 1) \leq -54 - (2^{10} - 1).$$

Case 1.2: *hugeNegScale* = 0 and *expDeficit* ≤ 54.

In this case,

$$shift = expDeficit = 1 - (e_A + e_B + 2^{10} - 1) \le 54.$$

Case 1.2.1: *prod*[105] = *shift* = 1.

By definition, *expInc* = 1, and we must establish (a). Clearly, *prod*[*shift* − 2 : 0] = 0, and hence *stkShft* = 0. Since

$$1 - (e_A + e_B + 2^{10} - 1) = shift = 1,$$

$e_A + e_B + 2^{10} - 1 = 0$, which implies $e_P = e_A + e_B + 1$. Furthermore, *prodShft* = *prod*, *frac105* = *prod*[104 : 0], and by Corollary 16.2,

$$
\begin{aligned}
|AB| &= 2^{e_A + e_B - 104} prod \\
&= 2^{e_A + e_B - 104}(2^{105} + frac105) \\
&= 2^{e_A + e_B + 1}(1 + 2^{-105} frac105) \\
&= 2^{e_P}(1 + 2^{-105} frac105).
\end{aligned}
$$

Case 1.2.2: *prod*[105] = 0 or *shift* > 1.

In this case, $e_P + (2^{10} - 1) = expInc = 0$, and we must establish the claims of (b). By Corollary 16.2,

$$
\begin{aligned}
|AB| &= 2^{e_A + e_B - 104} prod \\
&= 2^{e_A + e_B - 105} prod0 \\
&= 2^{e_A + e_B - 105} \left(2^{shift+52} prod0[106{:}shift+52] + prod0[shift+51{:}0]\right) \\
&= 2^{-52-(2^{10}-1)} \left(prod0[106{:}shift+52] + 2^{-(shift+52)} prod0[shift+51{:}0]\right).
\end{aligned}
$$

If *shift* > 1, then by Lemma 2.15,

$$frac105[104 : 52] = prod0[106 : shift][104 : 52] = prod0[106 : shift + 52],$$

and otherwise, *shift* = 1, *prod0*[106] = *prod*[105] = 0, and the same equation holds:

$$
\begin{aligned}
frac105[104 : 52] &= prod0[105 : shift][104 : 52] \\
&= prod0[105 : shift + 52] \\
&= prod0[106 : shift + 52].
\end{aligned}
$$

Thus

$$|AB| = 2^{-52-(2^{10}-1)} \left(frac105[104:52] + 2^{-(shift+52)} prod0[shift+51:0] \right),$$

where

$$0 \leq 2^{-(shift+52)} prod0[shift+51:0] < 2^{-(shift+52)} 2^{shift+52} = 1$$

and

$$|AB| = 2^{-52-(2^{10}-1)} frac105[104:52]$$

$$\Leftrightarrow prod0[shift+51:0] = 0$$

$$\Leftrightarrow prod0[shift+51:shift] = prod0[shift-1:0] = 0$$

$$\Leftrightarrow frac105[51:0] = stkShft = 0.$$

Case 2: $e_A + e_B + (2^{10} - 1) > 0$.

In this case, the function *leftShft* is called. As we shall see, the product is precisely represented by the data returned by this function, and therefore *stkShft* is set to 0. A shift is performed only when one of the operands is denormal. Note that they cannot both be denormal in this case since $expa = expb = 0$ implies

$$e_A + e_B + (2^{10} - 1) = 1 - (2^{10} - 1) + 1 - (2^{10} - 1) + (2^{10} - 1) = 3 - 2^{10} < 0.$$

Thus, the definition of *clz* may be reformulated as

$$clz = \begin{cases} CLZ53(mana) & \text{if } expa = 0 \\ CLZ53(manb) & \text{if } expb = 0 \\ 0 & \text{otherwise.} \end{cases}$$

It follows from Lemma 16.5 that $expo(s_A) + expo(s_B) = 104 - clz$, and hence $expo(prod)$ is either $104 - clz$ or $105 - clz$.

By Lemma 16.3, since $0 \leq clz \leq 52$,

$$expDiff = expBiased - clz = e_A + e_B + (2^{10} - 1) - clz.$$

Case 2.1: $e_A + e_B + (2^{10} - 1) > clz$.

In this case, $expDiff > 0$ and

$$e_P + (2^{10} - 1) = expShft + expInc$$

$$= expDiff + expInc$$

$$= e_A + e_B + (2^{10} - 1) - clz + expInc > 0,$$

and we must show that $|AB| = 2^{e_P}(1 + 2^{-105} frac105)$.

Since $shift = clz$,

$$expo(prodShft) = expo(prod) + clz \in \{104, 105\},$$

$ovfMask = 2^{63-clz}$, and

$$expInc = mulOvf = prod[105 - clz] = prodShft[105].$$

Suppose $expInc = 0$. Then $expo(prodShft) = 104$,

$$frac105 = (2prodShft)[104 : 0] = 2prodShft[103 : 0],$$

and

$$
\begin{aligned}
|AB| &= 2^{e_A+e_B-104}prod \\
&= 2^{e_A+e_B-clz-104}prodShft \\
&= 2^{e_A+e_B-clz-104}(2^{104} + prodShft[103 : 0]) \\
&= 2^{e_A+e_B-clz}(1 + 2^{-105}frac105) \\
&= 2^{e_P}(1 + 2^{-105}frac105).
\end{aligned}
$$

On the other hand, if $expInc = 1$, then $expo(prodShft) = 105$,

$$frac105 = prodShft[104 : 0],$$

and

$$
\begin{aligned}
|AB| &= 2^{e_A+e_B-104}prod \\
&= 2^{e_A+e_B-clz-104}prodShft \\
&= 2^{e_A+e_B-clz-104}(2^{105} + prodShft[104 : 0]) \\
&= 2^{e_A+e_B-clz+1}(1 + 2^{-105}frac105) \\
&= 2^{e_P}(1 + 2^{-105}frac105).
\end{aligned}
$$

Case 2.2: $e_A + e_B + (2^{10} - 1) = clz$.
   In this case, $clz > 0$, $expDiff = 0$, $shift = clz - 1$,

$$e_P + (2^{10} - 1) = expInc,$$

$$mulOvf = prod[105 - shift] = prod[106 - clz] = 0,$$

$$expo(prodShft) = expo(2^{clz-1}prod) \in \{103, 104\},$$

$$expInc = sub2Norm = prod[104 - shift] = prodShft[104],$$

$$|AB| = 2^{e_A + e_B - 104} prod = 2^{clz - (2^{10} - 1) - 104} 2^{1 - clz} prodShft = 2^{-102 - 2^{10}} prodShft,$$

and

$$frac105 = (2prodShft)[104:0] = 2prodShft[103:0].$$

Case 2.2.1: expInc = 1.
   Since $e_P + (2^{10} - 1) = expInc > 0$, we must show that

$$|AB| = 2^{e_P}(1 + 2^{-105} frac105).$$

But since $prodShft[104] = expInc = 1$, $expo(prodShft) = 104$ and

$$\begin{aligned}
|AB| &= 2^{-102 - 2^{10}} prodShft \\
&= 2^{-102 - 2^{10}}(2^{104} + prodShft[103:0]) \\
&= 2^{2 - 2^{10}}(1 + 2^{-104} prodShft[103:0]) \\
&= 2^{e_P}(1 + 2^{-105} frac105).
\end{aligned}$$

Case 2.2.2: expInc = 0.
   Now $e_P + (2^{10} - 1) = expInc = 0$, and we must establish the claims of (b). Since $prodShft[104] = expInc = 0$, $expo(prodShft) = 103$ and

$$frac105 = 2prodShft[103:0] = 2prodShft.$$

We have $shift = clz - 1 = e_A + e_B + 2^{10} - 2$, and hence

$$\begin{aligned}
|AB| &= 2^{e_A + e_B - 104} prod \\
&= 2^{shift + 2 - 2^{10} - 104} prod \\
&= 2^{-102 - 2^{10}}(2^{shift} prod) \\
&= 2^{-102 - 2^{10}} prodShft \\
&= 2^{-103 - 2^{10}} frac105 \\
&= 2^{-103 - 2^{10}}(2^{52} frac105[104:52] + frac105[51:0]) \\
&= 2^{-52 - (2^{10} - 1)}(frac105[104:52] + 2^{-52} frac105[51:0]).
\end{aligned}$$

Since $0 \le 2^{-52} frac105[51:0] < 1$ and $stkShft = 0$, the desired result follows.
Case 2.3: $e_A + e_B + (2^{10} - 1) < clz$.

In this case, $expDiff < 0$, $expShft = 0$, and

$$shift = expBiased - 1 \bmod 2^6 = e_A + e_B + 2^{10} - 2 \bmod 2^6.$$

By assumption, $e_A + e_B + 2^{10} - 1 > 0$ and $e_A + e_B + 2^{10} - 1 < clz \leq 52$. Therefore, $0 \leq e_A + e_B + 2^{10} - 2 < 64$, and as in Case 2.2.2, $shift = e_A + e_B + 2^{10} - 2$. Furthermore, since $shift < clz - 1$,

$$expo(prodShft) = shift + expo(prod) < (clz - 1) + (105 - clz) = 104,$$

$$expInc = mulOvfl = prod[105 - shift] = 0,$$

$$e_P + (2^{10} - 1) = expShft + expInc = 0,$$

and once again

$$frac105 = 2prodShft[103:0] = 2prodShft.$$

The proof is completed in Case 2.2.2. $\square$

The biased exponent of the product is the sum of $exp11 = expShft[10:0]$ and $expInc$ unless $expGTInf$ is asserted, indicating a product above the normal range:

Lemma 16.7

(a) If $expGTinf = 0$, then $e_P + (2^{10} - 1) = exp11 + expInc$;
(b) If $expGTinf = 1$, then $|AB| \geq 2^{2^{10}+1}$.

Proof If $expGTinf = 0$, then $expShft < 2^{11}$, $exp11 = expShft$, and the claim follows from (16.1). If $hugePosScale = 1$, then

$$|AB| \geq spd(DP)2^{2^{12}} > 2^{2^{10}+1}.$$

In the remaining case,

$$e_P + (2^{10} - 1) \geq expShft \geq 2^{11}$$

and by Lemma 16.6(a),

$$|AB| \geq 2^{e_P} \geq 2^{2^{11}-(2^{10}-1)} = 2^{2^{10}+1}.$$ $\square$

16.4   FMA Support

Collecting the results of the preceding sections, we have the following main result for the FMA case, which justifies the adder input assumptions postulated in Sect. 17.1.

Lemma 16.8 *Let opa, opb, scale, fz, dn, and rmode be bit vectors of widths 64, 64, 64, 1, 1, and 2, respectively. Assume that if fz* $= 0$, *then each of opa and opb is a normal or a denormal DP encoding, and if fz* $= 1$, *then each is a normal. Let*

$$A = decode(opa, DP),$$

$$B = decode(opa, DP),$$

and

$$\langle D, flags, piz, inz, expOvfl \rangle = fmul64(opa, opb, scale, fz, dn, rmode, 1, 0).$$

The following conditions hold:

(a) $piz = inz = 0$.
(b) *flags is an 8-bit vector with flags*$[k] = 0$ *for all* $k \neq IXC$.
(c) D *is a 117-bit vector with* $D[116] = 1 \Leftrightarrow AB < 0$.
(d) *If expOvfl* $= 1$, *then*

    (i) $|AB| \geq 2^{2^{10}+1}$;
    (ii) *flags*$[IXC] = 0$.

(e) *If expOvfl* $= 0$ *and* $D[115 : 105] > 0$, *then*

    (i) $|AB| = 2^{D[115:105]-(2^{10}-1)}(1 + 2^{-105}D[104 : 0])$;
    (ii) *flags*$[IXC] = 0$.

(f) *If expOvfl* $= D[115 : 105] = 0$, *then*

    (i) $2^{-52-(2^{10}-1)}D[104 : 52] \leq |AB| < 2^{-52-(2^{10}-1)}(D[104 : 52] + 1)$;
    (ii) $|AB| = 2^{-52-(2^{10}-1)}D[104 : 52] \Leftrightarrow D[51 : 0] = flags[IXC] = 0$.

Proof (a), (b), and (c) are trivial.

If *expOvfl* $= 1$, then either *expGTinf* $= 1$ or *expInf* $= expInc = 1$. In the former case, (d) follows from Lemma 16.7. In the latter case,

$$e_P = expInf = expInc - (2^{10} - 1) = (2^{11} - 1) + 1 - (2^{10} - 1) = 2^{10} + 1$$

and (d) follows from Lemma 16.6(a).

If *expOvfl* $= 0$, then *expGTinf* $= 0$ and

$$D[115 : 105] = exp11 + expInc = e_P + (2^{10} - 1).$$

Since $D[104 : 0] = frac105$ and $flags[IDC] = stkShft$, (e) and (f) follow from Lemma 16.6(a).   □

The remaining special cases are characterized by the following result. We omit the proof, which is a straightforward case analysis based on the definition of *fmul64* and has been mechanically checked.

Lemma 16.9 *Let opa, opb, fz, dn, and rmode be bit vectors of widths 64, 64, 1, 1, and 2, respectively, and let fma = 1. Assume that at least one of opa and opb is a NaN, an infinity, a zero, or a denormal with fz = 1. Let*

$$\langle D, flags, piz, inz, expOvfl \rangle = fmul64(opa, opb, fz, dn, rmode, fma).$$

The following conditions hold:

(a) piz = 1 ⇔ either opa and opb is an infinity and the other is either a zero or a denormal with fz = 1.

(b) inz = 1.

(c) expOvfl = 0.

(d) flags is an 8-bit vector with

  (i) flags[IOC] = 1 ⇔ either opa or opb is an SNaN or piz = 1;
  (ii) flags[IDC] = 1 ⇔ either opa or opb is a denormal and fz = 1;
  (iii) flags[k] = 0 for all k ∉ {IOC, IDC}.

(e) D is a 117-bit vector with D[52 : 0] = 0 and

  (i) If piz = 1, then D[116 : 53] is the real indefinite QNaN;
  (ii) If either opa or opb is an SNaN, then D[116 : 53] is the first SNaN;
  (iii) If neither opa nor opb is an SNaN but at least one is a QNaN, then D[116 : 53] is the first QNaN;
  (iv) If piz = 0, neither opa nor opb is a NaN, and at least one is an infinity, then D[116 : 53] is an infinity with sign D[116] = opa[63] ˆ opb[63];
  (v) In the remaining case D[116 : 53] is a zero with sign D[116] = opa[63]ˆ opb[63].

16.5   Rounded Product: FMUL and FSCALE

In the FMUL and FSCALE cases, rounding begins with the computation of the sticky, guard, and least significant bits of the shifted product. In the interest of timing, these bits are computed by *rightShft* or *leftShft* in advance of the shift, which complicates their definitions.

Lemma 16.10 *Assume hugePosScale* $= 0$.

(a) stk $= 0 \Leftrightarrow frac105[51:0] = stkShft = 0$;
(b) grd $= frac105[52]$;
(c) lsb $= frac105[53]$.

Proof

Case 1: $e_A + e_B + (2^{10} - 1) \leq 0$.
   (a) If *shift* ≤ 55, then *stkMask* $= 2^{52+shift} - 1$, *stkMask*$[106:1] = 2^{51+shift} - 1$, and

$$stk = 0 \Leftrightarrow prod[50 + shift : 0] = 0.$$

But if *shift* > 55, then *stkMask* $= 2^{107} - 1$, *stk* $= 1$, *prod*$[50+shift:0] = prod \neq 0$, and the same equivalence holds. Thus

$$stk = 0 \Leftrightarrow prod[50 + shift : 0] = 0$$

$$\Leftrightarrow prod0[51 + shift : 0] = 0$$

$$\Leftrightarrow prod0[51 + shift : shift] = prod0[shift-1:0] = 0$$

$$\Leftrightarrow frac105[51 : 0] = stkShft = 0.$$

   (b) For $k \geq 0$,

$$grdMask[k] = 1 \Leftrightarrow stkMask[106 : 52][k] = 0 \text{ and } stkMask[105 : 51][k] = 1$$

$$\Leftrightarrow stkMask[52 + k] = 0 \text{ and } stkMask[51 + k] = 1$$

$$\Leftrightarrow 52 + k = 52 + shift$$

$$\Leftrightarrow k = shift.$$

Thus, *grdMask* $= 2^{shift}$ and

$$grd = 1 \Leftrightarrow prod[105 : 51][shift] = 1$$

$$\Leftrightarrow prod[51 + shift] = 1$$

$$\Leftrightarrow prod0[52 + shift] = 1$$

$$\Leftrightarrow frac105[52] = 1.$$

   (c) is similar to (b).
Case 2: $e_A + e_B + (2^{10} - 1) > 0$.
   Recall that in this case, *stkShft* $= 0$.

(a) Clearly, $stkMask = 2^{52-shift} - 1$. If $mulOvf = 1$, then

$$stk = 0 \Leftrightarrow prod[51 - shift : 0] = 0$$
$$\Leftrightarrow prodShft[51 : 0] = 0$$
$$\Leftrightarrow frac105[51 : 0] = 0,$$

and if $mulOvf = 0$, then

$$stk = 0 \Leftrightarrow prod[50 - shift : 0] = 0$$
$$\Leftrightarrow prodShft[50 : 0] = 0$$
$$\Leftrightarrow frac105[51 : 0] = 0.$$

(b) We have $ovfMask = 2^{63-shift}$ and

$$grdMask = ovfMask[63 : 11] = 2^{52-shift}.$$

If $mulOvf = 1$, then

$$grd = 1 \Leftrightarrow prod[52 - shift] = 1$$
$$\Leftrightarrow prodShft[52] = 1$$
$$\Leftrightarrow frac105[52] = 1,$$

and if $mulOvf = 0$, then

$$grd = 1 \Leftrightarrow prod[51 - shift] = 1$$
$$\Leftrightarrow prodShft[51] = 1$$
$$\Leftrightarrow frac105[52] = 1.$$

(c) is similar to (b). □

The following allows us to replace the disjunction of $expInc$ and $expRndInc$, which appears in the RTL, with their sum:

Lemma 16.11 *Assume $hugePosScale = 0$. If $expInc = 1$, then $expRndInc = 0$.*

Proof Suppose $expInc = expRndInc = 1$. Then $fracUnrnd = 2^{52} - 1$. We again refer to the proof of Lemma 16.6.
Case 1: Note that

$$prod \le (2^{53} - 1)^2 = 2^{106} - 2^{54} + 1 < 2^{106} - 2^{53}.$$

Since $expInc = 1$, $prod[105] = shift = 1$. We have $frac105 = prod[104 : 0]$ and

$$fracUnrnd = frac105[104:53] = prod[104:53] = 2^{52} - 1.$$

Thus, $prod[105:53] = 2^{53} - 1$ and

$$prod \geq 2^{53}(2^{53} - 1) = 2^{106} - 2^{53},$$

a contradiction.

Case 2: Note that $prod \leq (2^{53} - 1)(2^{53-clz} - 1)$.

Case 2.1: In this case, $shift = clz$,

$$prodShft = 2^{clz}prod \leq 2^{clz}(2^{53} - 1)(2^{53-clz} - 1)$$
$$= 2^{106} - 2^{53+clz} - 2^{53} + 2^{clz} < 2^{106} - 2^{53},$$

$expInc = prodShft[105] = 1, frac105 = prodShft[104:0]$, and

$$fracUnrnd = frac105[104:53] = prodShft[104:53] = 2^{52} - 1.$$

It follows that $prodShft[105:53] = 2^{53} - 1$ and

$$prodShft \geq 2^{53}(2^{53} - 1) = 2^{106} - 2^{53},$$

a contradiction.

Case 2.2: In this case, $clz > 0$, $shift = clz - 1$,

$$prodShft = 2^{clz-1}prod \leq 2^{clz-1}(2^{53} - 1)(2^{53-clz} - 1)$$
$$= 2^{105} - 2^{52+clz} - 2^{52} + 2^{clz-1} < 2^{105} - 2^{52},$$

$expInc = prodShft[104] = 1, frac105 = (2prodShft)[104:0]$, and

$$fracUnrnd = frac105[104:53] = (2prodShft)[104:53]$$
$$= prodShft[103:52] = 2^{52} - 1.$$

It follows that $prodShft[104:52] = 2^{53} - 1$ and

$$prodShft \geq 2^{52}(2^{53} - 1) = 2^{105} - 2^{52},$$

a contradiction.

Case 2.3: In this case, $expo(prodShft) < 104$, which implies

$$mulOvf = prodShft[105] = 0$$

and

$$sub2Norm = prodShft[104] = 0,$$

contradicting $expInc = 1$.   □

We define

$$\mathcal{R}' = \begin{cases} RDN & \text{if } \mathcal{R} = RUP \text{ and } sign = 1 \\ RUP & \text{if } \mathcal{R} = RDN \text{ and } sign = 1 \\ \mathcal{R} & \text{otherwise.} \end{cases}$$

Then by Lemma 6.82,

$$\mathcal{R}(AB, 53) = \begin{cases} \mathcal{R}'(|AB|, 53) & \text{if } sign = 0 \\ -\mathcal{R}'(|AB|, 53) & \text{if } sign = 1. \end{cases} \tag{16.2}$$

If AB is subnormal, i.e., $|AB| < spn(DP)$, then denormal rounding is applied instead:

$$drnd(AB, \mathcal{R}, DP) = \begin{cases} drnd(AB, \mathcal{R}', DP) & \text{if } sign = 0 \\ -drnd(AB, \mathcal{R}', DP) & \text{if } sign = 1. \end{cases} \tag{16.3}$$

Lemma 16.12 $|AB| < spn(DP) \Leftrightarrow e_P + (2^{10} - 1) = 0.$

Proof This is an immediate consequence of Lemma 16.6.   □

The next two lemmas correspond to the normal and subnormal cases:

Lemma 16.13 *Assume that* $|AB| \geq spn(DP)$ *and let* $r = \mathcal{R}(AB, 53)$.

(a) $r = AB \Leftrightarrow stk = grd = 0;$
(b) $|r| = 2^{e_P - 52 + expRndInc}(2^{52} + fracRnd).$

Proof We shall invoke Lemma 6.97 with the substitutions $n = 53$:

$$x = 2^{53} + frac105[104 : 52],$$

$$z = 2^{53 - e_P}|AB|,$$

and replacing $\mathcal{R}$ with $\mathcal{R}'$. By Lemma 16.6(a),

$$
\begin{aligned}
z &= 2^{53}(1 + 2^{-105} frac105) \\
&= 2^{53} + 2^{-52} frac105 \\
&= 2^{53} + 2^{-52}(2^{52} frac105[104:52] + frac105[51:0]) \\
&= x + 2^{-52} frac105[51:0].
\end{aligned}
$$

Thus, $\lfloor z \rfloor = x$ and by Lemmas 16.6(a) and 16.10(a),

$$
z \in \mathbb{Z} \Leftrightarrow frac105[51:0] = 0 \Leftrightarrow stk = 0.
$$

Since $e = expo(x) = 53$, according to Definition 6.2,

$$
RTZ(x, 53) = 2\lfloor 2^{-1}x \rfloor = 2(2^{52} + frac105[104:53]) = 2(2^{52} + fracUnrnd)
$$

and by Definition 4.3,

$$
fp^{+}(RTZ(x, 53), 53) = 2(2^{52} + fracUnrnd) + 2 = 2(2^{52} + fracP1).
$$

By Lemma 16.10(b) and (c),

$$
x[e - n : 0] = x[e - n] = x[0] = frac105[52] = grd
$$

and

$$
x[e - n+1] = x[1] = frac105[53] = lsb.
$$

By a straightforward case analysis, the conditions under which $\mathcal{R}'(z, 53) = fp^{+}(RTZ(x, 53), 53)$, according to Lemma 6.97, are equivalent to the conditions for $rndUp = 1$.

Thus, (a) follows from Lemma 6.97. For the proof of (b), we consider the following cases.

If $rndUp = 0$, then $expRndInc = 0$ and

$$
\begin{aligned}
\mathcal{R}'(z, 53) &= RTZ(x, 53) \\
&= 2(2^{52} + fracUnrnd) \\
&= 2(2^{52} + fracRnd).
\end{aligned}
$$

If $rndUp = 1$, then

$$
\begin{aligned}
\mathcal{R}'(z, 53) &= fp^{+}(RTZ(x, 53), 53) \\
&= 2(2^{52} + fracP1).
\end{aligned}
$$

In the latter case, if $fracP1 < 2^{52}$, then $expRndInc = 0$ and

$$\mathcal{R}'(z, 53) = 2(2^{52} + fracRnd),$$

and otherwise, $fracP1 = 2^{52}$, $fracRnd = 0$, $expRndInc = 1$, and

$$\mathcal{R}'(z, 53) = 2(2^{52} + 2^{52}) = 4(2^{52}) = 4(2^{52} + fracRnd).$$

Thus, in all cases,

$$\mathcal{R}'(2^{53-e_P}|AB|, 53) = \mathcal{R}'(z, 53) = 2^{1+expRndInc}(2^{52} + fracRnd)$$

and by Eq. (16.2),

$$|r| = \mathcal{R}'(|AB|, 53) = 2^{e_P - 52 + expRndInc}(2^{52} + fracRnd). \qquad \square$$

Lemma 16.14 *Assume* $|AB| < spn(DP)$ *and let* $d = drnd(AB, \mathcal{R}, DP)$.

(a) $d = AB \Leftrightarrow stk = grd = 0$;

(b) $|d| = \begin{cases} 2^{1-(2^{10}-1)}(1 + 2^{-52}fracRnd) & \text{if } expRndInc = 1 \\ 2^{-51-(2^{10}-1)}fracRnd & \text{if } expRndInc = 0. \end{cases}$

Proof We shall invoke Lemma 6.97 with the substitutions $n = 53$,

$$x = 2^{53} + frac105[104 : 52],$$

and

$$z = 2^{52+(2^{10}-1)}(|AB| + 2^{1-(2^{10}-1)}),$$

and replacing $\mathcal{R}$ with $\mathcal{R}'$. By Lemma 16.6(b),

$$frac105[104 : 52] \leq 2^{52+(2^{10}-1)}|AB| < frac105[104 : 52] + 1,$$

and hence,

$$x = 2^{53} + \left\lfloor 2^{52+(2^{10}-1)}|AB| \right\rfloor = \left\lfloor 2^{53} + 2^{52+(2^{10}-1)}|AB| \right\rfloor = \lfloor z \rfloor.$$

By Lemmas 16.6(b) and 16.10(a),

$$z \in \mathbb{Z} \Leftrightarrow frac105[51 : 0] = stkShft = 0 \Leftrightarrow stk = 0.$$

As in the proof of Lemma 16.13,

$$\mathcal{R}'(z, 53) = z \Leftrightarrow stk = grd = 0$$

and

$$\mathcal{R}'(z, 53) = 2^{1+expRndInc}(2^{52} + fracRnd),$$

which implies

$$\mathcal{R}'(|AB| + 2^{1-(2^{10}-1)}, 53) = 2^{-52-(2^{10}-1)}\mathcal{R}'(z, 53)$$

$$= 2^{-52-(2^{10}-1)}2^{1+expRndInc}(2^{52} + fracRnd)$$

$$= 2^{-51-(2^{10}-1)+expRndInc}(2^{52} + fracRnd).$$

By Eq. (16.3) and Lemma 6.100,

$$|d| = drnd(|AB|, \mathcal{R}', DP) = \mathcal{R}'(|AB| + 2^{1-(2^{10}-1)}), 53) - 2^{1-(2^{10}-1)},$$

and (a) follows easily. To complete the proof of (b), we note that if $expRndInc = 1$, then $fracRnd = 0$ and

$$|d| = 2^{2-(2^{10}-1)} - 2^{1-(2^{10}-1)} = 2^{1-(2^{10}-1)} = 2^{1-(2^{10}-1)}(1 + 2^{-52}fracRnd),$$

and if $expRndInc = 0$, then

$$|d| = 2^{-51-(2^{10}-1)}(2^{52} + fracRnd) - 2^{1-(2^{10}-1)} = 2^{-51-(2^{10}-1)}fracRnd. \qquad \square$$

Lemma 16.15

(a) $|\mathcal{R}(AB, 53)| < spn(DP) \Leftrightarrow underflow = 1.$
(b) $|\mathcal{R}(AB, 53)| > lpn(DP) \Leftrightarrow overflow = 1.$

Proof
    (a) By Eq. (16.1),

$$e_P + (2^{10} - 1) = 0 \Leftrightarrow expShft = expInc = 0$$

$$\Leftrightarrow expZero = 1 \text{ and } expInc = 0$$

$$\Leftrightarrow underflow = 1.$$

(b) We may assume $|AB| > spn(DP)$, and hence $e_P + (2^{10} - 1) > 0$. By Lemma 16.13, $expo(\mathcal{R}(AB, 53)) = e_P + expRndInc$. The claim follows from Eq. (16.1), the definition of *overflow*, and Lemma 16.11. $\qquad \square$

Lemma 16.16 *Assume that* $|AB| \geq spn(DP)$ *and let* $r = \mathcal{R}(AB, 53) \leq lpn(DP)$. *Then*

$$D = nencode(r, DP).$$

Proof By Lemmas 16.13 and 16.7,

$$expo(r) = e_P + expRndInc$$
$$= exp11 + 2^{11} \cdot expGTinf + expInc - (2^{10} - 1) + expRndInc$$
$$\leq expo(lpn(DP))$$
$$= 2^{10} - 1,$$

which implies $expGTinf = 0$ and

$$|r| = 2^{e_P + expRndInc}(1 + 2^{-52}fracRnd)$$
$$= 2^{exp11 + +expInc - (2^{10} - 1) + expRndInc}(1 + 2^{-52}fracRnd)$$
$$= 2^{expRnd - (2^{10} - 1)}(1 + 2^{-52}fracRnd).$$

It is clear that

$$sign = \begin{cases} 0 \text{ if } r > 0 \\ 1 \text{ if } r < 0, \end{cases}$$

and the claim follows from Definition 5.8 and Lemma 5.2.   □

Lemma 16.17 *Assume that* $|AB| < spn(DP)$ *and* $fz = 0$. *Let* $d = drnd(AB, \mathcal{R}, DP)$.

(a) If $d = 0$, *then* $D = zencode(sign, DP)$;
(b) If $|d| = spn(DP)$, *then* $D = nencode(d, DP)$;
(c) If $0 < |d| < spn(DP)$, *then* $D = dencode(d, DP)$.

Proof By Lemma 16.7, $exp11 = expInc = 0$, and hence $expRnd = expRndInc$. If $d = 0$, then we must have $expRnd = fracRnd = 0$, and (a) follows trivially, while the cases $expRnd = 1$ and $expRnd = 0$ correspond (b) and (c), respectively.   □

Our correctness theorem for FMUL matches the behavior of the top-level function *fmul64* with the specification function *arm-binary-spec* of Chap. 14. The arguments of the latter are a binary operation (*ADD*, *SUB*, *MUL*, or *DIV*), two operands, the initial FPSCR, and a FP format, and its returned values are the data result and the updated FPSCR. To account for the different interface of *fmul64*, the *FZ*, *DN*, and rounding control bits are extracted from the FPSCR, which is ultimately combined with the *flags* output.

For the trivial cases involving a zero, infinity, or NaN operand, the proof is a straightforward comparison of the function *specialCase* of the model with the specification functions *arm-binary-pre-comp* and *arm-binary-comp*. For the remaining computational case, the theorem similarly follows from a simple comparison of *fmul64* and *arm-post-comp* and the application of Lemmas 16.13–16.17.

Theorem 16.1 *Let opa, opb, and scale be 64-bit vectors; let R_{in} be a 32-bit vector with $fz = R_{in}[24]$, $dn = R_{in}[25]$, and $rmode = R_{in}[23:22]$; and let $fma = fscale = 0$. Let*

$$\langle D_{spec}, R_{spec} \rangle = arm\text{-}binary\text{-}spec(MUL, opa, opb, R_{in}, DP)$$

and

$$\langle D, flags, piz, inz, expOvfl \rangle = fmul64(opa, opb, scale, fz, dn, rmode, fma, fscale).$$

Then $D[63:0] = D_{spec}$ and $R_{in} \mid flags = R_{spec}$.

Similarly, we have the following correctness theorem for FSCALE:

Theorem 16.2 *Let opa and scale be 64-bit vectors; let R_{in} be a 32-bit vector with $fz = R_{in}[24]$, $dn = R_{in}[25]$, and $rmode = R_{in}[23:22]$; and let $fma = 0$ and $fscale = 1$. Assume that $opb = nencode(1, DP)$. Let*

$$\langle D_{spec}, R_{spec} \rangle = arm\text{-}fscale\text{-}spec(opa, scale, R_{in}, DP)$$

and

$$\langle D, flags, piz, inz, expOvfl \rangle = fmul64(opa, opb, scale, fz, dn, rmode, fma, fscale).$$

Then $D[63:0] = D_{spec}$ and $R_{in} \mid flags = R_{spec}$.

Chapter 17
Double-Precision Addition and Fused Multiply-Add

The double-precision addition module `fadd64` supports both the binary FADD instruction and, in concert with the multiplier, the ternary FMA instruction. Thus, both operands may be simple DP encodings, or one of them may be an unrounded product in the format of the data output of `fmul64` as described in Chap. 16. As is evident in the pseudocode version, which is available online at

  https://go.sn.pub/fadd64,

the separation of *near* and *far* data paths and the technique of leading zero anticipation (LZA) described in Sect. 8.3 are central to the design.

As usual, the RAC model has been constructed to be as simple as possible while preserving the essential computations performed by the RTL. Consequently, the timing details as well as the divergence of the data paths are largely obscured in the model. We note, however, that the computation is performed in two cycles. In the first cycle, the LZA logic is executed concurrently with the right shift and the addition, which is followed by the normalizing left shift. (In this design, the left shift is not performed in advance of the addition as suggested by Fig. 8.13.) Exponent computation, rounding, and the detection of post-computation exceptions are done in the second cycle.

The trivial cases of NaN and infinite inputs and a zero sum are handled by separate logic, represented in the model by the function `checkSpecial`. We shall limit our attention here to the more interesting case of two numerical operands with a nonzero sum.

17.1 Preliminary Analysis

The inputs of `fadd64` include the following:

- `ui64 opa`: A double-precision encoding of the first operand.

- `ui117 opp`: A 117-bit representation of the second operand. In the FMA case, as seen in Chap. 16, the multiplier produces a 106-bit product, and therefore, the width of the mantissa field of this operand is 105 rather than 52. In the FADD case, the most significant 64 bits form a DP encoding of the operand and the low 53 bits are 0. The name of this operand is intended to suggest its relation to the product in the FMA case and to avoid name conflict with the operands of `fmul64`.
- `bool fz, dn`: The FZ and DN fields of the FPSCR (Sect. 14.1).
- `ui2 rmode`: The RC field of the FPSCR, a 2-bit encoding of an IEEE rounding mode (Table 14.1), which we shall denote as $\mathcal{R}$.
- `bool fma`: An indication of an FMA operation.

The remaining inputs are generated by the multiplier in the FMA case and are ignored in the FADD case:

- `bool expOvfl`: An indication that the exponent of the product exceeds the normal 11-bit range.
- `ui8 mulExcp`: A vector representing the exceptional conditions reported by the multiplier, corresponding to *FPSCR*[7 : 0]. This is the *flags* output of *fmul64*. We use the mnemonics defined in Fig. 14.1 to refer to the bits of this vector. Recall that the inexact indication *mulExcp[IXC]* does not indicate an exception but rather that the product is subnormal, the 105-bit mantissa field of `opp` has been right-shifted in order to produce a zero exponent field, and at least one nonzero bit has been shifted out.
- `bool inz`: An indication that the multiplier output is an infinity, a NaN, or a zero.
- `bool piz`: An indication that the multiplier operands were an infinity and a zero.

Two results are returned by `fadd64`:

- `ui64 D`: The double-precision data result.
- `ui8 flags`: The exception flags.

The local variables include the following:

- The sign, exponent, and mantissa fields of the operands, with the mantissa of *opa* zero-extended to 105 bits to match that of *opp*, the mantissa of *opa* is coerced to 0 if the exponent field is 0 and *fz* = 1, and the same holds for *opp* if *fma* = 0:

$$signa = opa[63], \ signp = opp[116]$$
$$expa = opa[62 : 52], \ expp = opp[115 : 105]$$
$$fraca = \begin{cases} 0 & \text{if } expa = 0 \text{ and } fz = 1 \\ 2^{53} \cdot opa[51 : 0] & \text{otherwise} \end{cases}$$
$$fracp = \begin{cases} 0 & \text{if } expp = 0, \ fz = 1, \text{ and } fma = 0 \\ opp[104 : 0] & \text{otherwise} \end{cases}$$

- The significand of each operand, formed by appending a zero bit to the mantissa and prepending an integer bit:

$$siga = \begin{cases} 2^{106} + 2 \cdot fraca & \text{if } expa > 0 \\ 2 \cdot fraca & \text{if } expa = 0 \end{cases}$$

$$sigp = \begin{cases} 2^{106} + 2 \cdot fracp & \text{if } expp > 0 \\ 2 \cdot fracp & \text{if } expp = 0 \end{cases}$$

The reason for appending the zero is to avoid loss of accuracy in the near case in the event of a 1-bit right shift (see the proof of Lemma 17.7).

- Qualified versions of *mulExcps[IXC]* and *expOvfl*:

$$mulStk = \begin{cases} mulExcps[IXC] & \text{if } fma = 1 \\ 0 & \text{if } fma = 0, \end{cases}$$

$$mulOvfl = \begin{cases} expOvfl & \text{if } fma = 1 \text{ and } inz = 0 \\ 0 & \text{otherwise.} \end{cases}$$

The operands are represented internally by a sign bit, an 11-bit biased exponent, and a 107-bit significand. Given bit vectors b, e, and s (representing sign, exponent, and significand, respectively), we define

$$\delta(e, s) = \begin{cases} 2^{e-(2^{10}-1)-106}s & \text{if } e > 0 \\ 2^{1-(2^{10}-1)-106}s & \text{if } e = 0 \end{cases}$$

and

$$\Delta(b, e, s) = \begin{cases} \delta(e, s) & \text{if } b = 0 \\ -\delta(e, s) & \text{if } b \neq 0. \end{cases}$$

In the event that *opa* is a numerical encoding, i.e., $expa < 2^{11} - 1$, we define A to be its value, possibly forced to zero:

$$A = \Delta(signa, expa, siga) = \begin{cases} 0 & \text{if } fz = 1 \text{ and } expa = 0 \\ decode(opa, DP) & \text{otherwise.} \end{cases}$$

P will denote the value represented by the second operand. In the FADD case, if $opp[116 : 53]$ is a numerical encoding, then we define P by the same formula as A:

$$P = \Delta(signp, expp, sigp)$$

$$= \begin{cases} 0 & \text{if } fz = 1 \text{ and } expp = 0 \\ decode(opp[116 : 53], DP) & \text{otherwise.} \end{cases}$$

In the FMA case, we assume that *opp* and related inputs are supplied by the multiplier, i.e.,

$$\langle opp, mulExcps, piz, inz, mulOvfl \rangle = fmul64(opb, opc, scale, fz, dn, rmode, 1, 0),$$

where the final two parameters indicate an FMA operation, *opb* and *opc* are *DP* encodings, and the irrelevant *scale* input is arbitrary. In the event that both *opb* and *opc* are numerical, we define

$$B = decode(opb, DP),$$

$$C = decode(opc, DP),$$

and

$$P = BC.$$

The following properties hold in both cases.

Lemma 17.1 *Assume that if fma* $= 0$, *then opp*$[116 : 53]$ *is a numerical encoding, and if fma* $= 1$, *then opb and opc are numerical.*

(a) *If* $P \neq 0$, *then signp* $= 1 \Leftrightarrow P < 0$.
(b) *If mulOvfl* $= 1$, *then*

$$|P| \geq 2^{2^{10}+1};$$
$$mulStk = 0.$$

(c) *If mulOvfl* $= 0$ *and expp* > 0, *then*

$$P = \Delta(signp, expp, sigp);$$
$$mulStk = 0.$$

(d) *If mulOvfl* $= 0$ *and expp* $= 0$, *then*

$$\delta(expp, sigp^{(-53)}) \leq |P| < \delta(expp, sigp^{(-53)} + 2^{53});$$
$$\delta(expp, sigp^{(-53)}) = |P| \Leftrightarrow sigp[52 : 0] = mulStk = 0;$$

Proof In the case *fma* $= 0$, (a) holds trivially and

$$mulOvfl = mulStk = opp[52 : 0] = 0.$$

Thus, (b) holds vacuously and (c) is true by definition. If *expp* $= 0$, then by Lemma 2.14

$$sigp[52 : 0] = (2fracp)[52 : 0] = 2fracp[51 : 0] = 2opp[51 : 0] = 0,$$

which implies $sigp = 2^{53} sigp[106 : 53]$, and by Definition 1.5

$$sigp^{(-53)} = 2^{53} \left\lfloor \frac{sigp}{2^{53}} \right\rfloor = sigp.$$

Thus, (d) reduces to $P = \delta(expp, sigp)$, which again holds by definition.

In the case $fma = 1$, we invoke Lemmas 16.8 and 16.9, substituting opb, opc, opp, $mulExcps$, and P for opa, opb, D, $flags$, and AB, respectively. First suppose $P = 0$. Lemma 16.9 yields

$$mulOvfl = expOvfl = 0,$$

$$mulStk = mulExcps[IXC] = 0,$$

$$opp[52 : 0] = 0,$$

and by Lemma 16.9(c)(v), $opp[115 : 53] = 0$. It follows that

$$expp = sigp = P = 0,$$

and (d) holds trivially.

If $P \neq 0$, then Lemma 16.8 applies and (a) and (b) follow immediately. If $expp > 0$, then by Lemma 16.8(e)

$$\begin{aligned} \delta(expp, sigp) &= 2^{expp-(2^{10}-1)-106} sigp \\ &= 2^{opp[115:105-(2^{10}-1)-106} (2^{106} + 2opp[104 : 0]) \\ &= 2^{opp[115:105-(2^{10}-1)} (1 + 2^{-105} opp[104 : 0]) \\ &= |P|, \end{aligned}$$

and (c) follows. If $expp = 0$, then

$$sigp^{(-53)} = 2^{53} \left\lfloor \frac{sigp}{2^{53}} \right\rfloor = 2^{53} \left\lfloor \frac{opp[104 : 0]}{2^{53}} \right\rfloor = 2^{53} opp[104 : 52],$$

and (d) similarly follows from Lemma 16.8(f). □

We shall focus on the case of numerical operands with a nonzero sum. This is precisely the complement of the case in which the function $checkSpecial$ sets the variable $isSpecial$:

Lemma 17.2 $isSpecial = 1 \Leftrightarrow$ *any of the following conditions holds:*

(a) *$fma = 0$ and either opa or $opp[116 : 53]$ is non-numerical;*
(b) *$fma = 1$ and opa, opb, or opc is non-numerical;*
(c) *All operands are numerical and $A + P = 0$.*

Proof It is clear by inspection of the definitions of *opaz* and the local variables of *checkSpecial* that *opa* is non-numerical iff any of the variables *opaInf*, *opaQnan*, and *opaSnan* is set and that $A = 0$ iff *opaZero* is set. In the case *fma* = 0, *opp*[116 : 53] and P are analogously related to *oppInf*, *oppQnan*, *oppSnan*, and *oppZero*. In the case *fma* = 1, by examining the same definitions and Lemma 16.9, it is easily seen that at least one of *opb* and *opc* is non-numerical iff any of the variables *oppInf*, *oppQnan*, and *oppSnan* is set and again that $P = 0$ iff *oppZero* is set.

Thus, we may assume that all operands are numerical and that A and P are not both 0, and we must show that $A + P = 0$ iff the following conditions hold: *expa* = *expp*, *fraca* = *fracp*, *signa* $\neq$ *signp*, and *mulovfl* = *mulStk* = 0. We shall derive this equivalence from Lemma 17.1.

By Lemma 17.1(a) and (b), we may assume that *signa* $\neq$ *signp* and *mulOvfl* = 0. If *expp* > 0, then by Lemma 17.1(c), *mulStk* = 0 and

$$A + P = 0 \Leftrightarrow \delta(expa, siga) = \delta(expp, sigp),$$

and it is easily shown that the latter equation holds iff *expa* = *expp* and *fraca* = *fracp*. Thus, we may assume that *expp* = 0 and appeal to Lemma 17.1(d).

If *sigp*[52 : 0] = *mulStk* = 0, then

$$|P| = \delta(expp, sigp^{(-53)}) = \delta(expp, sigp)$$

and the claim follows as in the case *expp* > 0. In the remaining case, either *mulStk* = 1 or *fraca* $\neq$ *fracp*, and

$$\delta(0, sigp^{(-53)}) < |P| < \delta(expp, sigp^{(-53)} + 2^{53}),$$

where

$$\delta(0, sigp^{(-53)}) = 2^{1-(2^{10}-1)-106} 2^{53} \left\lfloor \frac{sigp}{2^{53}} \right\rfloor = 2^{-52-(2^{10}-1)} \left\lfloor \frac{sigp}{2^{53}} \right\rfloor$$

and

$$\delta(0, sigp^{(-53)} + 2^{53}) = 2^{-52-(2^{10}-1)} \left(\left\lfloor \frac{sigp}{2^{53}} \right\rfloor + 1 \right).$$

This implies $2^{52+(2^{10}-1)}|P| \notin \mathbb{Z}$. On the other hand,

$$2^{52+(2^{10}-1)}|A| = 2^{52+(2^{10}-1)}\delta(0, siga) = 2^{-53} siga = 2opa[51 : 0] \in \mathbb{Z},$$

and hence $A + P \neq 0$.   □

The analysis to follow, through Lemma 17.22 of Sect. 17.7, will be based on the input conditions prescribed above along with the additional assumption that *isSpecial* $= 0$. In the lemmas of Sects. 17.2–17.6, we shall further assume that *mulOvfl* $= 0$. The case *mulOvfl* $= 1$ will be handled specially in the proof of the main result of Sect. 17.7.

17.2 Alignment

Prior to the addition, the significands are aligned by a right shift of the significand corresponding to the lesser exponent. The design follows the convention of employing two parallel data paths, *near* and *far*. The near path is used when the signs are opposite and the exponents differ by at most 1. This is the condition under which massive cancellation may occur.

Lemma 17.3 *near* $= 1 \Leftrightarrow$ *signa* $\neq$ *signp and* $|expa - expp| \leq 1$.

Proof Since

$$exppP1 = (expp + 1) \bmod 2^{12} = expp + 1,$$

$$expaEQexppP1 = 1 \Leftrightarrow expa = expp + 1.$$

Similarly

$$exppEQexpaP1 = 1 \Leftrightarrow expp = expa + 1,$$

and the claim follows.   □

By definition, *signl* and *expl* are the sign and exponent fields of the larger operand. We shall also define *signs* and *exps* to be the corresponding fields of the smaller operand. This selection is based on the variable *oppGEopa*, which requires justification:

Lemma 17.4 *oppGEopa* $= 1 \Leftrightarrow |P| \geq |A|$.

Proof If *expa* $\leq$ *expp* and *siga* $\leq$ *sigp*, then *oppGEopa* $= 1$ and

$$siga = siga^{(-53)} \leq sigp^{(-53)},$$

which implies

$$|A| = \delta(expa, siga) \leq \delta(expa, sigp^{(-53)}) \leq \delta(expp, sigp^{(-53)}) \leq |P|.$$

If *expp* $\leq$ *expa* and *sigp* $<$ *siga*, then *oppGEopa* $= 0$, and since

$$sigp^{(-53)} < siga = siga^{(-53)},$$

$$sigp^{(-53)} \leq siga - 2^{53},$$

and

$$|P| < \delta(expp, sigp^{(-53)} + 2^{53}) \le \delta(expp, siga) \le \delta(expa, siga) = |A|.$$

In the remaining case, $expa$ and $expp$ are positive and distinct. If $expa > expp$, then $oppGEopa = 0$ and

$$
\begin{aligned}
|P| &= \delta(expp, sigp) \\
&< \delta(expp, 2^{107}) \\
&\le \delta(expa - 1, 2^{107}) \\
&= \delta(expa, 2^{106}) \\
&\le \delta(expa, siga) \\
&= |A|,
\end{aligned}
$$

and similarly, if $expa < expp$, then $oppGEopa = 1$ and $|A| < |P|$.   □

As an immediate consequence, $signl$ represents the sign of the sum:

Lemma 17.5 $signl = \begin{cases} 0 \ if \ |A + P| > 0 \\ 1 \ if \ |A + P| < 0. \end{cases}$

In the far case with opposite signs, both significands are shifted left by 1 bit to facilitate rounding, and the exponent must be adjusted accordingly. We define

$$expl' = \begin{cases} expl - 1 \ \text{if } far = 1 \text{ and } signa \ne signp \\ expl \qquad \text{otherwise.} \end{cases}$$

The resulting significands are $sigaPrime$ and $sigpPrime$, which we shall denote as $siga'$ and $sigp'$. The one that corresponds to the larger operand is the value of $sigl$. The other is the value of $sigs$, which is shifted right to produce $sigShft$:

Lemma 17.6

(a) $sigShft = \lfloor 2^{-expDiff} sigs \rfloor$;
(b) $shiftOut = 1 \Leftrightarrow sigShft \ne 2^{-expDiff} sigs$.

Proof If $expDiff < 128$, then $rshift = expDiff$ and the lemma is trivial. If $expDiff \ge 128$, then $rshift \ge 112$,

$$sigShft = \lfloor 2^{-rshift} sigs \rfloor = 0 = \lfloor 2^{-expDiff} sigs \rfloor,$$

and

$$shiftOut = 1 \Leftrightarrow sigs \ne 0 \Leftrightarrow sigShft \ne 2^{-expDiff} sigs.$$ □

Lemma 17.7 *near* $= 1 \Rightarrow$ *shiftOut* $= 0$.

Proof By Lemma 17.3, *expdiff* $\in \{0, 1\}$. Since *sigs*[0] $= 0$, *sigs* is even and $2^{-expDiff}$ *sigs* $\in \mathbb{Z}$. The claim follows from Lemma 17.6. □

Lemma 17.8 *If shiftOut* $= 1$, *then*

$$sigShft + 2^{1-expDiff} \leq 2^{-expDiff} sigs \leq sigShft + 1 - 2^{1-expDiff}.$$

Proof Let $k = sigs \bmod 2^{expDiff} \neq 0$. Then

$$sigs = \lfloor 2^{-expDiff} sigs \rfloor 2^{expDiff} + k = sigShft \cdot 2^{expDiff} + k,$$

or

$$2^{-expDiff} sigs = sigShft + 2^{-expDiff} k.$$

Since *sigs* is even, k is even, and hence $2 \leq k \leq 2^{expDiff} - 2$, or

$$2^{1-expDiff} \leq 2^{-expDiff} k \leq 1 - 2^{1-expDiff},$$

and the lemma follows. □

17.3 Addition

Next we establish error bounds for the sum. We first consider the case in which the product is represented precisely:

Lemma 17.9 *Assume that either expp* > 0 *or mulStk* $= sigp[52 : 0] = 0$.

(a) $\delta(expl', sum) \leq |A + P| < \delta(expl', sum + 1)$;
(b) $\delta(expl', sum) = |A + P| \Leftrightarrow stk = 0$.

Proof First suppose *signa* $=$ *signp*. Then *usa* $= 0$, *sum* $= sigl + sigShft$, and

$$|A + P| = \delta(expl', sigl) + \delta(expl', 2^{-expDiff} sigs)$$
$$= \delta(expl', sigl + 2^{-expDiff} sigs).$$

If *shiftOut* $= 0$, then *stk* $=$ *mulStk* $= 0$, and by Lemma 17.6, $\delta(expl', sum) = |A + P|$. On the other hand, if *shiftOut* $= 1$, then *stk* $= 1$, and by Lemma 17.8,

$$sum < sigl + 2^{-expDiff} sigs \leq sum + 1 - 2^{1-expDiff},$$

which implies

$$\delta(expl', sum) < |A + P| \leq \delta(expl', sum + 1 - 2^{1-expDiff}) < \delta(expl', sum + 1).$$

In the remaining case, $signa \neq signp$ and

$$|A + P| = \delta(expl', sigl - 2^{-expDiff} sigs).$$

Suppose $shiftOut = 1$. Then $stk = 1, far = 1, cin = 0$, and

$$sum = sigl - sigShft - 1.$$

By Lemma 17.8,

$$sum + 2^{1-expDiff} \leq sigl - 2^{-expDiff} sigs \leq sum + 1 - 2^{1-expDiff}.$$

Thus

$$\delta(expl', sum) + \delta(expl', 2^{1-expDiff}) \leq |A + P| \leq \delta(expl', sum + 1) - \delta(expl', 2^{1-expDiff})$$

and the lemma follows trivially.

But if $shiftOut = 0$, then $stk = 0$ and $cin = 1$. By Lemma 17.6,

$$sum = sigl - sigShft = sigl - 2^{-expDiff} sigs,$$

and hence

$$\delta(expl', sum) = \delta(expl', sigl - 2^{-expDiff} sigs) = |A + P|. \qquad \square$$

In the case of an approximation error, there is a loss of precision of the sum:

Lemma 17.10 *Assume that* $expp = 0$ *and either* $mulStk = 1$ *or* $sigp[52 : 0] \neq 0$.

(a) $\delta(expl', sum^{(-53)}) < |A + P| < \delta(expl', sum^{(-53)} + 2^{53})$;
(b) *Either* $stk = 1$ *or* $sum[52 : 0] \neq 0$.

Proof

Case 1: $oppGEopa = 1$.

In this case, $expa = expp = expp' = expl = expl' = 0$, $sigl = sigp' = sigp$, $sigs = siga' = siga$, and $shiftOut = 0$.

If $signa = signp$, then $|A + P| = |A| + |P|$ and $sum = siga + sigp$. Since $siga[53 : 0] = 0$,

$$\delta(expl', sum^{(-53)}) = \delta(0, siga) + \delta(0, sigp^{(-53)})$$

$$< |A| + |P|$$

$$< \delta(0, siga) + \delta(0, sigp^{(-53)} + 2^{53})$$
$$= \delta(expl', sum^{(-53)} + 2^{53}).$$

On the other hand, if $signa \neq signp$, then $sum = sigp - siga$, $sum^{(-53)} = sigp^{(-53)} - siga$, and the same results follow similarly.

Thus, $sum = sigp \pm siga$. If $stk = 0$, then $mulStk = 0$, $sigp[52 : 0] \neq 0$ and $siga[52 : 0] = 0$, and it follows that $sum[52 : 0] \neq 0$.

Case 2 oppGEopa = 0.

In this case, $sigl = siga'$ $sigs = sigp'$, and $sigShft = \lfloor 2^{-expDiff} sigp' \rfloor$.

Case 1.2: signa = signp.

We have $|A + P| = |A| + |P|$, $expl' = expl = expa$, $siga' = siga$, $sigp' = sigp$, and $sum = siga + \lfloor 2^{-expDiff} sigp \rfloor$. We invoke Lemma 1.28 with the substitutions $k = 53$, $n = expDiff$, and $x = sigp$. Lemma 1.28(a) yields

$$\lfloor 2^{-expDiff} sigp \rfloor^{(-53)} = \left\lfloor \frac{x}{2^n} \right\rfloor^{(-k)} \leq \frac{x^{(-k)}}{2^n} = 2^{-expDiff} sigp^{(-53)},$$

which implies

$$\delta(expl', sum^{(-53)}) = \delta(expa, siga) + \delta(expa, \lfloor 2^{-expDiff} sigp \rfloor^{(-53)})$$
$$\leq \delta(expa, siga) + \delta(expa, 2^{-expDiff} sigp^{(-53)})$$
$$= |A| + \delta(0, sigp^{(-53)})$$
$$< |A| + |P|.$$

Lemma 1.28(b) yields

$$2^{-expDiff}\left(sigp^{(-53)} + 2^{53}\right) = \frac{x^{(-k)} + 2^k}{2^n}$$
$$\leq \left\lfloor \frac{x}{2^n} \right\rfloor^{(-k)} + 2^k$$
$$= \lfloor 2^{-expDiff} sigp \rfloor^{(-53)} + 2^{53},$$

which implies part (a) of the present lemma:

$$\delta(expl', sum^{(-53)} + 2^{53}) = \delta(expa, siga) + \delta(expa, \lfloor 2^{-expDiff} sigp \rfloor^{(-53)} + 2^{53})$$
$$\geq \delta(expa, siga) + \delta(expa, 2^{-expDiff}(sigp^{(-53)} + 2^{53}))$$
$$= |A| + \delta(0, sigp^{(-53)} + 2^{53})$$
$$> |A| + |P|.$$

For the proof of (b), suppose $stk = 0$. Then $mulStk = shiftOut = 0$, and by hypothesis, $sigp[52 : 0] \neq 0$. Thus

$$sum = siga' + sigShft = siga' + 2^{-expDiff} sigp.$$

Since $siga'[52 : 0] = 0$ and

$$(2^{-expDiff} sigp)[52 : 0] = 2^{-expDiff} sigp[52 + expDiff : 0] \neq 0,$$

$sum[52 : 0] \neq 0$.

Case 2.2: signa $\neq$ signp.

In this case, $|A + P| = |A| - |P|$. Let $x = 2^{106+(2^{10}-1)-1}|P|$, so that $\delta(0, x) = |P|$, and let

$$n = \begin{cases} expa - 2 & \text{if } expa > 1 \\ 0 & \text{if } expa \leq 1. \end{cases}$$

A straightforward case analysis shows that

$$2^{-expDiff} sigp' = 2^{-n} sigp$$

and

$$\delta(expl', 2^{-n}x) = \delta(0, x).$$

Since

$$\delta(0, sigp^{(-53)}) < \delta(0, x) < \delta(0, sigp^{(-53)} + 2^{53}),$$

$$2^{53} \left\lfloor \frac{sigp}{2^{53}} \right\rfloor = sigp^{(-53)} < x < sigp^{(-53)} + 2^{53} = 2^{53} \left(\left\lfloor \frac{sigp}{2^{53}} \right\rfloor + 1 \right)$$

and

$$\left\lfloor \frac{sigp}{2^{53}} \right\rfloor < \frac{x}{2^{53}} < \left\lfloor \frac{sigp}{2^{53}} \right\rfloor + 1,$$

which implies $2^{-53}x \notin \mathbb{Z}$ and $\lfloor 2^{-53}x \rfloor = \lfloor 2^{-53} sigp \rfloor$.

We shall show that

$$sum^{(-53)} = (siga' - 2^{-n}x)^{(-53)}.$$

If $cin = 1$, then $shiftOut = mulStk = 0$. It follows that $sigp[52 : 0] \neq 0$, i.e., $2^{-53} sigp \notin \mathbb{Z}$. Consequently

$$\left\lfloor -\frac{sigp}{2^{n+53}} \right\rfloor = -\left\lfloor \frac{sigp}{2^{n+53}} \right\rfloor - 1 = -\left\lfloor \frac{x}{2^{n+53}} \right\rfloor - 1 = \left\lfloor -\frac{x}{2^{n+53}} \right\rfloor,$$

which implies $(-2^{-n} sigp)^{(-53)} = (-2^{-n} x)^{(-53)}$. Since $shiftOut = 0$,

$$sum = siga' - \lfloor 2^{-expDiff} sigp' \rfloor = siga' - 2^{-expDiff} sigp' = siga' - 2^{-n} sigp,$$

and

$$
\begin{aligned}
sum^{(-53)} &= (siga' - 2^{-n} sigp)^{(-53)} \\
&= siga' + (-2^{-n} sigp)^{(-53)} \\
&= siga' + (-2^{-n} x)^{(-53)} \\
&= (siga' - 2^{-n} x)^{(-53)}.
\end{aligned}
$$

For the case $cin = 0$, we invoke Lemma 1.29 with n and x as specified above, $y = sigp$, and $k = 53$. This yields

$$(-\lfloor 2^{-n} sigp \rfloor - 1)^{(-53)} = (-2^{-n} x)^{(-53)}.$$

Thus

$$sum = siga' - \lfloor 2^{-expDiff} sigp' \rfloor - 1 = siga' - \lfloor 2^{-n} sigp \rfloor - 1$$

and once again

$$
\begin{aligned}
sum^{(-53)} &= siga'^{(-53)} + (-\lfloor 2^{-n} sigp \rfloor - 1)^{(-53)} \\
&= siga' + (-2^{-n} x)^{(-53)} \\
&= (siga' - 2^{-n} x)^{(-53)}.
\end{aligned}
$$

This equation is equivalent to

$$sum^{(-53)} \leq siga' - 2^{-n} x < sum^{(-53)} + 2^{53}.$$

In fact, both inequalities are strict, for otherwise

$$2^{-n} x = siga' - sum^{(-53)}$$

and

$$\frac{x}{2^{53}} = 2^n \left(\frac{siga'}{2^{53}} - \left\lfloor \frac{sum}{2^{53}} \right\rfloor \right) \in \mathbb{Z}.$$

Thus

$$\delta(expl', sum^{(-53)}) < \delta(expl', siga') - \delta(expl', 2^{-n}x) < \delta(expl', sum^{(-53)} + 2^{53}),$$

where

$$\delta(expl', siga') = \delta(expa, siga) = |A|$$

and

$$\delta(expl', 2^{-n}x) = \delta(0, x) = |P|.$$

Thus, (a) holds in this case as well.

The proof of (b) is similar to Case 2.1. Suppose $stk = 0$. Then $mulStk = shiftOut = 0$, $sigp[52:0] \neq 0$, $cin = 1$, and

$$sum = siga' - sigShft = siga' - 2^{-n}sigp.$$

Since $siga'[52:0] = 0$ and

$$(2^{-n}sigp)[52:0] = 2^{-n}sigp[52+n:0] \neq 0,$$

$sum[52:0] \neq 0$. $\square$

17.4 Leading Zero Anticipation

The near case may involve cancellation, requiring a normalizing left shift of the sum. The exponent of the sum is estimated in advance of the addition by the method of leading zero anticipation described in Sect. 8.3. Given two 128-bit vectors, the function *LZA128* computes a vector with an exponent that is either equal to or one less than that of the 128-bit sum and passes that vector to the leading zero counter *CLZ128*, which is a straightforward implementation of the algorithm of Definition 8.9.

Lemma 17.11 *If near $= 1$ and exps $\neq 0$, then lza > 0 and*

$$106 - lza \leq expo(sum) \leq 107 - lza.$$

Proof The hypothesis implies that

$$rshift = expDiff = |expa - expp| \in \{0, 1\}$$

and

$$sigShft = 2^{-rshift}sigs = \begin{cases} sigs & \text{if } expa[0] = expp[0] \\ \frac{1}{2}sigs & \text{if } expa[0] \neq expp[0]. \end{cases}$$

It also implies $stk = 0$, and therefore $\delta(expl', sum) = |A + P| \neq 0$, which implies $sum \neq 0$. Furthermore, $cin = 1$, which implies $sum = sigl - sigShft$.

We shall apply Lemma 8.27 with $n = 128$, $a = in1LZA$, and $b = in2LZA$. Note that the vector w defined in the body of $LZA128$ is as specified in the lemma. By inspection, $a = 2^{21}sigl$, $b = 2^{128} - 2^{21}sigShft - 1 = ops$,

$$s = a + b = 2^{128} + 2^{21}(sigl - sigShft) - 1 = 2^{128} + 2^{21}sum - 1,$$

$s' = s + 1 = 2^{128} + 2^{21}sum$, and $s'[127 : 0] = 2^{21}sum$. Thus, the lemma yields $w \geq 2$ and

$$expo(w) - 1 \leq expo(2^{21}sum) = expo(sum) + 21 \leq expo(w).$$

By Lemma 8.29,

$$lza = CLZ128\left(\left\lfloor \frac{w}{2} \right\rfloor\right) = 127 - expo\left(\left\lfloor \frac{w}{2} \right\rfloor\right) = 128 - expo(w) > 0$$

and the lemma follows. $\qquad\qquad\square$

17.5 Normalization

In the near case, if the anticipated number of leading zeroes is less than the exponent of the larger operand, then it becomes the shift amount, and the exponent is adjusted accordingly. Otherwise, the shift is limited by the exponent, which is then reduced to 0. We have estimates for the exponent of the shifted sum in both cases. The estimates also apply to the far case.

Lemma 17.12

(a) $sumShft = 2^{lshift}sum$;

(b) $expShft > 0 \Rightarrow expo(sumShft) \geq 106$;

(c) $expShft = 0 \Rightarrow expo(sumShft) \leq 106$.

Proof

(a) Since $sumShft = (2^{lshift} sum) \bmod 2^{108}$, we need only show that $2^{lshift} sum < 2^{108}$, i.e., $lshift + expo(sum) < 108$. This holds trivially if $far = 1$, since $lshift = 0$. If $far = 0$, then $lshift \leq lza$, and Lemma 17.11 applies.

(b) First suppose $far = 1$. If $signa = signp$, then

$$sumShft = sum = sigl + sigShft \geq sigl \geq 2^{106}$$

and

$$expo(sumShft) \geq 106.$$

But if $signa \neq signp$, then

$$sumShft = sum = sigl - sigShft - 1 + cin,$$

where $sigl \geq 2^{107}$ and $sigShft = \lfloor 2^{-expDiff} sigs \rfloor$. If $exps > 0$, then $expDiff = expl - exps \geq 2$, $sigs < 2^{108}$, and $sigShft < 2^{106}$. If $exps = 0$, then $expDiff = expl - exps - 1 \geq 1$, $sigs < 2^{107}$, and again $sigShft < 2^{106}$. Thus,

$$sumShft > 2^{107} - 2^{106} - 1 = 2^{106} - 1,$$

i.e., $sumShft \geq 2^{106}$.

In the remaining case, $far = 0$ and $lshft = lza < expl$. By Lemma 17.11, $expl > 1$, which implies $exps > 0$, and Lemma 17.11 yields

$$expo(sumShft) = lshift + expo(sum) = lza + expo(sum) \geq 106.$$

(c) If $far = 1$, then $expShft = 0$ implies $usa = expl = 0$ and

$$sumShft = sum = sigs + sigl < 2^{106} + 2^{106} = 2^{107}.$$

Thus, we may assume $far = 0$, and hence $lza \geq expl$. If $expl \leq 1$, then $lshift = 0$ and

$$sumShft = sum \leq sigl < 2^{107}.$$

But if $expl > 1$, then $exps > 0$, $lshift = expl - 1 \leq lza - 1$, and by Lemma 17.11,

$$expo(sumShft) = lshift + expo(sum) \leq lza - 1 + expo(sum) \leq 106. \qquad \square$$

The accuracy of the sum, as given by Lemmas 17.9 and 17.10, is preserved by the normalization.

Lemma 17.13 *Assume that either $expp > 0$ or $mulStk = sigp[52 : 0] = 0$.*

(a) $\delta(expShft, sumShft) \le |A + P| < \delta(expShft, sumShft + 1)$;
(b) $\delta(expShft, sumShft) = |A + P| \Leftrightarrow stk = 0$.

Proof If $far = 1$, then $expShft = expl'$, $sumShft = sum$, and the lemma follows from Lemma 17.9. Thus, we may assume $far = 0$, which implies $expl' = expl$, and the lemma will follow from Lemma 17.9 once we prove the following two claims:

(1) $\delta(expShft, sumShft) = \delta(expl, sum)$.
(2) If $stk = 1$, then $\delta(expShft, sumShft + 1) = \delta(expl, sum + 1)$.

Note that if $stk = 1$, then since $expDiff \le 1$ and $sigs[0] = 0$, Lemma 17.6 implies

$$sigShft = \lfloor 2^{-expDiff} sigs \rfloor = 2^{-expDiff} sigs,$$

and hence $shiftOut = 0$, which implies $mulStk = 1$, $expp = 0$, and $expl = expa \le 1$.

Suppose $lza < expl$. Then $lshift = lza$, $expShft = expl - lza > 0$, $sumShft = 2^{lza} sum$, and

$$\delta(expl, sum) = 2^{expl - (2^{10}-1) - 106} sum$$

$$= 2^{expl - lza - (2^{10}-1) - 106} 2^{lza} sum$$

$$= 2^{expShft - (2^{10}-1) - 106} sumShft$$

$$= \delta(expShft, sumShft).$$

If $stk = 1$, then $lza < expl \le 1$, which implies $= lshift = lza = 0$, $expShft = expl$, and $sumShft = sum$.

Finally, suppose $lza \ge expl$. Then $expShft = 0$. We may assume $expl > 0$, for otherwise $lshift = 0$, $expShft = expl$, and $sumShft = sum$. Thus, $lshift = expl - 1$ and

$$\delta(expShft, sumShft) = \delta(0, 2^{expl-1} sum)$$

$$= 2^{1 - (2^{10}-1) - 106} 2^{expl-1} sum$$

$$= 2^{expl - (2^{10}-1) - 106} sum$$

$$= \delta(expl, sum).$$

if $stk = 1$, then $lshift = expl - 1 = 0$, which implies $sumShft = sum$ and

$$\delta(expShft, sumShft + 1) = \delta(0, sum + 1) = \delta(1, sum + 1) = \delta(expl, sum + 1). \quad \Box$$

The following estimate holds in all cases:

Lemma 17.14

(a) $\delta(expShft, sumShft^{(-53)}) \le |A + P| < \delta(expShft, sumShft^{(-53)} + 2^{53})$;
(b) $\delta(expShft, sumShft^{(-53)}) = |A + P| \Leftrightarrow sumShft[52 : 0] = stk = 0$.

Proof If $expp > 0$ or $mulStk = sigp[52 : 0] = 0$, then the lemma is a weakening of Lemma 17.13. In the remaining case, $lshift = 0$, and therefore $sum = sumShft$. If $far = 1$, then $expShft = expl'$; if $far = 0$, then $expl' = expl \leq 1$ and $expShft \leq expl$. In either case, the lemma follows from Lemma 17.10. $\square$

17.6 Rounding

For the purpose of rounding, we distinguish between the *overflow* case, in which the exponent of the shifted sum has the maximum value of 107, and the *normal* case. Rounding is performed for both cases before that exponent is known, and the appropriate result is selected later.

Rounding involves a possible increment of *sumShft* at the index of the least significant bit of the rounded result. For both the overflow and normal cases, we compute a set of masks for extracting the lsb and the guard and sticky bits, which determine whether the increment occurs. These masks are actually applied to the unshifted sum as the shift is being performed.

Lemma 17.15

(a) $lOvfl = sumShft[55]$;

(b) $gOvfl = sumShft[54]$;

(c) $sOvfl = 0 \Leftrightarrow sumShft[53 : 0] = stk = 0$;

(d) $lNorm = sumShft[54]$;

(e) $gNorm = sumShft[53]$;

(f) $sNorm = 0 \Leftrightarrow sumShft[52 : 0] = stk = 0$.

Proof We shall prove (a) and (c); the proofs of the other claims are similar.

(a) If $lshift \leq 55$, then

$$lOvfl = sum \; \& \; lOvflMask = sum \; \& \; 2^{55-lshift} = sum[55 - lshift] = sumShft[55].$$

On the other hand, if $lshift > 55$, then

$$lOvfl = 0 = sumShft[55].$$

(c) We shall show that $sum \; \& \; sOvflMask = 0 \Leftrightarrow sumShft[53 : 0] = 0$. If $lshift \leq 53$, then

$$sum \; \& \; sOvflMask = 0 \Leftrightarrow sum \; \& \; \left\lfloor \frac{2^{54} - 1}{2^{lshift}} \right\rfloor = 0$$

$$\Leftrightarrow sum \; \& \; (2^{54-lshift} - 1) = 0$$

$$\Leftrightarrow sum[53 - lshift : 0] = 0$$

$$\Leftrightarrow sumShft[53 : 0] = 0.$$

But if $lshift > 53$, then

$$sum \ \& \ sOvflMask = sum \ \& \ 0 = 0 = sumShft[53:0]. \qquad \square$$

Since the rounder operates on the absolute value of the sum, the rounding mode $\mathcal{R}$ must be adjusted. We define

$$\mathcal{R}' = \begin{cases} RDN & \text{if } \mathcal{R} = RUP \text{ and } signOut = 1 \\ RUP & \text{if } \mathcal{R} = RDN \text{ and } signOut = 1 \\ \mathcal{R} & \text{otherwise.} \end{cases}$$

Lemma 17.16

(a) $\mathcal{R}(A + P, 53) = \begin{cases} \mathcal{R}'(|A + P|, 53) & \text{if } signl = 0 \\ -\mathcal{R}'(|A + P|, 53) & \text{if } signl = 1; \end{cases}$

(b) $drnd(A + P, \mathcal{R}, DP) = \begin{cases} drnd(A + P, \mathcal{R}', DP) & \text{if } signl = 0 \\ -drnd(A + P, \mathcal{R}', DP) & \text{if } signl = 1. \end{cases}$

Proof This follows from the above definition, the definition of *signOut*, and Lemma 17.5. $\qquad \square$

We begin with the case $expShft > 0$. By Lemma 17.12, $expo(sumShft)$ is either 107 or 106. For both cases, we apply Lemma 6.97 to show that $|A + P|$ is correctly rounded.

Lemma 17.17 *Assume* $expShft > 0$.

(a) *If* $expo(sumShft) = 107$, *then* $|A + P| \geq spn(DP)$,

$$\mathcal{R}'(|A + P|, 53) = 2^{expShft-(2^{10}-1)-51}sumOvfl$$

and

$$\mathcal{R}'(|A + P|, 53) = |A + P| \Leftrightarrow gOvfl = sOvfl = 0.$$

(b) *If* $expo(sumShft) = 106$, *then* $|A + P| \geq spn(DP)$,

$$\mathcal{R}'(|A + P|, 53) = 2^{expShft-(2^{10}-1)-52}sumNorm$$

and

$$\mathcal{R}'(|A + P|, 53) = |A + P| \Leftrightarrow gNorm = sNorm = 0.$$

Proof We instantiate Lemma 6.97 with $n = 53$:

$$z = 2^{53+(2^{10}-1)-expShft}|A + P|,$$

and

$$x = \left\lfloor \frac{sumShft}{2^{53}} \right\rfloor.$$

Upon expanding the definition of δ, Lemma 17.14 yields

$$x \le z < x + 1$$

and

$$x = z \Leftrightarrow sumShft[52 : 0] = stk = 0.$$

Thus, $x = \lfloor z \rfloor$ and $z \in \mathbb{Z} \Leftrightarrow sumShft[52 : 0] = stk = 0$.

Let $e = expo(x) = expo(RTZ(x, n)) = expo(sumShft) - 53 \in \{53, 54\}$. Then

$$RTZ(x, 53) = 2^{e-52}x[e : e - 52]$$

and by definition,

$$fp^+(RTZ(x, 53), 53) = RTZ(x, 53) + 2^{e-52} = 2^{e-52}(x[e : e - 52] + 1).$$

Consider the case $expo(sumShft) = 107$, $e = 54$. The lower bound on $|A + P|$ follows from Lemma 17.14. According to Lemmas 6.97(b) and 17.15 and the definition of $incOvfl$,

$$
\begin{aligned}
\mathcal{R}'(z, 53) &= \begin{cases} RTZ(x, 53) & \text{if } incOvfl = 0 \\ fp^+(RTZ(x, 53), 53) & \text{if } incOvfl = 1 \end{cases} \\
&= 2^{e-52}(x[e : e - 52] + incOvfl) \\
&= 2^2(sumShft[107 : 55] + incOvfl) \\
&= 2^2(sumUnrnd[53 : 1] + incOvfl) \\
&= 2^2 sumOvfl,
\end{aligned}
$$

or

$$\mathcal{R}'(|A + P|, 53) = 2^{expShft - (2^{10} - 1) - 51} sumOvfl.$$

Moreover, by Lemmas 6.97(a) and 17.15,

$\mathcal{R}'(|A + P|, 53) = |A + P|$

$\Leftrightarrow z$ is 53-exact

$\Leftrightarrow x[1 : 0] = sumShft[54 : 53] = 0$ and $sumShft[52 : 0] = stk = 0$

$\Leftrightarrow gOvfl = sOvfl = 0.$

The case $e = 106$ is similar, with *incOvfl*, *sumOvfl*, *gOvfl*, and *sOvfl* replaced by *incNorm*, *sumNorm*, *gNorm*, and *sNorm*. $\qquad\square$

If *expShft* = 0, then *expo*(*sumShft*) ≤ 106. We distinguish between the subcases *expo*(*sumShft*) = 106, which indicates that the sum has overflowed to the normal range, and *expo*(*sumShft*) < 106:

Lemma 17.18 *Assume expShft = 0.*

(a) If expo(sumShft) = 106, then $|A + P| \geq spn(DP)$,

$$\mathcal{R}'(|A + P|, 53) = 2^{-2^{10}-50} sumNorm,$$

and

$$\mathcal{R}'(|A + P|, 53) = |A + P| \Leftrightarrow gNorm = sNorm = 0.$$

(b) If expo(sumShft) < 106, then $|A + P| < spn(DP)$,

$$drnd(|A + P|, \mathcal{R}', DP) = 2^{-2^{10}-50} sumNorm$$

and

$$drnd(|A + P|, \mathcal{R}', DP) = |A + P| \Leftrightarrow gNorm = sNorm = 0.$$

Proof Following the proof of Lemma 17.17 with *expShft* replaced by 1, let

$$z = 2^{51+2^{10}} |A + P|.$$

Once again, $x = \lfloor 2^{-53} sumShft \rfloor = \lfloor z \rfloor$, $z \in \mathbb{Z} \Leftrightarrow sumShft[52 : 0] = stk = 0$, and $e = expo(x) = expo(sumShft) - 53$. The proof for the case *expo*(*sumShft*) = 106 is the same as that of Lemma 17.17. For *expo*(*sumShft*) < 106, we consider the subcases *expo*(*sumShft*) ≥ 54 and *expo*(*sumShft*) < 54 separately.

For *expo*(*sumShft*) ≥ 54, we have $e \geq 1$, and we repeat the proof for *expo*(*sumShft*) = 106 with $n = e$ instead of $n = 53$. This yields

$$\mathcal{R}'(|A + P|, e) = 2^{-2^{10}-50} sumNorm,$$

and

$$\mathcal{R}'(|A + P|, e) = |A + P| \Leftrightarrow gNorm = sNorm = 0.$$

But by definition,

$$drnd(|A + P|, \mathcal{R}, DP) = \mathcal{R}(|A + P|, 53 + expo(|A + P|) - expo(spn(DP))),$$

where

$$53 + expo(|A + P|) - expo(spn(DP)) = 53 + (e - 2^{10} - 51) - (2 - 2^{10}) = e.$$

For the case $expo(sumShft) < 54$, since

$$expo(|A + P|) = e - 2^{10} - 51 < -2^{10} - 50,$$

$$|A + P| < 2^{-2^{10}-50} = spd(DP)$$

and we may invoke Lemma 6.101. Note that since $expo(sumShft) = e + 53 < 54$, $sumUnrnd = sumShft[107 : 54] = 0$ and $sumNorm = incNorm$. Thus

$$2^{-2^{10}-50} sumNorm = \begin{cases} spd(DP) & \text{if } incNorm = 1 \\ 0 & \text{if } incNorm = 0. \end{cases}$$

The proof is completed by a straightforward case analysis. Suppose, for example,

$$|A + P| > \frac{1}{2} spd(DP) = 2^{-2^{10}-51}.$$

Then $e = 0$, $expo(sumShft) = 53$, and $gNorm = sumShft[53] = 1$. Furthermore, since

$$|A + P| > 2^{-2^{10}-51} = \delta(0, 2^{53}) = \delta(expShft, sumShft^{(-53)}),$$

Lemma 17.14(b) implies $sNorm = 1$. According to the definition of $incNorm$, it follows that $incNorm = 1 \Leftrightarrow \mathcal{R}' = RNE$ or $\mathcal{R}' = RUP$, and the desired result follows from Lemma 6.101(c). The cases $|A + P| = \frac{1}{2} spd(DP)$ and $0 < |A + P| < \frac{1}{2} spd(DP)$ are similar.   □

In the overflow case, *double overflow* occurs if rounding results in an additional exponent increment. The next three lemmas pertain to the implications of overflow and double overflow, which are indicated by the variables $ovfl$ and $ovfl2$.

Lemma 17.19 If $expo(sumShft) = 107$, then

(a) $ovfl2 = 0 \Rightarrow expo(sumOvfl) = 52$;
(b) $ovfl2 = 1 \Rightarrow sumOvfl = 2^{53}$;
(c) Either $ovfl2 = 1$ or $ovfl = 1$.

Proof (a) and (b) are trivial. To prove (c), suppose $ovfl2 = ovfl = 0$. Since $sumUnrnd[53] = sumShft[107] = 1$ and $sumNorm[53] = ovfl = 0$, we must have $sumUnrnd[53 : 0] = 2^{54} - 1$ and $incNorm = 1$. But then $sumUnrnd[53 : 1] = 2^{53} - 1$, and since $ovfl2 = 0$, we must also have $incOvfl = 0$.

Suppose $rndDir = rndInf$. Since $incNorm = 1$, either $gNorm = 1$ or $sNorm = 1$. But this implies $sOvfl = 1$, contradicting $incOvfl = 0$.

In the remaining case, $rndDir = rndNear$. Since $incNorm = 1$, $gNorm = 1$, which implies $sOvfl = 1$. Furthermore, $gOvfl = sumShft[54] = sumUnrnd[0] = 1$, again contradicting $incOvfl = 0$. $\qquad\square$

Lemma 17.20 *If* $expo(sumShft) = 106$, *then*

(a) $ovfl2 = 0$;
(b) $ovfl = 0 \Rightarrow expo(sumNorm) = 52$;
(c) $ovfl = 1 \Rightarrow sumNorm = 2^{53}$, $sumOvfl = 2^{52}$, $gOvfl = 1$, *and either* $gNorm = 1$
   or $sNorm = 1$.

Proof (a) and (b) are trivial. To prove (c), suppose $ovfl = sumNorm[53] = 1$. Then $sumUnrnd = 2^{53} - 1$, $incNorm = 1$, and $sumNorm = 2^{53}$. Since $sumUnrnd = 2^{53} - 1$, $gOvfl = sumUnrnd[0] = 1$. Since $incNorm = 1$, $rndDir$ is either $rndNear$ or $rndInf$, and either $gNorm = 1$ or $sNorm = 1$. It follows that $sOvfl = 1$, which implies $incOvfl = 1$, and hence $sumOvfl = 2^{52}$. $\qquad\square$

Lemma 17.21 *If* $expo(sumShft) < 106$, *then*

(a) $ovfl2 = ovfl = 0$;
(b) $sumNorm \le 2^{52}$.

Proof Both claims are trivial. $\qquad\square$

17.7 Correctness Theorems

Our first objective is to relate the function *fadd64* with the specification function *arm-post-comp* of Sect. 14.3, which takes an unrounded value, an initial FPSCR state, and a floating-point format and returns a rounded result and the updated FPSCR. We shall show that the data value D computed by *fadd64*, which is the concatenation of the fields *signOut*, *expOut*, and *fracOut*, and the post-computation exception flags, $R[IXC]$, $R[UFC]$, and $R[OFC]$, match the corresponding values returned by *arm-post-comp*. The two correctness theorems corresponding to FADD and FMA follow easily from this result:

Lemma 17.22 *Let* R_{in} *be a 32-bit vector with* $R_{in}[12 : 10] = 0$ *and let* $fz = R_{in}[24]$, $dn = R_{in}[25]$, *and* $rmode = R_{in}[23 : 22]$. *Let*

$$\langle D, flags \rangle = fadd64(opa, opp, fz, dn, rmode, fma, inz, piz, expOvfl, mulExcps)$$

and

$$\langle D_{spec}, R_{spec} \rangle = arm\text{-}post\text{-}comp(A + P, R_{in}, DP).$$

Then $D = D_{spec}$ *and for* $k = IXC$, *UFC, and OFC,* $(flags \mid R_{in})[k] = R_{spec}[k]$.

Proof Note that the two functions *fadd64* and *arm-post-comp* are similarly structured, treating the overflow, underflow, and normal cases in that order.

Case 1: Overflow

The *arm-post-comp* overflow condition is $|\mathcal{R}(A+P, 53)| > lpn(DP)$, which may be expressed as $expo(\mathcal{R}(A + P, 53)) \geq 2^{10}$, and that of *fadd64* is *infOrMax* $= 1$. Once these conditions are shown to be equivalent, the comparison for this case is trivial.

Subcase 1.1: *mulOvfl* $= 1$.

In this case, *infOrMax* $= 1$. Since *expp*[11] $= 1$, *expp* $\geq 2^{11}$, which implies *sigp* $\geq 2^{106}$ and

$$|P| \geq \delta(expp, sigp^{(-53)}) \geq 2^{2^{11}-(2^{10}-1)-106+106} = 2^{2^{10}+1}.$$

On the other hand, since $expa \leq 2^{11} - 2$ and $siga < 2^{107}$,

$$|A| = \delta(expa, siga) < 2^{2^{11}-2-(2^{10}-1)-106+107} = 2^{2^{10}}.$$

Consequently, $|A + P| \geq |P| - |A| > 2^{2^{10}+1} - 2^{2^{10}} = 2^{2^{10}}$ and

$$expo(\mathcal{R}(A + P, 53)) \geq expo(A + P) \geq 2^{10}.$$

Thus, we may assume *mulOvfl* $= 0$, and Lemmas 17.3–17.21 apply. We may also assume that *expShft* > 0; otherwise, both conditions are clearly false. By Lemma 17.12, $expo(sumShft) \geq 106$.

In order to simplify the definition of *infOrMax*, we note that if *expShft* $= 2047$, then since

$$expShft \leq expl = \max(expa, expp),$$

our assumptions imply *fma* $= 1$ and *expp* $= 2047$, and therefore *opplong* $= 1$.

Subcase 1.2: $expo(sumShft) = 107$.

If *ovfl2* $= 1$, then by Lemmas 17.17 and 17.19,

$$expo(\mathcal{R}(A + P, 53)) = expShft - (2^{10} - 1) - 51 + 53 = expShft - 2^{10} + 3$$

and

$$expo(\mathcal{R}(A + P, 53)) \geq 2^{10} \Leftrightarrow expShft \geq 2^{11} - 3 \Leftrightarrow infOrMax = 1.$$

Similarly, if *ovfl2* $= 0$, then *ovfl1* $= 1$,

$$expo(\mathcal{R}(A + P, 53)) = expShft - (2^{10} - 1) - 51 + 52 = expShft - 2^{10} + 2,$$

and

$$expo(\mathcal{R}(A + P, 53)) \geq 2^{10} \Leftrightarrow expShft \geq 2^{11} - 2 \Leftrightarrow infOrMax = 1.$$

Subcase 1.3: $expo(sumShft) = 106$.
    If $ovfl = 1$, then by Lemmas 17.17 and 17.20,

$$expo(\mathcal{R}(A + P, 53)) = expShft - (2^{10} - 1) - 52 + 53 = expShft - 2^{10} + 2$$

and

$$expo(\mathcal{R}(A + P, 53)) \geq 2^{10} \Leftrightarrow expShft \geq 2^{11} - 2 \Leftrightarrow infOrMax = 1.$$

But if $ovfl = 0$, then

$$expo(\mathcal{R}(A + P, 53)) = expShft - (2^{10} - 1) - 52 + 52 = expShft - 2^{10} + 1$$

and

$$expo(\mathcal{R}(A + P, 53)) \geq 2^{10} \Leftrightarrow expShft \geq 2^{11} - 1 \Leftrightarrow infOrMax = 1.$$

Case 2: Underflow
    The *arm-post-comp* underflow condition is $|A + P| < spn(DP)$, or

$$expo(A + P) < 2 - 2^{10},$$

and that of *fadd64* is $tiny = 1$. Once again, we must establish the equivalence of
these conditions. By Lemma 17.14

$$expo(A + P) = expo(\delta(expShft, sumShft)),$$

and it follows from Lemma 17.12 that

$$expo(A + P) < 2 - 2^{10} \Leftrightarrow expo(sumShft) < 106$$

$$\Leftrightarrow expo(sumUnrnd) < 52$$

$$\Leftrightarrow tiny = 1.$$

If $fz = 1$, then $R[UFC] = 1$, the data result is a zero with the sign of $A + P$, and
the theorem holds trivially. In the remaining case, according to the specification, the
sum is rounded to $d = drnd(A + P, \mathcal{R}, DP)$, and if $d \neq A + P$, then $R[UFC] =
R[IXC] = 1$. Since the sign of the result is correctly represented by *signl*, we need
only show that $|d|$ is correctly encoded by *exp* and *frac*. By Lemma 17.18, $|d| =
2^{-2^{10}-50} sumNorm$.

Subcase 2.1: $sumNorm[52] = 1$.
    By Lemma 17.21, $sumNorm = 2^{52}$, which implies $|d| = 2^{2-2^{10}} = spn(DP)$,
$expOut = 1$, and $fracOut = 0$. Thus, D is a normal encoding, and the encoded

absolute value is

$$2^{expOut-(2^{10}-1)}(1 + 2^{-52}fracOut) = 2^{1-(2^{10}-1)} = 2^{2-2^{10}} = |d|.$$

By Lemma 17.18, $|d| = |A + P| \Leftrightarrow gNorm = sNorm = 0$, which is the condition under which $R[UFC]$ and $R[IXC]$ are not set.

Subcase 2.2: sumNorm[52] = 0.
In this case, $sumNorm < 2^{52}$, $expOut = 0$, and $fracOut = sumNorm$. Thus, D is a denormal encoding, and the encoded absolute value is

$$2^{1-(2^{10}-1)-52}fracOut = 2^{-2^{10}-50}sumNorm = |d|.$$

Correctness of the flags follows as in Subcase 2.1.

Case 3: $A + P$ and $\mathcal{R}(A + P)$ are both within the normal range.
In this case, $expShft \geq 106$, the specified value of D is the normal encoding of $r = \mathcal{R}(A + P, 53)$, and $R[IXC]$ is set when $r \neq A + P$. Again, since it is clear that the sign of the sum is correctly represented by $signl$, we need only show that $|r|$ is correctly encoded by the exponent and mantissa fields, i.e.,

$$|r| = 2^{expOut-(2^{10}-1)}(1 + 2^{-52}fracOut).$$

Subcase 3.1: ovfl2 = 1.
In this case, $expOut = expShft+2$ and $fracOut = 0$. By Lemmas 17.19 and 17.20, $expo(sumShft) = 107$ and $sumOvfl = 2^{53}$. By Lemma 17.17,

$$|r| = 2^{expShft-(2^{10}-1)-51}2^{53}$$
$$= 2^{expShft+2-(2^{10}-1)}$$
$$= 2^{expOut-(2^{10}-1)}(1 + 2^{-52}fracOut)$$

and $r = A + P \Leftrightarrow gOvfl = sOvfl = 0$, which is the condition under which $R[IXC]$ is not set.

Subcase 3.2: ovfl2 = 0 and ovfl = 1.
If $expo(sumShft) = 107$, then by Lemma 17.19, $expo(sumOvfl) = 52$ and

$$sumOvfl = 2^{52} + sumOvfl[51 : 0] = 2^{52} + fracOut.$$

Since $expShft > 0$, $expOut = expShft + 1$, and by Lemma 17.17,

$$|r| = 2^{expShft-(2^{10}-1)-51}sumOvfl$$
$$= 2^{expShft+1-(2^{10}-1)}(2^{-52}sumOvfl)$$
$$= 2^{expOut-(2^{10}-1)}(1 + 2^{-52}fracOut)$$

and $r = A + P \Leftrightarrow gOvfl = sOvfl = 0$, which is the condition under which $R[IXC]$ is not set.

On the other hand, if $expo(sumShft) = 106$, then by Lemma 17.20, $sumNorm = 2^{53}$, $gOvfl = 1$, and either $gNorm = 1$ or $sNorm = 1$. If $expShft > 0$, then $expOut = expShft + 1$, and by Lemma 17.17,

$$|r| = 2^{expShft-(2^{10}-1)-52} sumNorm$$

$$= 2^{expShft+1-(2^{10}-1)}$$

$$= 2^{expOut-(2^{10}-1)}(1 + 2^{-52} fracOut).$$

Similarly, if $expShft = 0$, then by Lemma 17.18,

$$|r| = 2^{1-(2^{10}-1)-52} sumNorm$$

$$= 2^{expShft-(2^{10}-1)-52} 2^{53}$$

$$= 2^{2-(2^{10}-1)}$$

$$= 2^{expOut-(2^{10}-1)}(1 + 2^{-52} fracOut).$$

In either case, since either $gNorm = 1$ or $sNorm = 1$, $r \neq A + P$, and since $gOvfl = 1$, $R[IXC]$ is set.

Subcase 3.3: $ovfl2 = ovfl = 0$.

By Lemmas 17.19 and 17.20, $expo(sumShft) = 106$ and $expo(sumNorm) = 52$. Since $fracOut = sumNorm[51 : 0]$,

$$sumNorm = 2^{52} + sumNorm[51 : 0] = 2^{52}(1 + 2^{-52} fracOut).$$

If $expo(expShft) > 0$, then $expOut = expShft$, and by Lemma 17.17,

$$|r| = 2^{expShft-(2^{10}-1)-52} sumNorm = 2^{expOut-(2^{10}-1)}(1 + 2^{-52} fracOut).$$

Similarly, if $expo(expShft) = 0$, then $expOut = 1$, and by Lemma 17.18,

$$|r| = 2^{1-(2^{10}-1)-52} sumNorm = 2^{expOut-(2^{10}-1)}(1 + 2^{-52} fracOut).$$

In either case, $r = A + P \Leftrightarrow gNorm = sNorm = 0$, which is the condition under which $R[IXC]$ is not set.   □

We have the following correctness theorem for FADD. For the case of numerical operands with a nonzero sum, the behavior of the data result and the post-computation exception flags follow from Lemma 17.22. The behavior of the pre-computation flags and the remaining special cases are readily verified by a straightforward case analysis comparing *fadd64* with the specification function *arm-binary-spec*:

Theorem 17.1 *Let*

$$\langle D_{spec}, R_{spec} \rangle = arm\text{-}binary\text{-}spec(ADD, opa, opp[116:53], R_{in}, DP),$$

where opa is a 64-bit vector, opp is a 117-bit vector with opp[52 : 0] = 0, and R_{in} is a 32-bit vector. Let

$$\langle D, R \rangle = fadd64(opa, opp, fz, dn, rmode, 0, inz, piz, mulOvfl, mulExcps),$$

where $fz = R_{in}[24]$, $dn = R_{in}[25]$, $rmode = R_{in}[23:22]$, and inz, piz, mulOvfl, and mulExcps are arbitrary.
   Then $D = D_{spec}$ and $R_{in} \mid flags = R_{spec}$.

According to the following theorem, FMA is correctly implemented by a combination of *fadd64* and *fmul64*. This similarly follows from Lemmas 17.22, 16.8, and 16.9 and a comparison of these functions with *arm-fma-spec*:

Theorem 17.2 *Let*

$$\langle D_{spec}, R_{spec} \rangle = arm\text{-}fma\text{-}spec(opa, opb, opc, R_{in}, DP),$$

where opa, opb, and opc are 64-bit vectors and R_{in} is a 32-bit vector. Let

$$\langle opp, mulExcps, piz, inz, mulOvfl \rangle = fmul64(opb, opc, mathitscale, fz, dn, rmode, 1, 0),$$

where $fz = R_{in}[24]$, $dn = R_{in}[25]$, $rmode = R_{in}[23:22]$, and scale is arbitrary. Let

$$\langle D, flags \rangle = fadd64(opa, opp, fz, dn, rmode, 1, inz, piz, mulOvfl, mulExcps).$$

Then $D = D_{spec}$ and $R_{in} \mid flags = R_{spec}$.

Chapter 18
Multi-precision Radix-8 SRT Division

Unlike the dedicated double-precision multiplier and adder described in the preceding two chapters, a single module of our FPU performs floating-point division and square root extraction at all three precisions: double, single, and half, as well as 32-bit and 64-bit integer division. For the sake of clarity, however, this module is modeled by three separate functions, `fdiv8`, `idiv8`, and `fsqrt4`, the first of which is the subject of this chapter. This function is based on an implementation of the minimally redundant radix-8 case of SRT division that is addressed by Lemma 10.8 of Sect. 10.3. The pseudocode version is available online at

https://go.sn.pub/fdiv8.

18.1 Overview

The input parameters of the top-level function `fdiv8` are as follows:

- `ui64 opa, opb`: Encodings of the dividend and divisor, respectively. For formats *SP* and *HP*, the operands reside in the low-order bits.
- `ui2 fmt`: A 2-bit encoding of a floating-point format ($DP = 2$, $SP = 1$, $HP = 0$).
- `bool fz, dn`: The FZ and DN fields of the FPSCR (Sect. 14.1).
- `ui2 rmode`: The RC field of the FPSCR, a 2-bit encoding of an IEEE rounding mode (Table 14.1), which we shall denote as $\mathcal{R}$.

Two results are returned:

- `ui64 D`: The data result. For *fmt* = *SP* (resp., *HP*), $D[31:0]$ (resp., $D[15:0]$) holds the data, and the higher bits are 0.
- `ui8 flags`: The exception flags, *FPSCR*[7 : 0]. We use the mnemonics defined in Fig. 14.1 to refer to the bits of this vector.

© The Author(s), under exclusive license to Springer Nature Switzerland AG 2022   325
D. M. Russinoff, *Formal Verification of Floating-Point Hardware Design*,
https://doi.org/10.1007/978-3-030-87181-9_18

We also define the following values:

- The precision of the operation, $p = \begin{cases} 11 \text{ if } fmt = HP \\ 24 \text{ if } fmt = SP \\ 53 \text{ if } fmt = DP; \end{cases}$

- The exponent width, $e = \begin{cases} 5 \ \ \text{if } fmt = HP \\ 8 \ \ \text{if } fmt = SP \\ 11 \text{ if } fmt = DP; \end{cases}$

- The exponent $bias = 2^{e-1} - 1$;
- The operand values, $A = decode(opa, fmt)$ and $B = decode(opb, fmt)$;
- The divisor,

$$d = \frac{sig(B)}{2} \tag{18.1}$$

- The dividend,

$$x = \begin{cases} \frac{sig(A)}{2} & \text{if } sig(A) \geq sig(B) \\ sig(A) & \text{if } sig(A) < sig(B), \end{cases} \tag{18.2}$$

Note that the definitions (18.1) and (18.2) ensure that the bounds (10.11) and (10.1) are satisfied.

The initial phase of fdiv8 handles the early termination cases, in which at least one operand is a zero, a NaN, or an infinity. Since these cases are trivial, we shall focus on the remaining computational case, in which each operand is either a denormal that is not forced to 0 or normal.

The iterative phase of the algorithm is naturally modeled in C as a for loop. As usual, successful equivalence checking between the model and the RTL requires faithful modeling of the essential computations, which in this case means that the partial remainder and quotient must be replicated precisely at each iteration. With the goal of minimizing latency, the RTL executes two iterations on each cycle, and in order to achieve this, different approximations are used for the iterations within a cycle. Consequently, each iteration of the for loop corresponds to a cycle, i.e., two iterations of the algorithm of Sect. 10.1 rather than one.

Before entering the for loop, the operands are normalized if necessary, and the first iteration of the algorithm of Sect. 10.1 is executed. Let C be the number of these iterative cycles. Then the number of iterations of the algorithm is $N = 2C + 1$.

Notation For each variable that is updated within the for loop, we shall use the subscript j to denote the value produced by the jth iteration of the algorithm.

In particular, the value of the jth quotient digit is q_j. Now that d, x, and q_j have been specified, the partial quotients Q_j and R_j are also defined, according to (10.3) and (10.4).

The partial remainder is represented in redundant signed-digit form by RP and RN, 71-bit vectors with 4 implicit integer bits. That is, we shall show that

$$2^{67} R_j \equiv RP_j - RN_j \pmod{2^{71}}. \tag{18.3}$$

To see that 4 integer bits are sufficient to represent the full range of remainders, note that the bound on d (10.11) together with Lemma 10.8(a) yields

$$|R_j| \le \frac{4}{7} d \le \frac{4}{7} \tag{18.4}$$

and hence $|8R_j| \le \frac{32}{7} < 8$.

The partial quotient is represented explicitly by the vector $quot$, consisting of 54 fractional bits. The integer bits of the quotient are not explicitly stored. Thus, we shall show that

$$quot_j = 8^{j-1} Q_j.$$

This value is computed "on the fly", which requires that the decremented value

$$quotM1_j = quot_j - 1$$

is also maintained.

As specified in Lemma 10.8, the derivation of the quotient digit q_{j+1} is based on an approximation A_j of the remainder R_j, which is compared with the constants $m_k(i)$ of Fig. 10.3. These constants are encoded as 10-bit signed integers in the array $computeCmpConst(i)$. They include 6 implicit fractional bits and are negated for efficient comparison. Thus, we have the following, which may be verified by direct computation:

Lemma 18.1 *For $0 \le i < 64$ and $-3 \le k < \ell \le 4$,*

$$m_k(i) = -\frac{1}{64} si(computeCmpConst(i)[k + 3], 10),$$

where $m_k(i)$ is defined in Fig. 10.3.

The approximation of R_j is defined by

$$A_j = \frac{1}{64} si(RS10_j, 10).$$

Note that the requirement $64A_j \in \mathbb{Z}$ of Lemma 10.8 holds trivially. As is typical of SRT designs, when the $(j + 1)^{\text{st}}$ iteration is the first iteration of a cycle, this approximation is derived from the leading bits of RP_j and RN_j, in this case by means of a 10-bit adder. In order to satisfy timing constraints, however, when the

$(j+1)^{\text{st}}$ iteration is the second iteration of a cycle, $RS10_j$ is computed into two steps, beginning in the preceding cycle, before RP_j and RN_j are available. The analysis of this critical feature of the design appears in the proof of Lemma 18.6.

18.2   Pre-processing

The function *normalize* performs a mantissa shift for each denormal operand, using the same auxiliary function *CLZ53* as the multiplier of Chap. 16. The values computed by normalize, representing the operand significands and the predicted exponent of the result, satisfy the following:

Lemma 18.2

(a) $siga = 2^{52} sig(A)$;

(b) $sigb = 2^{52} sig(B)$;

(c) $\left| \frac{A}{B} \right| = 2^{expDiff - bias} \left(\frac{siga}{sigb} \right) = 2^{expQ - bias} \left(\frac{x}{d} \right)$.

Proof It is easily shown that $2^{52} \leq siga < 2^{53}$ and

$$|A| = 2^{expaShft - bias - 52} siga :$$

if *opa* is normal, this is trivial, and the denormal case follows from Lemma 16.5. It follows that $siga = 2^{52} sig(A)$. The analogous results hold for B. Thus

$$\left| \frac{A}{B} \right| = 2^{expaShft - expbShft} \left(\frac{siga}{sigb} \right),$$

where

$$expaShft - expbShft = (expaShft - expbShft + bias) - bias = expDiff - bias.$$

If $siga < sigb$, then $sigLTsigb = 1$ and

$$\left| \frac{A}{B} \right| = 2^{expDiff - bias} \left(\frac{sig(A)}{sig(B)} \right) = 2^{expQ - 1 - bias} \left(\frac{x}{2d} \right) = 2^{expQ - bias} \left(\frac{x}{d} \right);$$

if $siga \geq sigb$, then $sigLTsigb = 0$ and

$$\left| \frac{A}{B} \right| = 2^{expDiff - bias} \left(\frac{sig(A)}{sig(B)} \right) = 2^{expQ - bias} \left(\frac{2x}{2d} \right) = 2^{expQ - bias} \left(\frac{x}{d} \right). \qquad \square$$

The divisor is represented in the same 71-bit format as the remainder:

Lemma 18.3

(a) $div = 2^{67}d$;
(b) $div2 = 2^{68}d$;
(c) $div3 = 2^{67}3d$.

Proof This follows trivially from Lemma 18.2 and the definitions of div and d. □

Next, the partial remainder is initialized and the biased exponent $expQ$ of the unrounded quotient is computed:

Lemma 18.4 $RP_1 = 2^{67}x$.

Proof If $siga < sigb$, then $sigLTsigb = 1$ and

$$RP_1 = 2^{15}siga = 2^{67}sig(A) = 2^{67}x;$$

if $siga \geq sigb$, then $sigLTsigb = 0$ and

$$RP_1 = 2^{14}siga = 2^{66}sig(A) = 2^{67}x.$$ □

Prior to entering the loop, the first iteration of the algorithm is performed, producing the first quotient digit q_1 and a signed-digit redundant representation of the partial remainder R_1:

Lemma 18.5 Let $i = \left\lfloor 128\left(d - \frac{1}{2}\right)\right\rfloor$.

(a) q_1 is the greatest $k \in \{-4, \ldots, 4\}$ such that $m_k(i) \leq A_0$;
(b) $|A_0 - 8R_0| < \frac{1}{64}$.
(c) $|R_1| \leq \frac{4}{7}d$;
(d) $2^{67}R_1 = RP_1 - RN_1$ and $RP_1[66 - p : 0] = RN_1[66 - p : 0] = 0$.

Proof
(a) By (10.11) and Lemma 18.3(a), $2^{66} \leq div < 2^{67}$. Thus

$$i = \left\lfloor 128\left(d - \frac{1}{2}\right)\right\rfloor = \left\lfloor 2^{-60}(div - 2^{66})\right\rfloor = \left\lfloor 2^{-60}div \bmod 2^{66}\right\rfloor = div[65 : 60],$$

$$computeCmpConst(i) = computeCmpConst(div[65 : 60]) = cmpConst,$$

and for $-3 \leq k \leq 4$,

$$m_k(i) = -\frac{1}{64}si(computeCmpConst(i)[k + 3], 10) = -\frac{1}{64}si(cmpConst[k + 3], 10).$$

Since $RS10_0 = RP_1[70 : 61] < 2^{10}$, $A_0 = \frac{1}{64}RS10_0$. We must show that q_1 is the greatest $k \in \{-4, \ldots, 4\}$ such that $64m_k(i) \leq RS10_0$. Since $q_1 \in \{1, 2\}$, this is a consequence of the following three observations:

(i) $64m_1(i) \leq RS10_0$:

Since $RP_1 \geq 2^{14}siga \geq 2^{66}$, $RS10_0 = RP_1[70:61] \geq 2^5$. Since $cmpConst[4] \geq 992$ (hexadecimal 3e0),

$$64m_1(i) = -si(cmpConst[4], 10)$$
$$= 2^{10} - cmpConst[4] \leq 2^{10} - 992$$
$$= 2^5 \leq RS10_0.$$

(ii) $64m_3(i) > RS10_0$:

Since $x < 2d$, $RP_1 = 2^{67}x < 2^{68}d = 2div$, and it follows that

$$RS10_0 = RP_1[70:61]$$
$$\leq div[69:60]$$
$$= 2^6 div[69:66] + div[65:60]$$
$$= 2^6 + i.$$

Since $computeCmpConst(i)[6] > 2^9$ for all i,

$$64m_3(i) = -si(computeCmpConst(i)[6], 10)$$
$$= 2^{10} - computeCmpConst(i)[6],$$

and it suffices to show that

$$2^{10} - computeCmpConst(i)[6] > 2^6 + i$$

for $0 \leq i < 64$, which may be verified by exhaustive computation.

(iii) $q_1 = 2 \Leftrightarrow 64m_2(i) \leq RS10_0$:

Since $computeCmpConst(i)[5] > 2^9$ for all i,

$$64m_2(i) = -si(computeCmpConst(i)[6], 10)$$
$$= 2^{10} - computeCmpConst(i)[6]$$

and

$$q_1 = 2 \Leftrightarrow geP2[10] = 1$$
$$\Leftrightarrow RS10_0 + computeCmpConst(i)[6] \geq 2^{10}$$
$$\Leftrightarrow 64m_2(i) \leq RS10_0.$$

(b) We must show that $|A_0 - 8R_0| = |A_0 - x| < \frac{1}{64}$, i.e., $|RS10_0 - 64x| < 1$. By Lemma 18.4,

$$2^{67}x = RP_1 = 2^{61}RP_1[70:61] + RP_1[60:0] = 2^{61}RS10_0 + RP_1[60:0],$$

and hence

$$|RS10_0 - 64x| = 2^{-61}|2^{61}RS10_0 - 2^{67}x| = 2^{-61}RP_1[60:0] < 1.$$

(c) This follows from (a), (b), and Lemmas 10.2 and 10.8.

(d) $RP_1 - RN_1 = 2^{67}x - q_1 div = 2^{67}(8R_0 - q_1d) = 2^{67}R_1$, and it follows from $siga[52 - p:0] = sigb[52 - p:0] = 0$ that

$$RP_1[66 - p:0] = RN_1[66 - p:0] = 0.$$

$\square$

18.3 Iterative Phase

Lemma 18.6 *The following hold for all j, $1 \le j \le N = 2C + 1$:*

(a) q_j *is the greatest $k \in \{-4, \ldots, 4\}$ such that $m_k(i) \le A_{j-1}$;*
(b) $|A_{j-1} - 8R_{j-1}| < \frac{1}{64}$;*
(c) $|R_j| \le \frac{4}{7}d$;*
(d) $2^{67}R_j \equiv RP_j - RN_j \pmod{2^{71}}$ *and $RP_j[63 - p:0] = RN_j[63 - p:0] = 0$.*

Proof The case $j = 1$ is given by Lemma 18.5. Let $1 \le j \le N - 1$. We assume the lemma holds for all ℓ such that $1 \le \ell \le j$ and show that it holds for $j + 1$.

(a) The quotient digit q_{j+1} is computed as

$$nextDigit(RS10_j, computeCmpConst(i)),$$

where $0 \le RS10_j < 1024$ and $0 \le i < 64$. It may be verified by direct computation that for all values of these arguments, q_{j+1} is the greatest $k \in \{-4, \ldots, 4\}$ such that $64m_k(i) \le si(RS10_j, 10) = 64A_j$.

(b) We must show that $|A_j - 8R_j| < \frac{1}{64}$.

When either $j = 1$ or j is even (in the latter case the jth iteration occurs as the first iteration of a cycle),

$$RS10_j = (RP_j[67:58] - RN_j[67:58]) \bmod 2^{10}.$$

We invoke Lemma 2.37 with

$$X = RP_j[67:0] - RN_j[67:0] \equiv 2^{67} R_j \pmod{2^{68}},$$

$$Y = 2^{58}(RP_j[67:58] - RN_j[67:58]) \equiv 2^{58} RS10_j \pmod{2^{68}},$$

and $n = 68$. Note that $|X - Y| = |RP_j[57:0] - RN_j[57:0]| < 2^{58}$ and

$$|2^{67} R_j| \leq 2^{67} \cdot \frac{4}{7} < 2^{67} - 2^{58} < 2^{67} - |X - Y|.$$

Thus, by Lemma 2.32,

$$si(X \bmod 2^{68}, 68) = 2^{67} R_j,$$

and by Lemmas 2.34 and 2.37,

$$\begin{aligned}
|si(X \bmod 2^{68}, 68) - si(Y \bmod 2^{68}, 68)| &= |2^{67} R_j - si(2^{58} RS10_j, 68)| \\
&= |2^{67} R_j - 2^{58} si(RS10_j, 10)| \\
&= |X - Y| \\
&< 2^{58}.
\end{aligned}$$

Finally, division by 2^{64} yields

$$|8R_j - A_j| = |8R_j - \frac{1}{64} si(RS10_j, 10)| < \frac{1}{64}.$$

When $j > 1$ and j is odd, i.e., the jth iteration is the second iteration of a cycle, the computation of $RS10_j$ is performed in two steps. First, the preceding iteration (the first iteration of the same cycle) computes an 11-bit approximation $RS11_{j-1}$ of R_{j-1}, which may be described as follows:

We begin by defining three 65-bit vectors, which may be related to both $RS11_{j-1}$ and R_{j-1}:

$$\begin{aligned}
\pi &= \quad RP_{j-2}[61] \quad RP_{j-2}[60] \ldots \quad RP_{j-2}[0] \quad 1 \quad 0 \quad \alpha \\
\nu &= {\sim}RN_{j-2}[61] \ {\sim}RN_{j-2}[60] \ldots \ {\sim}RN_{j-2}[0] \quad 1 \quad 0 \quad 0 \\
\delta &= \quad\quad \Delta[64] \quad\quad\quad \Delta[63] \ldots \quad\quad\quad \Delta[3] \ \Delta[2] \ \Delta[1] \ \Delta[0],
\end{aligned}$$

where

$$\alpha = \begin{cases} 0 \text{ if } q_{j-1} \leq 0 \\ 1 \text{ if } q_{j-1} > 0 \end{cases}$$

and

$$\Delta = -q_{j-1} div - \alpha.$$

Note that the two's complement completion of v is handled by the two 1's at index 2. Thus

$$\pi = 8RP_{j-2} \bmod 2^{65} + 4 + \alpha,$$

$$v = -8(RN_{j-2} + 1) \bmod 2^{65} + 4,$$

$$\delta = \Delta \bmod 2^{65} = (-q_{j-1} div - \alpha) \bmod 2^{65},$$

and

$$\begin{aligned} \pi + v + \delta &\equiv 8RP_{j-2} + 4 + \alpha - 8(RN_{j-2} + 1) + 4 - q_{j-1} div - \alpha \\ &\equiv 8(RP_{j-2} - RN_{j-2}) - q_{j-1} div \\ &\equiv 8 \cdot 2^{67} R_{j-2} - 2^{67} q_{j-1} d \\ &\equiv 2^{67} R_{j-1} \pmod{2^{65}}. \end{aligned}$$

Following the definition of *ComputeRS11*, we distinguish the case $q_{j-1} = 0$. Let

$$\sigma = \begin{cases} \pi & \text{if } q_{j-1} = 0 \\ \pi \ \char94\ v \ \char94\ \delta & \text{if } q_{j-1} \neq 0, \end{cases}$$

$$\chi = \begin{cases} v & \text{if } q_{j-1} = 0 \\ 2(\pi \ \& \ v \mid (\pi \mid v) \ \& \ \delta) & \text{if } q_{j-1} \neq 0, \end{cases}$$

and

$$R = \sigma + \chi[64:0].$$

By Lemma 8.7,

$$R \equiv \pi + v + \delta \equiv 2^{67} R_{j-1} \pmod{2^{65}}.$$

Let

$$\tilde{R} = 2^{54}(\sigma[64:54] + \chi[64:54] + \epsilon),$$

where

$$\epsilon = \begin{cases} 0 \text{ if } \sigma[53] = \chi[53] = 0 \\ 1 \text{ otherwise.} \end{cases}$$

Then

$$\begin{aligned} |R - \tilde{R}| &= |\sigma[53:0] + \chi[53:0] - 2^{54}\epsilon| \\ &= |2^{53}(\sigma[53] + \chi[53] - 2\epsilon) + \sigma[52:0] + \chi[52:0]| \\ &\leq \max\left(2^{53}(2\epsilon - (\sigma[53] + \chi[53])), \sigma[52:0] + \chi[52:0]\right) \\ &< 2^{54}. \end{aligned}$$

By definition, $RS11_{j-1} = sum12[11:1]$, where

$$\begin{aligned} sum12 &= (\sigma[64:53] + \chi[64:53] + 1) \bmod 2^{12} \\ &= (2\sigma[64:54] + 2\chi[64:54] + \sigma[53] + \chi[53] + 1) \bmod 2^{12}. \end{aligned}$$

Thus, by Lemma 1.21,

$$\begin{aligned} RS11_{j-1} &= \left\lfloor \frac{1}{2} sum12_{j-1} \right\rfloor \\ &= \left(\sigma[64:54] + \chi[64:54] + \left\lfloor \frac{1}{2}(\sigma[53] + \chi[53] + 1) \right\rfloor \right) \bmod 2^{11} \\ &= (\sigma[64:54] + \chi[64:54] + \epsilon) \bmod 2^{11} \\ &= \left(2^{-54} \tilde{R} \right) \bmod 2^{11}. \end{aligned}$$

In the second iteration, an 11-bit sum $sum11$ is derived from $RS11_{j-1}$ and q_j, and the 10-bit approximation of R_j is computed as $RS10_j = sum11[10:1]$.
Let

$$D = -q_j div$$

and

$$\tilde{D} = \begin{cases} -q_j div & \text{if } q_j \leq 0 \\ -q_j div - 1 & \text{if } q_j > 0 \end{cases}$$

Let $S = \lfloor 2^{-55} \tilde{R} \rfloor + \lfloor 2^{-58} \tilde{D} \rfloor + c$, where

$$c = \begin{cases} 0 \text{ if } \tilde{R}[54] = \tilde{D}[57] = 0 \\ 1 \text{ otherwise.} \end{cases}$$

Now

$$\begin{aligned} sum11 &= (RS11_{j-1} + \lfloor 2^{-57} \tilde{D} \rfloor + 1) \bmod 2^{11} \\ &= (2^{-54} \tilde{R} + \lfloor 2^{-57} \tilde{D} \rfloor + 1) \bmod 2^{11} \\ &= \left((2\lfloor 2^{-55} \tilde{R} \rfloor + \tilde{R}[54]) + (2\lfloor 2^{-58} \tilde{D} \rfloor + \tilde{D}[57]) + 1 \right) \bmod 2^{11} \end{aligned}$$

and therefore, by Lemma 1.21,

$$\begin{aligned} RS10_j &= sum11[10 : 1] \\ &= \left\lfloor \frac{1}{2} sum11 \right\rfloor \\ &= \left(\lfloor 2^{-55} \tilde{R} \rfloor + \lfloor 2^{-58} \tilde{D} \rfloor + \left\lfloor \frac{1}{2}(\tilde{R}[54] + \tilde{D}[57] + 1) \right\rfloor \right) \bmod 2^{10} \\ &= S \bmod 2^{10}. \end{aligned}$$

We shall invoke Lemma 2.37 with $X = 8R + D$, $Y = 2^{58} S$, and $n = 68$. Thus

$$X \equiv 2^{67} \cdot 8R_{j-1} - q_j div = 2^{67}(8R_{j-1} - q_j d) = 2^{67} R_j \pmod{2^{68}}$$

so that once again, $si(X \bmod 2^{68}, 68) = 2^{67} R_j$, and

$$\begin{aligned} si(Y \bmod 2^{68}, 68) &= si(2^{58} S \bmod 2^{68}, 68) \\ &= 2^{58} si(S \bmod 2^{10}, 10) \\ &= 2^{58} si(RS10_j, 10). \end{aligned}$$

Since

$$\tilde{R} = 2^{55} \lfloor 2^{-55} \tilde{R} \rfloor + \tilde{R}[54 : 0] = 2^{55} \lfloor 2^{-55} \tilde{R} \rfloor + 2^{54} \tilde{R}[54]$$

and

$$\tilde{D} = 2^{58} \lfloor 2^{-58} \tilde{D} \rfloor + \tilde{D}[57 : 0] = 2^{58} \lfloor 2^{-58} \tilde{D} \rfloor + 2^{57} \tilde{D}[57] + \tilde{D}[56 : 0],$$

$$8\tilde{R} + \tilde{D} - 2^{58} S = 2^{57}(\tilde{R}[54] + \tilde{D}[57] - 2c) + \tilde{D}[56 : 0],$$

and it is readily seen that $|8\tilde{R} + \tilde{D} - 2^{58}S| \le 2^{57}$. Thus

$$
\begin{aligned}
|X - Y| &= |8R + D - 2^{58}S| \\
&\le |8\tilde{R} + \tilde{D} - 2^{58}S| + 8|R - \tilde{R}| + |D - \tilde{D}| \\
&\le 2^{57} + 8(2^{54} - 1) + 1 \\
&< 2^{58},
\end{aligned}
$$

which implies

$$
|si(X \bmod 2^{68}, 68) - si(Y \bmod 2^{68}, 68)| = |2^{67}R_j - 2^{58}si(RS10_j, 10)| < 2^{58},
$$

and division by 2^{64} yields

$$
|8R_j - A_j| = \left|8R_j - \frac{1}{64}si(RS10_j, 10)\right| < \frac{1}{64}.
$$

(c) This follows from (a), (b), and Lemma 10.7.
(d) By induction and the definition of *nextRem*,

$$
RP_{j+1}[63 - p : 0] = RN_{j+1}[63 - p : 0] = 0;
$$

we must show that $2^{67}R_{j+1} \equiv RP_{j+1} - RN_{j+1} \pmod{2^{71}}$. We refer to the definition of *nextRem* and its local variables.

If $q_{j+1} = 0$, then $RP_{j+1} = RP8_{j+1}, RN_{j+1} = RN8_{j+1}, R_{j+1} = 8R_j$, and

$$
\begin{aligned}
RP_{j+1} - RN_{j+1} &\equiv RP8_{j+1} - RN8_{j+1} \\
&\equiv 8(RP_j - RN_j) \\
&\equiv 8 \cdot 2^{67}R_j \\
&\equiv 2^{67}R_{j+1} \pmod{2^{71}}.
\end{aligned}
$$

Thus, we may assume $q_{j+1} \ne 0$. Let

$$
\epsilon = \begin{cases} 0 \text{ if } q_{j+1} \le 0 \\ 1 \text{ if } q_{j+1} > 0. \end{cases}
$$

Since

$$
div[70 : 64 - p] = div[67 : 64 - p] = 2^{p+3}d,
$$

it is clear that for all values of q_{j+1},

$$
divMult_{j+1}[70 : 64 - p] \equiv -q_{j+1}div[70 : 64 - p] - \epsilon = -2^{p+3}q_{j+1}d - \epsilon \pmod{2^{p+7}}.
$$

Since $RP_j[63 - p : 0] = RN_j[63 - p : 0] = 0$, Lemma 1.20 implies

$$2^{64-p} \left((RP8_{j+1}[70 : 64 - p] - RN8_{j+1}[70 : 64 - p]) \bmod 2^{p+7} \right)$$

$$= \left(2^{64-p}(RP8_{j+1}[70 : 64 - p] - RN8_{j+1}[70 : 64 - p]) \right) \bmod 2^{71}$$

$$= (RP8_{j+1} - RN8_{j+1}) \bmod 2^{71}$$

$$= \left(8(RP_j - RN_j) \right) \bmod 2^{71}$$

$$= (2^{70} R_j) \bmod 2^{71}$$

$$= 2^{64-p} \left((2^{p+6} R_j) \bmod 2^{p+7} \right),$$

or

$$RP8_{j+1}[70 : 64 - p] - RN8_{j+1}[70 : 64 - p] \equiv 2^{p+6} R_j \pmod{2^{p+7}}.$$

By Lemma 8.7,

$$\tilde{sum}_{j+1}[70 : 64 - p] + 2car_{j+1}[70 : 64 - p]$$

$$= RP8_{j+1}[70 : 64 - p] + {\sim}RN8_{j+1}[70 : 64 - p] + divMult_{j+1}[70 : 64 - p]$$

$$\equiv RP8_{j+1}[70 : 64 - p] - RN8_{j+1}[70 : 64 - p] - 1 + divMult_{j+1}[70 : 64 - p]$$

$$\equiv 2^{p+6} R_j - 1 - 2^{p+3} q_{j+1} d - \epsilon$$

$$\equiv 2^{p+3}(8R_j - q_{j+1}d) - 1 - \epsilon$$

$$\equiv 2^{p+3} R_{j+1} - 1 - \epsilon \pmod{2^{p+7}}.$$

Now

$$(RP_{j+1} - RN_{j+1}) \bmod 2^{71}$$

$$= \left(2^{64-p}(RP_{j+1}[70 : 64 - p] - RN_{j+1}[70 : 64 - p]) \right) \bmod 2^{71}$$

$$= 2^{64-p} \left((RP_{j+1}[70 : 64 - p] - RN_{j+1}[70 : 64 - p]) \bmod 2^{p+7} \right),$$

where

$$RP_{j+1}[70 : 64 - p] = 2RP_{j+1}[70 : 65 - p] + RP_{j+1}[64 - p]$$

$$= 2car_{j+1}[69 : 64 - p] + \epsilon$$

$$= 2(car_{j+1}[70 : 64 - p] \bmod 2^{p+6}) + \epsilon$$

$$= (2car_{j+1}[70 : 64 - p]) \bmod 2^{p+7}) + \epsilon$$

and

$$-RN_{j+1}[70:64-p] = -sum_{j+1}[70:64-p] \equiv \tilde{\ }sum_{j+1}[70:64-p]+1 \pmod{2^{p+7}}.$$

Thus

$$(RP_{j+1} - RN_{j+1}) \bmod 2^{71}$$

$$= 2^{64-p}\left((2car_{j+1}[70:64-p]+\epsilon+\tilde{\ }sum_{j+1}[70:64-p]+1) \bmod 2^{p+7}\right)$$

$$= 2^{64-p}\left((2^{p+3}R_{j+1}-1-\epsilon+\epsilon+1) \bmod 2^{p+7}\right)$$

$$= 2^{64-p}\left((2^{p+3}R_{j+1}) \bmod 2^{p+7}\right)$$

$$= (2^{67}R_{j+1}) \bmod 2^{71}. \qquad \square$$

Lemma 18.7 *For all* j, $1 \le j \le N$, *quot$_j$ and quotM1$_j$ are* $(3j-1)$-*bit vectors,*

$$quot_j = 8^{j-1}Q_j,$$

and

$$quotM1_j = quot_j - 1.$$

Proof This is a simple consequence of (10.3), the definition of *nextQuot*, and induction. The claim holds trivially for $j = 1$. Assume that it holds for some $j \le 2N$. if $q_{j+1} \ge 0$, then

$$quot_{j+1} = 8quot_j + q_{j+1}$$

$$= 8^j Q_j + q_{j+1}$$

$$= 8^j(Q_j + 8^{-j}q_{j+1})$$

$$= 8^j Q_{j+1},$$

and if $q_{j+1} < 0$, then

$$quot_{j+1} = 8quotM1_j + q_{j+1} \bmod 8$$

$$= 8(quot_j - 1) + q_{j+1} + 8$$

$$= 8quot_j + q_{j+1}$$

$$= 8^j Q_{j+1}.$$

The computation of *quotM1$_{j+1}$* is similar. $\qquad \square$

On the final iteration, incremented versions of *quot* and *quotM1* are computed, to be used by the rounder:

Lemma 18.8 *Let* $inc = \begin{cases} 2 & \text{if } \mathit{fmt} = SP \\ 4 & \text{if } \mathit{fmt} \neq SP. \end{cases}$

(a) $quotP_N = quot_N + inc;$
(b) $quotM1P_N = quotM1_N + inc.$

Proof We refer to the function *incQuot*. Note that $lsbIs2 = 1 \Leftrightarrow \mathit{fmt} \neq SP$. The proof is based on Lemma 18.7 and the recurrence relation

$$Q_j = Q_{j-1} + 8^{1-j} q_j,$$

which follows from (10.3).

We present the proof for the case $\mathit{fmt} \neq SP$; the other case is similar. Suppose first that $q_N = 4$. Since *qLast*, *quotLast*, and *quotM1Last* are computed during iteration $2N$ after q_{2N} is derived, we have $qLast_N = q_{2N}$, $quotLast_N = quot_{N-2}$, and $quotM1Last_N = quotM1_{N-2}$. If $q_{2N} \geq -1$, then

$$\begin{aligned} quotP_N &= 64quot_{N-2} + 8(q_{2N} + 1)[2:0] \\ &= 8^2 quot_{N-2} + 8(q_{2N} + 1) \\ &= 8^2 8^{2N-2} Q_{N-2} + 8q_{2N} + q_N + inc \\ &= 8^{2N}(Q_{N-2} + 8^{-2N} q_{2N} + 8^{-(N)} q_N) + inc \\ &= 8^{2N} Q_N + inc \\ &= quot_N + inc, \end{aligned}$$

and if $q_{2N} < -1$, then

$$\begin{aligned} quotP_N &= 64quotM1_{N-2} + 8(q_{2N} + 1)[2:0] \\ &= 8^2(quot_{N-2} - 1) + 8(q_{2N} + 1 + 8) \\ &= 8^2 quot_{N-2} + 8(q_{2N} + 1) \\ &= quot_N + inc. \end{aligned}$$

If $q_{2N} \geq 0$, then

$$\begin{aligned} quotM1P_N &= 64quot_{N-2} + 8q_{2N}[2:0] + 7 \\ &= 64quot_{N-2} + 8q_{2N} + 7 \\ &= quotP_{N-2} - 1 \\ &= quot_N + inc - 1 \\ &= quotM1_N + inc, \end{aligned}$$

and if $q_{2N} < 0$, then

$$
\begin{aligned}
quotM1P_N &= 64quotM1_{N-2} + 8q_{N-1}[2:0] + 7 \\
&= 64(quot_{N-2} - 1) + 8(q_{N-1} + 8) + 7 \\
&= 64quot_{N-2} + 8q_{N-1} + 7 \\
&= quotP_{N-2} - 1 \\
&= quotM1_N + inc.
\end{aligned}
$$

Now suppose $q_N < 4$. Then

$$
\begin{aligned}
quotP_N &= 8quot_{N-1} + q_N + 4 \\
&= 8^{N-1}Q_{N-1} + q_N + 4 \\
&= 8^{N-1}(Q_{N-1} + 8^{-(N)}q_N) + 4 \\
&= 8^{N-1}Q_N + 4 \\
&= quot_N + inc.
\end{aligned}
$$

If $q_N = -4$, then

$$
\begin{aligned}
quotM1P_N &= 8quotM1_{N-1} + (q_N + 3) \bmod 8 \\
&= 8(quot_{N-1} - 1) + 7 \\
&= 8quot_{N-1} + q_N - 1 + 4 \\
&= quot_N - 1 + 4 \\
&= quotM1_N + inc,
\end{aligned}
$$

and otherwise,

$$
\begin{aligned}
quotM1P_N &= 8quot_{N-1} + q_N + 3 \\
&= quot_N - 1 + 4 \\
&= quotM1_N + inc. \qquad \square
\end{aligned}
$$

18.4   Post-processing and Rounding

Notation In discussing the final loop variable values RP_N, RN_N, $quot_N$, $quotM1_N$, $quotP_N$, and $quotM1P_N$, the subscript may be omitted. We shall similarly abbreviate the final quotient Q_N and remainder R_N as Q and R.

We have the following error bound for the final quotient:

Lemma 18.9 $\left| Q - \frac{x}{d} \right| \le \frac{4}{7} \cdot 2^{-6C}$.

Proof This is an immediate consequence of (10.4) and Lemma 18.6(c) with $j = N = 2C + 1$.   □

The rounder selects one of two values, the truncated quotient $Qtrunc$ or the incremented truncated quotient $Qinc$. The choice depends on the variable stk, which indicates inexactness.

In the trivial case of division by a power of 2, the quotient is exact and $Qtrunc$ is selected:

Lemma 18.10 *Assume divPow2* $= 1$.

(a) $2^p \cdot \frac{x}{d} \in \mathbb{Z}$;
(b) $Qtrunc \equiv 2^p \cdot \frac{x}{d} \pmod{2^p}$;
(c) $stk = 0$.

Proof In this case, $sig(B) = 1$ and by Definitions (18.1) and (18.2),

$$2^p \cdot \frac{x}{d} = 2^p sig(A) \in \mathbb{Z}.$$

By Lemma 5.1(c),

$$sig(A) = 1 + 2^{1-p} mana,$$

and hence,

$$2^p \cdot \frac{x}{d} = 2^p sig(A) = 2^p + 2mana = 2^p + Qtrunc \equiv Qtrunc \pmod{2^p}.$$

By definition, $stk = 0$.   □

In the usual case $divPow2 = 0$, $Qtrunc$ and $Qinc$ are computed by the function $computeQ$.

Lemma 18.11

(a) $remZero = 1 \Leftrightarrow R = 0$;
(b) $remSign = 1 \Leftrightarrow R < 0$;

Proof By Lemma 18.6(d), $rem = 2^{67} R \bmod 2^{71}$. The first claim follows from Lemma 18.6(c), and by Lemma 2.32, $2^{67} R = si(rem, 71)$, which implies the second.   □

Lemma 18.12 *Assume divPow2* $= 0$.

(a) $Qtrunc \equiv \left\lfloor 2^p \cdot \frac{x}{d} \right\rfloor \pmod{2^p}$;
(b) $Qinc \equiv \left\lfloor 2^p \cdot \frac{x}{d} \right\rfloor + 2 \pmod{2^p}$.

Proof According to (10.3), $2^{6C} Q = 8^{2C} Q \in \mathbb{Z}$. If *remSign* $= 0$, then by Lemmas 18.9 and 18.11,

$$Q \le \frac{x}{d} < Q + 2^{-6C},$$

or

$$2^{6C} Q \le 2^{6C} \frac{x}{d} < 2^{6C} Q + 1,$$

which implies

$$\left\lfloor 2^{6C} \frac{x}{d} \right\rfloor = 2^{6C} Q.$$

Let *inc* be defined as in Lemma 18.8. Then

$$quotLo = quot = 2^{6C} Q = \left\lfloor 2^{6C} \frac{x}{d} \right\rfloor$$

and

$$quotLoP = quotP = quot + inc = \left\lfloor 2^{6C} \frac{x}{d} \right\rfloor + inc.$$

But if *remSign* $= 1$, then we have the same expressions for *quotLo* and *quotLoP*:

$$Q - 2^{-6C} < \frac{x}{d} < Q,$$

$$2^{6C} Q - 1 < 2^{6C} \frac{x}{d} < 2^{6C} Q,$$

$$\left\lfloor 2^{6C} \frac{x}{d} \right\rfloor = 2^{6C} Q - 1,$$

$$quotLo = quotM1 = quot - 1 = 2^{6C} Q - 1 = \left\lfloor 2^{6C} \frac{x}{d} \right\rfloor,$$

and

$$quotLoP = quotM1P = quotM1 + inc = quot - 1 + inc = \left\lfloor 2^{6C} \frac{x}{d} \right\rfloor + inc.$$

Suppose *lsbIs2* $= 0$. Then $inc = 2, fmt = SP, 6C = 24 = p,$

$$Qtrunc = quotLo \bmod 2^{53} = \left\lfloor 2^p \frac{x}{d} \right\rfloor \bmod 2^{53},$$

and

$$Qinc = quotLoP \bmod 2^{53} = \left(\left\lfloor 2^p \frac{x}{d} \right\rfloor + 2 \right) \bmod 2^{53}.$$

On the other hand, if $lsbIs2 = 1$, then $inc = 4$, $6C = p + 1$,

$$Qtrunc = \left\lfloor \frac{quotLo}{2} \right\rfloor \bmod 2^{53} = \left\lfloor \frac{\left\lfloor 2^{p+1} \frac{x}{d} \right\rfloor}{2} \right\rfloor \bmod 2^{53} = \left\lfloor 2^p \frac{x}{d} \right\rfloor \bmod 2^{53},$$

and

$$Qinc = \left\lfloor \frac{quotLoP}{2} \right\rfloor \bmod 2^{53} = \left\lfloor \frac{\left\lfloor 2^{p+1} \frac{x}{d} \right\rfloor + 4}{2} \right\rfloor \bmod 2^{53}$$

$$= \left(\left\lfloor 2^p \frac{x}{d} \right\rfloor + 2 \right) \bmod 2^{53}. \qquad \square$$

Inexactness is also determined by *computeQ*.

Lemma 18.13 $stk = 0 \Leftrightarrow 2^p \cdot \frac{x}{d} \in \mathbb{Z}$.

Proof We may assume $divPow2 = 0$. Lemma 18.11, $remZero = 1 \Leftrightarrow R = 0$. By Lemma 18.11, if $remSign = 0$, then since

$$\left\lfloor 2^{6C} \frac{x}{d} \right\rfloor = 2^{6C} Q,$$

this condition is equivalent to $2^{6C} \frac{x}{d} \in \mathbb{Z}$. On the other hand, if $remSign = 1$, then $R < 0$, which implies $remZero = 0$, and since

$$2^{6C} Q - 1 < 2^{6C} \frac{x}{d} < 2^{6C} Q,$$

$2^{6C} \frac{x}{d} \notin \mathbb{Z}$. Thus, in general,

$$remZero = 1 \Leftrightarrow 2^{6C} \frac{x}{d} \in \mathbb{Z}.$$

If $lsbIs2 = 0$, then $6C = p$ and

$$stk = 0 \Leftrightarrow remSign = 1 \Leftrightarrow 2^{6C} \frac{x}{d} = 2^p \cdot \frac{x}{d} \in \mathbb{Z}.$$

On the other hand, if $lsbIs2 = 1$, then $6C = p + 1$,

$$quotLo = \left\lfloor 2^{p+1}\frac{x}{d} \right\rfloor,$$

and

$$stk = 0 \Leftrightarrow remZero = 1 \text{ and } rootLo \text{ is even}$$

$$\Leftrightarrow 2^{p+1}\frac{x}{d} \in Z \text{ and } 2^{p+1}\frac{x}{d} \text{ is even}$$

$$\Leftrightarrow 2^{p}\frac{x}{d} \in Z. \qquad\qquad \square$$

Since the rounder operates on the absolute value of the quotient, the rounding mode $\mathcal{R}$ must be adjusted. We define

$$\mathcal{R}' = \begin{cases} RDN & \text{if } \mathcal{R} = RUP \text{ and } \frac{A}{B} < 0 \\ RUP & \text{if } \mathcal{R} = RDN \text{ and } \frac{A}{B} < 0 \\ \mathcal{R} & \text{otherwise.} \end{cases}$$

Lemma 18.14

(a) $\frac{x}{d}$ *is p-exact* $\Leftrightarrow inx = 0$;
(b) $\mathcal{R}'\left(2^{p}\frac{x}{d}, p\right) \equiv 2Qrnd[p-2 : 0] \pmod{2^{p}}$.

Proof We instantiate Lemma 6.97 with $z = 2^{p}\frac{x}{d}$ and $n = e = expo(z) = p$. Note that by Lemma 18.12, $Qtrunc[p-1 : 0] = \lfloor 2^{p}\frac{x}{d} \rfloor[p-1 : 0]$.

(a) By Lemma 18.13, $stk = 0 \Leftrightarrow 2^{p}\frac{x}{d} \in \mathbb{Z}$. Thus, by Lemma 6.97(a) and the definition of *rounder*,

$$2^{p}\frac{x}{d} \text{ is p-exact} \Leftrightarrow Qtrunc[0] = \left\lfloor 2^{p}\frac{x}{d} \right\rfloor[0] = stk = 0 \Leftrightarrow inx = 0.$$

(b) First suppose that none of the conditions listed in Lemma 6.97(b) holds. It is clear from the definition of *rounder* that in this case, $Qrnd = Qtrunc[53 : 1]$. Thus

$$\mathcal{R}'\left(2^{p}\frac{x}{d}, p\right) = RTZ\left(\left\lfloor 2^{p}\frac{x}{d} \right\rfloor, p\right)$$

$$= 2\left\lfloor 2^{p}\frac{x}{d} \right\rfloor[p : 1]$$

$$\equiv 2\left\lfloor 2^{p}\frac{x}{d} \right\rfloor[p-1 : 1]$$

$$= 2Qtrunc[p-1 : 1]$$

$$= 2Qrnd[p-2 : 0] \pmod{2^{p}}.$$

In the remaining case, $Qrnd = Qinc[53 : 1]$. Furthermore, either $stk = 1$ or $Qtrunc[0] = 1$, which implies $divPow2 = 0$, and therefore Lemma 18.12 applies. Thus

$$\mathcal{R}' \left(2^p \frac{x}{d}, p \right) = fp^+ \left(RTZ \left(\left\lfloor 2^p \frac{x}{d} \right\rfloor, p \right), p \right)$$

$$= RTZ \left(\left\lfloor 2^p \frac{x}{d} \right\rfloor, p \right) + 2$$

$$= 2 \left\lfloor 2^p \frac{x}{d} \right\rfloor [p : 1] + 2$$

$$\equiv \left\lfloor 2^p \frac{x}{d} \right\rfloor - \left\lfloor 2^p \frac{x}{d} \right\rfloor [0] + 2$$

$$\equiv Qinc - Qinc[0]$$

$$\equiv 2Qinc[p-1 : 1]$$

$$= 2Qrnd[p-2 : 0] \pmod{2^p}. \qquad \square$$

The design is simplified by the observation that the quotient is never rounded up to a power of 2:

Lemma 18.15 $\mathcal{R}' \left(2^p \frac{x}{d}, p \right) \leq 2^{p+1} - 2.$

Proof It will suffice to show that $2^p \frac{x}{d} \leq 2^{p+1} - 2$, or $\frac{x}{d} \leq 2 - 2^{1-p}$.
   If $sig(A) \geq sig(B)$, then by Eqs. (18.1) and (18.2),

$$\frac{x}{d} = \frac{sig(A)}{sig(B)} \leq \frac{2 - 2^{1-p}}{1} = 2 - 2^{1-p}.$$

If $sig(A) < sig(B)$, then

$$\frac{x}{d} = \frac{2sig(A)}{sig(B)} \leq \frac{2(sig(B) - 2^{1-p})}{sig(B)} = 2 - \frac{2^{2-p}}{sig(B)} < 2 - 2^{1-p}. \qquad \square$$

We consider the normal and subnormal cases separately:

Lemma 18.16 If $\left| \frac{A}{B} \right| \geq spn(fmt)$, then

(a) $\mathcal{R}' \left(\left| \frac{A}{B} \right|, p \right) = \left| \frac{A}{B} \right| \Leftrightarrow inx = 0$;
(b) $\mathcal{R}' \left(\left| \frac{A}{B} \right|, p \right) = 2^{expQ - bias - (p-1)}(2^{p-1} + Qrnd[p-2 : 0])$.

Proof
   (a) This follows from Lemma 18.14(a)

$$\mathcal{R}' \left(\left| \frac{A}{B} \right|, p \right) = \left| \frac{A}{B} \right| \Leftrightarrow \left| \frac{A}{B} \right| \text{ is } p\text{-exact} \Leftrightarrow \frac{x}{d} \text{ is } p\text{-exact} \Leftrightarrow inx = 0.$$

(b) By Lemma 18.15, $expo\left(\mathcal{R}'\left(2^{p}\frac{x}{d}, p\right)\right) = p$, and by Lemma 18.14(b),

$$\mathcal{R}'\left(2^{p}\frac{x}{d}, p\right) = 2^{p} + \mathcal{R}'\left(2^{p}\frac{x}{d}, p\right) \bmod 2^{p} = 2^{p} + 2Qrnd[p-2:0]$$

and by Lemma 18.2(c),

$$\mathcal{R}'\left(\left|\frac{A}{B}\right|, p\right) = 2^{expQ-bias-p}\mathcal{R}'\left(2^{p}\frac{x}{d}, p\right)$$

$$= 2^{expQ-bias-(p-1)}(2^{p-1} + Qrnd[p-2:0]). \qquad \square$$

For the subnormal case, we appeal to Lemmas 6.97 and 6.101:

Lemma 18.17 *If* $\left|\frac{A}{B}\right| < spn(fmt)$, *then*

(a) $drnd\left(\left|\frac{A}{B}\right|, \mathcal{R}', fmt\right) = \left|\frac{A}{B}\right| \Leftrightarrow inxDen = 0$;
(b) $drnd\left(\left|\frac{A}{B}\right|, \mathcal{R}', fmt\right) = 2^{1-bias-(p-1)}QrndDen[p-1:0]$.

Proof According to the definition of *rounder*,

$$QDen = 2^{p} + \left\lfloor 2^{p}\frac{x}{d}\right\rfloor \bmod 2^{p} = \left\lfloor 2^{p}\frac{x}{d}\right\rfloor,$$

$$Qshft = \lfloor 2^{-shft}QDen\rfloor = \left\lfloor 2^{p-shft}\frac{x}{d}\right\rfloor,$$

and

$$stkDen = 0 \Leftrightarrow Qshft = 2^{-shft}QDen \text{ and } QDen = 2^{p}\frac{x}{d}$$

$$\Leftrightarrow Qshft = 2^{p-shft}\frac{x}{d}.$$

By Lemma 18.2(c),

$$expo\left(\frac{A}{B}\right) = expQ - bias < expo(spn(fmt)) = 1 - bias,$$

and hence $expQ \le 0$.

Case 1: $expQ > 1 - p$.
   In this case, $shft = 1 - expQ < p$. We shall invoke Lemma 6.97 with

$$n = p - shft = p + expQ - 1 > 0$$

and

$$z = 2^{p-shft}\frac{x}{d} = 2^{p-(1-expQ)}\frac{x}{d} = 2^{n}\frac{x}{d}.$$

Note that $Qshft = \lfloor z \rfloor$. By Definition 6.9,

$$drnd\left(\left|\frac{A}{B}\right|, \mathcal{R}', fmt\right) = \mathcal{R}'\left(\left|\frac{A}{B}\right|, p + expo\left(\frac{A}{B}\right) - expo(spn(fmt))\right)$$

$$= \mathcal{R}'\left(\left|\frac{A}{B}\right|, p + (expQ - bias) - (1 - bias)\right)$$

$$= \mathcal{R}'\left(\left|\frac{A}{B}\right|, n\right).$$

(a) This now follows from Lemmas 6.97(a) and 18.2(c)

$$drnd\left(\left|\frac{A}{B}\right|, \mathcal{R}', fmt\right) = \left|\frac{A}{B}\right| \Leftrightarrow \mathcal{R}'\left(\left|\frac{A}{B}\right|, n\right) = \left|\frac{A}{B}\right|$$

$$\Leftrightarrow 2^n \frac{x}{d} \in \mathbb{Z} \text{ and } Qshft[0] = \left\lfloor 2^n \frac{x}{d} \right\rfloor [0] = 0$$

$$\Leftrightarrow stkDen = grdDen = 0$$

$$\Leftrightarrow inxDen = 0.$$

(b) If none of the conditions listed in Lemma 6.97(b) holds, then it is clear from the definition of *rounder* that $QrndDen = Qshft[53 : 1]$, and therefore

$$\mathcal{R}'\left(2^n \frac{x}{d}, n\right) = RTZ(Qshft, n) = 2Qshft[n : 1] = 2QrndDen[p-1 : 0].$$

Otherwise, $QrndDen = Qshft[53 : 1] + 1$ and

$$\mathcal{R}'\left(2^n \frac{x}{d}, n\right) = fp^+(RTZ(Qshft, n), n)$$

$$= RTZ(Qshft, n) + 2$$

$$= 2(Qshft[n : 1] + 1)$$

$$= 2QrndDen[p-1 : 0].$$

Thus, by Lemma 18.2(c),

$$\mathcal{R}'\left(\left|\frac{A}{B}\right|, n\right) = 2^{expQ-bias-n}\mathcal{R}'\left(2^n \frac{x}{d}, n\right)$$

$$= 2^{expQ-bias-(p+expQ-1)}2QrndDen[p-1 : 0]$$

$$= 2^{1-bias-(p-1)}QrndDen[p-1 : 0].$$

Case 2: $expQ \leq 1 - p$.

In this case, $p - shft \leq 0$, and therefore $Qshft < 2$. By Lemma 18.2(c),

$$\left|\frac{A}{B}\right| < 2^{expQ-bias+1} \leq 2^{2-p-bias} = spd(fmt).$$

(a) Either $drnd\left(\left|\frac{A}{B}\right|, \mathcal{R}', fmt\right) \geq spd(fmt)$ or $drnd\left(\left|\frac{A}{B}\right|, \mathcal{R}', fmt\right) = 0$. Thus

$$drnd\left(\left|\frac{A}{B}\right|, \mathcal{R}', fmt\right) \neq \left|\frac{A}{B}\right|$$

and we must show $inxDen = 1$.

If $inxDen = 0$, then $stkDen = Qshft[0] = 0$. But if $stkDen = 0$, then as noted above, $Qshft = 2^{p-shft}\frac{x}{d} \neq 0$, which implies $Qshft = Qshft[0] = 1$.

(b) Since $Qshft \leq 1$, $lsbDen = Qshft[1] = 0$ and

$$grdDen = Qshft[0] = 1 \Leftrightarrow Qshft = \left\lfloor 2^{p-shft}\frac{x}{d} \right\rfloor \geq 1$$

$$\Leftrightarrow p - shft = 0$$

$$\Leftrightarrow \left|\frac{A}{B}\right| \geq \frac{1}{2}spd(fmt).$$

We invoke Lemma 6.101, considering the three cases of the lemma separately.

Suppose, for example, $\left|\frac{A}{B}\right| > \frac{1}{2}spd(fmt)$. Then $grdDen = stkDen = 1$. If $\mathcal{R}' = RNE$ or $\mathcal{R}' = RUP$, then by Lemma 6.101(c), $drnd\left(\left|\frac{A}{B}\right|, \mathcal{R}', fmt\right) = spd(fmt)$, and according to the definition of *round*, $QrndDen = Qshft[53 : 1] + 1 = 1$, which implies

$$2^{1-bias-(p-1)}QrndDen[p-1 : 0] = 2^{1-bias-(p-1)} = spd(fmt)$$

as well. But if $\mathcal{R}' = RTZ$ or $\mathcal{R}' = RDN$, then $drnd\left(\left|\frac{A}{B}\right|, \mathcal{R}', fmt\right) = 0$, $QrndDen = Qshft[53 : 1] = 0$, and the claim again holds.

The other two cases are similar.   □

Our correctness theorem for division is similar to that of multiplication (Theorem 16.1), matching the behavior of the top-level function *fdiv8* with the specification function *arm-binary-spec*. The proof is an extensive but entirely straightforward case analysis involving nothing more than inspection of the two functions and Lemmas 18.16 and 18.17.

Theorem 18.1 *Let*

$$\langle D_{spec}, R_{spec} \rangle = arm\text{-}binary\text{-}spec(DIV, opa, opb, R_{in}, fmt),$$

where fmt $\in \{DP, SP, HP\}$, *opa and opb are 64-bit vectors and* R_{in} *is a 32-bit vector, and let*

$$\langle D, flags \rangle = fdiv8(opa, opb, fmt, R_{in}[24], R_{in}[25], R_{in}[23:22]).$$

Then $D = D_{spec}$ *and* $R_{in} \mid flags = R_{spec}$.

Chapter 19
64-Bit Integer Division

The function `idiv8`, available online https://go.sn.pub/idiv8, performs 64-bit and 32-bit signed and unsigned integer division. While this function is based on the same SRT algorithm and has much in common with the corresponding floating-point function of Chap. 18 (reflecting the sharing of hardware), we note several differences:

(1) The analysis of operands is generally simpler in the integer case, except that a sophisticated algorithm is used in the detection of division by a power of 2.
(2) The number of bits of the quotient to be computed during the iterative phase is the difference in leading zero counts of the operands, resulting in a variable latency that is unknown until after normalization. In the 64-bit case, this difference may be as large as 62, requiring 11 iterative cycles (again, 6 bits per cycle), as compared to 9 cycles for a double-precision floating-point division.
(3) The integer quotient is returned in two's complement rather than sign-magnitude form. Therefore, in the case of opposite-signed operands, each quotient digit must be negated for the purpose of on-the-fly computation of the quotient and decremented quotient.
(4) Post-processing is much simpler in the integer case: the quotient is simply rounded toward 0, with no exceptional conditions to be noted.

In spite of these differences, the same essential computations are performed in the main iterative phase, and consequently (as noted in Sect. 19.3) many of the results of Chap. 18 remain valid.

19.1 Preliminary Analysis and Early Exit

The input parameters of the top-level function `idiv8` are as follows:

© The Author(s), under exclusive license to Springer Nature Switzerland AG 2022
D. M. Russinoff, *Formal Verification of Floating-Point Hardware Design*,
https://doi.org/10.1007/978-3-030-87181-9_19

- $ui64$ opa, opb: In the 64-bit case, the integer dividend and divisor are *opa* and *opb*, respectively; in the 32-bit case, they are *opa*[63 : 32] and *opb*[63 : 32].
- bool int32: An indication of 32-bit division.
- bool sgnd: An indication of signed division, i.e., that the operands are to be interpreted as signed rather than unsigned integers.

A single result is returned:

- $ui64$ D: The integer quotient, computed by rounding the rational quotient toward 0, is represented by D or $D[31 : 0]$, depending on *int32*, interpreted as a signed or unsigned integer according to *sgnd*.

Thus, the integer dividend and divisor represented by the operands are

$$
A = \begin{cases} opa & \text{if } int32 = 0 \text{ and } sgnd = 0 \\ si(opa, 64) & \text{if } int32 = 0 \text{ and } sgnd = 1 \\ opa[63 : 32] & \text{if } int32 = 1 \text{ and } sgnd = 0 \\ si(opa[63 : 32], 32) & \text{if } int32 = 1 \text{ and } sgnd = 1 \end{cases} \tag{19.1}
$$

and

$$
B = \begin{cases} opb & \text{if } int32 = 0 \text{ and } sgnd = 0 \\ si(opb, 64) & \text{if } int32 = 0 \text{ and } sgnd = 1 \\ opb[63 : 32] & \text{if } int32 = 1 \text{ and } sgnd = 0 \\ si(opb[63 : 32], 32) & \text{if } int32 = 1 \text{ and } sgnd = 1. \end{cases} \tag{19.2}
$$

In the exceptional case $B = 0$, the value returned is easily seen to be $D = 0$, which is the architecturally prescribed result. Henceforth, we shall assume that $B \neq 0$. Clearly, the signs of A, B, and A/B are represented by the local variables *sgnA*, *sgnB*, and *sgnQ*, respectively. Our objective is to compute the result of truncating A/B to an integer, which may be expressed as

$$
I = (-1)^{sgnQ} \left\lfloor \left| \frac{A}{B} \right| \right\rfloor \begin{cases} \left\lfloor \frac{A}{B} \right\rfloor & \text{if } sgnQ = 0 \\ \left\lceil \frac{A}{B} \right\rceil & \text{if } sgnQ = 1. \end{cases} \tag{19.3}
$$

Thus, we would like to show that the data result is the signed or unsigned integer encoding of I, i.e.

$$
I = \begin{cases} D & \text{if } int32 = 0 \text{ and } sgnd = 0 \\ si(D, 64) & \text{if } int32 = 0 \text{ and } sgnd = 1 \\ D[32 : 0] & \text{if } int32 = 1 \text{ and } sgnd = 0 \\ si(D[32 : 0], 32) & \text{if } int32 = 1 \text{ and } sgnd = 1. \end{cases} \tag{19.4}
$$

When $sgnd = 1$, according to Lemma 2.32, Eq. (19.4) is equivalent to

$$D = I[63 : 0] \tag{19.5}$$

if $int32 = 0$ and

$$D[32 : 0] = I[32 : 0] \tag{19.6}$$

if $int32 = 1$, and clearly, the same is true when $sgnd = 0$.

There is one case of signed division, however, in which I is not representable: if A is the minimal representable integer (-2^{63} if $int32 = 0$, -2^{31} if $int32 = 1$) and $B = -1$, then the quotient $I = A/B = |A|$ lies above the representable range. In this case of "overflow", the prescribed result is the encoding of $A = -I$ instead of I. But this is consistent with (19.5) and (19.6). Also note that (19.6) is a trivial consequence of (19.5). In fact, for all formats, the implementation is designed to satisfy (19.5), which will therefore be our objective in all cases other than $B = 0$.

In the remainder of this section, we examine those cases that do not require significant computation. The case $I = 0$ is detected by the function $compareOps$:

Lemma 19.1 $BgtA = 1 \Leftrightarrow |B| > |A|$.

Proof First suppose $int32 = 0$. The proof is a case analysis based on the signs of the operands. In each case, we shall show that

$$diff = 2^{64} + |A| - |B|,$$

so that

$$BgtA = 1 \Leftrightarrow diff[64] = 0 \Leftrightarrow diff < 2^{64} \Leftrightarrow |B| > |A|. \tag{19.7}$$

We first consider the case $A < 0$ and $B \geq 0$. We have

$$argA = sum =\sim opa[63 : 0] \mathbin{\char`\^} \sim opb[63 : 0]$$

and

$$argB = car = 2(\sim opa[63 : 0] \mathbin{\&} \sim opb[63 : 0])[62 : 0] + 1,$$

where

$$\sim opa[63 : 0] = 2^{64} - opa - 1 = 2^{64} - (2^{64} - |A|) - 1 = |A| - 1$$

and

$$\sim opb[63 : 0] = 2^{64} - opb - 1 = 2^{64} - |B| - 1.$$

Since $opa[63] = sgna = 1$, $\sim opa[63] = 0$, and the second equation above may be written as

$$argB = 2(\sim opa[63:0] \;\&\; \sim opb[63:0]) + 1.$$

By Lemma 8.2,

$$\begin{aligned} argA + argB &= \sim opa[63:0] + \sim opb[63:0] + 1 \\ &= (|A| - 1) + (2^{64} - |B| - 1) + 1 \\ &= 2^{64} + |A| - |B| - 1, \end{aligned}$$

and hence

$$diff = argA + argB + 1 = 2^{64} + |A| - |B|.$$

In all remaining cases,

$$argA = \begin{cases} \sim opa[63:0] = |A| - 1 & \text{if } A < 0 \\ opa = |A| & \text{if } A > 0 \end{cases}$$

and

$$argB = \begin{cases} opb = 2^{64} - |B| & \text{if } B < 0 \\ \sim opb[63:0] = 2^{64} - |B| - 1 & \text{if } B > 0 \end{cases}$$

Thus, if $A > 0$ and $B > 0$, then

$$diff = argA + argB + 1 = |A| + (2^{64} - |B| - 1) + 1 = 2^{64} + |A| - |B|;$$

if $A > 0$ and $B < 0$, then

$$diff = argA + argB + 0 = |A| + (2^{64} - |B|) = 2^{64} + |A| - |B|;$$

and if $A < 0$ and $B < 0$, then

$$diff = argA + argB + 1 = (|A| - 1) + (2^{64} - |B|) + 1 = 2^{64} + |A| - |B|.$$

In the 32-bit case, a similar argument with opa, opb, $argA$, $argB$, $diff$, and 2^{64} replaced with $opa[63:32]$, $opb[63:32]$, $argA[63:32]$, $argB[63:32]$, $diff[64:32]$, and 2^{32} leads to

$$diff = \begin{cases} 2^{64} + 2^{32}(|A| - |B|) & \text{if } cin = 1 \\ 2^{64} + 2^{32}(|A| - |B|) + (2^{32} - 1) & \text{if } cin = 0 \end{cases}$$

and the equivalence (19.7) again follows. □

The desired Eq. (19.5) now follows easily:

Corollary 19.2 *If* $|B| > |A|$, *then* $D = 0 = I$.

The rest of the computation is based on *masked* versions of the operands,

$$mskA = \begin{cases} opa & \text{if } int32 = 0 \\ 2^{32}opa[63:32] & \text{if } int32 = 1 \end{cases}$$

and

$$mskB = \begin{cases} opb & \text{if } int32 = 0 \\ 2^{32}opb[63:32] & \text{if } int32 = 1. \end{cases}$$

Let

$$\hat{A} = \begin{cases} mskA & \text{if } sgnd = 0 \\ si(mskA, 64) & \text{if } sgnd = 1 \end{cases}$$

and

$$\hat{B} = \begin{cases} mskB & \text{if } sgnd = 0 \\ si(mskB, 64) & \text{if } sgnd = 1. \end{cases}$$

In the 64-bit case, $\hat{A} = A$ and $\hat{B} = B$, and in the 32-bit case, it follows from Lemma 2.32 that $\hat{A} = 2^{32}A$ and $\hat{B} = 2^{32}B$. Thus, in either case,

$$\frac{\hat{A}}{\hat{B}} = \frac{A}{B}.$$

The variables *mskA*, *negA*, *absA*, *mskB*, *negB*, and *absB* are conveniently expressed in terms of $|\hat{A}|$ and $|\hat{B}|$. The following expressions are direct consequences of the definitions:

Lemma 19.3 *Assume* $A \neq 0$ *and* $B \neq 0$.

(a) $mskA = \begin{cases} |\hat{A}| & \text{if } A \geq 0 \\ 2^{64} - |\hat{A}| & \text{if } A < 0 \end{cases}$ *and* $mskB = \begin{cases} |\hat{B}| & \text{if } B \geq 0 \\ 2^{64} - |\hat{B}| & \text{if } B < 0; \end{cases}$

(b) $negA = 2^{64} - mskA$ *and* $negB = 2^{64} - mskB$;

(c) $absA = |\hat{A}|$ *and* $absB = |\hat{B}|$.

The case of division by a power of 2 is detected by the function *isPow2*. The proof of the following is nontrivial and is deferred to Sect. 19.2:

Lemma 19.4 *If $B \neq 0$, then*

$$isPow2(mskB, sgnB) = 1 \Leftrightarrow |\hat{B}| \text{ is a power of } 2.$$

We shall also require the following characterization of the values computed by *CLZ64*:

Lemma 19.5 *If $A \neq 0$ and $B \neq 0$, then*

$$clzA = 63 - expo(\hat{A})$$

and

$$clzB = 63 - expo(\hat{B}).$$

Proof This is guaranteed by Lemma 8.29.   □

Pending the proof of Lemma 19.4, we may now establish the following:

Lemma 19.6 *If $|B|$ is a power of 2, then $D = I[63 : 0]$.*

Proof If $|B|$ is a power of 2, then so is $|\hat{B}|$, and in this case,

$$D = divPow2(arg, sgnQ, shft),$$

where

$$arg = \begin{cases} negA & \text{if } B < 0 \\ mskA & \text{if } B > 0 \end{cases}$$

and

$$shft = 2^6 - clzB - 1 = 64 - (63 - expo(\hat{B})) - 1 = expo(\hat{B}).$$

Suppose $sgnQ = 0$. Then $A > 0 \Leftrightarrow B > 0$. It follows from Lemma 19.3 that $arg = |\hat{A}|$ and

$$D = \lfloor 2^{-shft} arg \rfloor = \lfloor 2^{-expo(\hat{B})} |\hat{A}| \rfloor = \left\lfloor \left| \frac{\hat{A}}{\hat{B}} \right| \right\rfloor = \left\lfloor \frac{A}{B} \right\rfloor = \left\lfloor \frac{A}{B} \right\rfloor [63 : 0] = I[63 : 0].$$

Now suppose $sgnQ = 1$. Then $A > 0 \Leftrightarrow B < 0$, and $arg = 2^{64} - |\hat{A}|$. Referring to the local variables of *divPow2*, we have

$$padA = si(2^{64} arg, 128)$$

$$= si(2^{64}(2^{64} - |\hat{A}|), 128)$$

$$= si(2^{128} - 2^{64}|\hat{A}|, 128)$$

$$= -2^{64}|\hat{A}|$$

and

$$shftA = \lfloor 2^{-shft} padA \rfloor$$

$$= \lfloor -2^{64-expo(\hat{B})}|\hat{A}| \rfloor$$

$$= -2^{64-expo(\hat{B})}|\hat{A}|$$

$$= -2^{64} \left| \frac{\hat{A}}{\hat{B}} \right|$$

$$= 2^{64} \frac{A}{B}.$$

Thus

$$shftA[63:0] = 0 \Leftrightarrow 2^{-64} shftA \in \mathbb{Z} \Leftrightarrow \frac{A}{B} \in \mathbb{Z}$$

and

$$shftA[127:64] = \lfloor 2^{-64} shftA \rfloor [63:0] = \left\lfloor \frac{A}{B} \right\rfloor [63:0].$$

If $\frac{A}{B} \in \mathbb{Z}$, then

$$D = shftA[127:64] = \left\lfloor \frac{A}{B} \right\rfloor [63:0] = \left\lceil \frac{A}{B} \right\rceil [63:0] = I[63:0],$$

and if $\frac{A}{B} \notin \mathbb{Z}$, then

$$D = (shftA[127:64] + 1)[63:0]$$

$$= \left(\left\lfloor \frac{A}{B} \right\rfloor [63:0] + 1 \right) [63:0]$$

$$= \left(\left\lfloor \frac{A}{B} \right\rfloor + 1 \right) [63:0]$$

$$= \left\lceil \frac{A}{B} \right\rceil [63:0]$$

$$= I[63:0]. \qquad \square$$

One additional case is handled by preliminary analysis:

Lemma 19.7 *If* $|A| \geq |B| > 0$ *and* $expo(A) = expo(B)$, *then* $D = I[63 : 0]$.

Proof The hypothesis implies $\lfloor |A/B| \rfloor = 1$, and therefore $I = (-1)^{sgnQ}$. It also implies $expo(\hat{A}) = expo(\hat{B})$, and therefore $clzA = clzB$. In light of Lemma 19.6, we may assume that $|B|$ is not a power of 2, and hence, by Lemma 19.5 and the definition of *idiv8*,

$$D == \begin{cases} 1 = 1[63 : 0] = I[63 : 0] & \text{if } \frac{A}{B} > 0 \\ 2^{64} - 1 = (-1)[64 : 0] = I[63 : 0] & \text{if } \frac{A}{B} < 0. \end{cases} \qquad \square$$

Combining the above results, we have the following characterization of the "early exit" case:

Lemma 19.8 *Assume that at least one of the following conditions holds:*

(1) $B = 0$;
(2) $|B|$ is a power of 2;
(3) $|A| < |B|$;
(4) $expo(A) = expo(B)$.

If $B = 0$, then $D = 0$, and otherwise $D = I[63 : 0]$.

19.2   Detecting Powers of 2

The subject of this section is the function *isPow2*, which determines whether a given 64-bit integer is a power of 2. The function takes two arguments: a 64-bit vector x and a boolean *isNeg*, which indicates whether x is to be interpreted as an unsigned or a signed integer. If *isNeg* $= 1$, the function determines whether the absolute value of the represented integer, $|x - 2^{64}| = 2^{64} - x$, is a power of 2. In this case, the computation makes use of the following:

Lemma 19.9 *If x is an n-bit vector, $x \neq 0$, and $k \in \mathbb{Z}$, then*

$$2^n - x = 2^k \Leftrightarrow (2x \mathbin{\char94} x)[n - 1 : 0] = 2^k.$$

Proof We may assume that $0 \leq k < n$, since this is a consequence of either of the two conditions.

If $0 \leq j < n$, then by Lemma 2.16 (h),

$$2^k[j] = 1[j - k] = \begin{cases} 1 \text{ if } j = k. \\ 0 \text{ if } j \neq k; \end{cases}$$

by Lemma 2.16 (l),

$$(2^n - 2^k)[j] = \left\lfloor \frac{2^n - 2^k}{2^j} \right\rfloor \bmod 2$$

$$= \lfloor 2^{n-j} - 2^{k-j} \rfloor \bmod 2 \rfloor$$

$$= (2^{n-j} + \lfloor -2^{k-j} \rfloor) \bmod 2 \rfloor$$

$$= \lfloor -2^{k-j} \rfloor \bmod 2$$

$$= \begin{cases} 1 \text{ if } j \geq k. \\ 0 \text{ if } j < k; \end{cases}$$

and by Lemma 2.16 (h),

$$(2x \wedge x)[j] = (2x)[j] \wedge x[j] = x[j-1] \wedge x[j] = \begin{cases} 1 \text{ if } x[j] \neq x[j-1]. \\ 0 \text{ if } x[j] = x[j-1]. \end{cases}$$

Suppose $x = 2^n - 2^k$. We shall show that $(2x \wedge x)[n-1 : 0][j] = 2^k[j]$ for all $j \in \mathbb{N}$ and apply Lemma 2.22. By Lemma 2.16 (d), this holds for $j \geq n$, and for $j < n$,

$$(2x \wedge x)[n-1 : 0][j] = (2x \wedge x)[j] = 1 \Leftrightarrow x[j] \neq x[j-1]$$

$$\Leftrightarrow j \geq k \text{ and } j - 1 < k$$

$$\Leftrightarrow j = k$$

$$\Leftrightarrow 2^k[j] = 1.$$

The proof of the converse is similarly based on Lemma 2.22. Suppose $(2x \wedge x)[n-1 : 0] = 2^k$. We wish to show that for all $j \in \mathbb{N}$, $x[j] = (2^n - 2^k)[n-1 : 0][j]$. Again, the case $j \geq n$ follows from Lemma 2.16 (d). If $j < n$, then $(2x \wedge x)[j] = (2x \wedge x)[n-1 : 0][j]$ and

$$x[j] = x[j-1] \Leftrightarrow (2x \wedge x)[j] = 0 \Leftrightarrow 2^k[j] = 0 \Leftrightarrow j \neq k.$$

It follows by induction that $x[j] = 1 \Leftrightarrow j \geq k$, and therefore $x[j] = (2^n - 2^k)[j]$.

$\square$

The main loop of *isPow2* operates on a bit vector v, an array of bit vectors A, and an integer w. For $0 \leq i \leq 6$, let v_i, A_i, and w_i denote the values of these variables after i iterations of the loop. Since $w_0 = 64$ and w is divided by 2 on each iteration, $w_i = 2^{6-i}$. While v_i, $A_i[0], \ldots, A_i[5]$ are all 64-bit vectors, the only valid bits are the least significant w_i bits of v_i and $A_i[0], \ldots, A_i[i-1]$.

For $0 \le i \le 6$ and $0 \le b < w_i$, let $C_i(b)$ denote the number of indices j, $0 \le j < 64$, such that $j \bmod w_i = b$ and $z[j] = 1$. This count may alternatively be defined by the following recurrence relation:

Lemma 19.10 *Let* $0 \le i \le 6$ *and* $0 \le b < w_i$.

(a) $i = 0 \Rightarrow C_i(b) = z[b]$;
(b) $i > 0 \Rightarrow C_i(b) = C_{i-1}(b) + C_{i-1}(b + w_i)$.

Proof Let $0 \le j < 64$. The case $i = 0$ follows the observation that $w_0 = 64$ and hence

$$j \bmod w_i = b \Leftrightarrow j = b.$$

For the case $i > 0$, it suffices to show that

$$j \bmod w_i = b \Leftrightarrow j \bmod w_{i-1} = b \text{ or } j \bmod w_{i-1} = b + w_i,$$

where $w_{i-1} = 2w_i$. By Lemmas 1.16 and 1.12,

$$j \bmod w_i = b \Leftrightarrow (j \bmod w_{i-1}) \bmod w_i = b \Leftrightarrow \frac{j \bmod w_{i-1} - b}{w_i} \in \mathbb{Z}.$$

Since $0 \le j \bmod w_{i-1} < w_{i-1} = 2w_i$ and $0 \le b < w_i$,

$$-1 = \frac{-w_i}{w_i} < \frac{j \bmod w_{i-1} - b}{w_i} < \frac{2w_i}{w_i} = 2.$$

Thus

$$\frac{j \bmod w_{i-1} - b}{w_i} \in \mathbb{Z} \Leftrightarrow \frac{j \bmod w_{i-1} - b}{w_i} \in \{0, 1\}$$

$$\Leftrightarrow j \bmod w_{i-1} \in \{b, b + w_i\}. \qquad \square$$

The key to understanding the underlying algorithm is the identification of the relevant loop invariant $\Phi(i)$. For $1 \le i \le 6$, we define $\Phi(i)$ to be the conjunction of the following conditions over all b such that $0 \le b < w_i$:

(1) $C_i(b) > 0 \Leftrightarrow v_i[b] = 1$;
(2) $C_i(b) = 0 \Rightarrow A_i[0][b] = \ldots = A_i[i-1][b] = 0$;
(3) $C_i(b) = 1 \Leftrightarrow A_i[0][b] = \ldots = A_i[i-1][b] = 1$.

The next two lemmas establish the invariance of Φ.

Lemma 19.11 $\Phi(1)$ *holds.*

Proof By inspection of *isPow2*, $v_0 = z$, $w_1 = 32$,

$$A_1[0][31 : 0] = z[31 : 0] \;\char`\^\; z[63 : 32],$$

and

$$v_1[31:0] = z[31:0] \mid z[63:32].$$

Let $0 \le b < 32$. Then

$$A_1[0][b] = A_1[0][31:0][b] = z[31:0][b] \;\hat{}\; z[63:32][b] = z[b] \;\hat{}\; z[b+32]$$

and

$$v_1[b] = v_1[31:0][b] = z[31:0][b] \mid z[63:32][b] = z[b] \mid z[b+32].$$

By Lemma 19.10,

$$C_1(b) = C_0(b) + C_0(b+32) = \begin{cases} 0 \text{ if } z[b] = z[b+32] = 0 \\ 2 \text{ if } z[b] = z[b+32] = 1 \\ 1 \text{ if } z[b] \ne z[b+32]. \end{cases}$$

Thus

(1) $C_1(b) > 0 \Leftrightarrow z[b] = 1$ or $z[b+32] = 1 \Leftrightarrow v_1[b] = 1$;
(2) $C_1(b) = 0 \Rightarrow z[b] = z[b+32] = 0 \Rightarrow A_1[0][b] = 0$;
(3) $C_1(b) = 1 \Leftrightarrow z[b] \ne z[b+32] \Leftrightarrow A_1[0][b] = 1$.

$\square$

Lemma 19.12 *For $1 \le i \le 6$, $\Phi(i-1) \Rightarrow \Phi(i)$.*

Proof Let $0 \le b < w_i$. First we consider the case $C_{i-1}(b + w_i) = 0$. By Lemma 19.10, $C_i(b) = C_{i-1}(b)$. By hypothesis,

$$v_{i-1}[b + w_i] = A_{i-1}[0][b + w_i] = \ldots = A_{i-1}[i-2][b + w_i] = 0.$$

According to the definition of *isPow2*,

$$v_i[b] = v_{i-1}[b] \mid v_{i-1}[b + w_i] = v_{i-1}[b],$$

$$A_i[i-1][b] = v_{i-1}[b] \;\hat{}\; v_{i-1}[b + w_i] = v_{i-1}[b]$$

and for $0 \le k < i-1$,

$$A_i[k][b] = A_{i-1}[k][b] \mid A_{i-1}[k][b + w_i] = A_{i-1}[k][b].$$

We must establish the three conditions defining $\Phi(i)$:
(1) $C_i(b) > 0 \Leftrightarrow C_{i-1}(b) > 0 \Leftrightarrow v_{i-1}[b] = 1 \Leftrightarrow v_i[b] = 1$.

(2) If $C_i(b) = 0$, then $C_{i-1}(b) = 0$, which implies

$$A_i[i-1][b] = v_{i-1}[b] = 0$$

and for $0 \le k < i - 1$,

$$A_i[k][b] = A_{i-1}[k][b] = 0.$$

(3) We have

$$C_i(b) = 1 \Leftrightarrow C_{i-1}(b) = 1$$
$$\Leftrightarrow A_{i-1}[0][b] = \ldots = A_{i-1}[i-2][b] = 1$$
$$\Leftrightarrow A_i[0][b] = \ldots = A_i[i-2][b] = 1,$$

and we need only show that $C_i(b) = 1 \Rightarrow A_i[i-1][b] = 1$. But as noted in (1) above, if $C_i(b) = 1$, then $v_{i-1}[b] = 1$, which implies

$$A_i[i-1][b] = v_{i-1}[b] \text{ } \hat{} \text{ } v_{i-1}[b+w_i] = 1 \text{ } \hat{} \text{ } 0 = 1.$$

The case $C_{i-1}(b) = 0$ is similar. Thus, we may assume that $C_{i-1}(b) > 0$ and $C_{i-1}(b + w_i) > 0$. In this case, $v_{i-1}[b] = v_{i-1}[b + w_i] = 1$ and $C_i(b) \ge 2$. Again, we must verify the three conditions comprised by $\Phi(i)$:
(1) $C_i(b) > 0$ and

$$v_i[b] = v_{i-1}[b] \mid v_{i-1}[b+w_i] = 1 \mid 1 = 1.$$

(2) Since $C_i(b) \ne 0$, this holds vacuously.
(3) Since $C_i(b) \ne 1$, we need only show that $A_i[k][b] = 0$ for some $k < i$. But

$$A_i[i-1][b] = v_{i-1}[b] \text{ } \hat{} \text{ } v_{i-1}[b+w_i] = 1 \text{ } \hat{} \text{ } 1 = 0. \qquad \square$$

Lemma 19.4 is an immediate consequence of Lemma 19.3 and the following statement of the correctness of *isPow2*:

Lemma 19.13 *Let x be a 64-bit vector.*

(a) $isPow2(x, 0) = 1 \Leftrightarrow x$ *is a power of 2.*
(b) *If* $x \ne 0$, *then* $isPow2(x, 1) = 1 \Leftrightarrow 2^{64} - x$ *is a power of 2.*

Proof We have

$$z = \begin{cases} x & \text{if } isNeg = 0 \\ (2x \text{ } \hat{} \text{ } x)[63:0] & \text{if } isNeg = 1, \end{cases}$$

and in the latter case, according to Lemma 19.9, z is a power of 2 iff $2^{64} - x$ is a power of 2. Thus, we need only show that

$$isPow2(x, isNeg) = 1 \Leftrightarrow z \text{ is a power of 2.}$$

By Lemmas 19.11 and 19.12 and induction, $\Phi(6)$ holds. The only relevant component of $\Phi(6)$ is (2), which, since $w_6 = 1$, applies only to $b = 0$:

$$C_6(0) = 1 \Leftrightarrow A_i[0][0] = \ldots = A_6[5][0] = 1.$$

Upon inspection of the final loop of the function, it is clear that

$$isPow2(x, isNeg) = 1 \Leftrightarrow A_i[0][0] = \ldots = A_6[5][0] = 1 \Leftrightarrow C_6(0) = 1.$$

By definition, $C_6(0)$ is the number of indices j, $0 \le j < 64$, such that $j \bmod 1 = 0$ and $z[j] = 1$. But $j \bmod 1 = 0$ for all $j \in \mathbb{Z}$, and therefore $C_6(0) = 1$ iff exactly 1 bit of z is set, i.e., z is a power of 2. $\qquad \square$

19.3 Instantiating the SRT Algorithm

We may now assume that $|A| \ge |B| > 0$, $expo(A) > expo(B)$, and $|B|$ is not a power of 2. For this case, the design is based on the same minimally redundant radix-8 SRT algorithm and performs the same essential computations as the floating-point divider *fdiv8* of Chap. 18. This requires that the operands be scaled to satisfy the requirements of Sect. 10.1. Thus, we define

$$x = 2^{clzA-64}|\hat{A}| = 2^{-expo(\hat{A})-1}|\hat{A}| = \frac{1}{2}sig(\hat{A}) = \frac{1}{2}sig(A)$$

and

$$d = 2^{clzB-64}|\hat{B}| = 2^{-expo(\hat{B})-1}|\hat{B}| = \frac{1}{2}sig(\hat{B}) = \frac{1}{2}sig(B),$$

so that $\frac{1}{2} \le x < 1, \frac{1}{2} \le d < 1$, and

$$\frac{1}{2} < \frac{x}{d} < 2.$$

In particular, (10.2) holds.

In order to facilitate a development consistent with that of Chap. 18, in place of the floating-point precisions, we define

$$p = \begin{cases} 32 \text{ if } int32 = 1 \\ 64 \text{ if } int32 = 0. \end{cases}$$

Since $2^{p-64}\hat{A} \in \mathbb{Z}$ and $2^{p-64}\hat{B} \in \mathbb{Z}$, we have $2^p x \in \mathbb{Z}$ and $2^p d \in \mathbb{Z}$.

An important difference, however, between floating-point and integer division is that the latter exhibits a variable latency, determined by the relative magnitudes of the dividend and divisor. In this integer divider, the number of SRT iterations is determined by the local variable *delta*, which we denote as Δ

$$\Delta = clzB - clzA = expo(\hat{A}) - expo(\hat{B}) = expo(A) - expo(B).$$

We have the following obvious analog of Lemma 18.2 (c):

Lemma 19.14 $\left|\frac{A}{B}\right| = 2^\Delta \left(\frac{x}{d}\right).$

Thus, the quotient x/d will be computed to Δ fractional bits, where $1 \le \Delta \le 62$. With 3 bits computed on each iteration, the number of iterations is specified to be

$$N = \left\lceil \frac{\Delta}{3} \right\rceil + 1,$$

which lies in the range $2 \le N \le 12$.

As in the floating-point divider, the first iteration is performed before entering the `for` loop, each iteration of which corresponds to a cycle in which two SRT iterations are executed (with possibly one exception, depending on the parity on N). Thus, the number of such cycles is

$$C = \left\lceil \frac{N-1}{2} \right\rceil = \left\lceil \frac{\Delta}{6} \right\rceil. \tag{19.8}$$

As in Chap. 18, the values of the local variables q, $quot$, $quotM$, RP, and RN produced by the jth iteration will be denoted with subscript j. The partial quotients and remainders Q_j and R_j are defined in terms of x, d, and q_j according to Eqs. (10.3) and (10.4). In addition to the values $quot_N$ and $quotM_N$, the final iteration computes an incremented version of the quotient, $quotP_N$.

While only Δ fractional bits of the quotient are required, the number of fractional bits produced is $3(N-1)$, which may exceed Δ by 1 or 2. These extra bits will ultimately be discarded; their number is given by the variable

$$K = 3(N-1) - \Delta = 3\left\lceil \frac{\Delta}{3} \right\rceil - \Delta = \begin{cases} 0 \text{ if } \Delta \bmod 3 = 0 \\ 2 \text{ if } \Delta \bmod 3 = 1 \\ 1 \text{ if } \Delta \bmod 3 = 2. \end{cases} \tag{19.9}$$

The design distinguishes three cases:

(1) $N = 2$: There is only one iteration of the loop ($C = 1$), in which there is only
one SRT iteration, producing $quot_2$, $quotM_2$, RP_2, and RN_2. The incremented
quotient is computed as $quotP_2 = quot_2 + 2^K$. Note that in this case, this
computation is feasible because it requires only a 3-bit adder.
(2) N is odd: Since

$$2C + 1 = 2 \left\lceil \frac{N-1}{2} \right\rceil + 1 = 2 \cdot \frac{N-1}{2} + 1 = N,$$

two SRT iterations are executed in each iteration of the loop. The incremented
quotient $quotP_N$ is computed during the final SRT iteration. In general, $quot_N$
is too wide to allow $quotP_N$ to be derived directly from $quot_N$ as in (1). Instead,
it is computed from q_{N-1}, $quot_{N-2}$, $quotM_{N-2}$, q_N, $quot_{N-1}$, and $quotM_{N-1}$
by the function $incQuot$, which requires only concatenation and no addition.
(3) N is even and $N > 2$: In this case, since

$$2C + 1 = 2 \left\lceil \frac{N-1}{2} \right\rceil + 1 = 2 \cdot \frac{N}{2} + 1 = N + 1,$$

one of the iterative cycles must execute one SRT iteration rather than two.
Instead of omitting the second iteration of the last cycle, the second iteration
of the preceding cycle is skipped. This allows the computation of $quotP_N$ in the
final cycle to use the same hardware as in (2).

The computations of q_j, RP_j, and RN_j are the same as in the floating-point
divider, with the minor exception that in the above case (3), skipping the second
iteration of the penultimate cycle affects whether each of the last two iterations
occurs as the first or second iteration of a cycle. This determines which of the two
alternative methods is used to compute the remainder approximation $RS10_j$ for $j =
N - 1$ and $j = N$, but it does not affect the proofs of Lemmas 18.3–18.6, all of
which, therefore, remain valid in the present context.

The principal difference between the floating-point and integer quotient computations is that in the integer case $A/B < 0$, $-Q_j$ is required instead of Q_j:

Lemma 19.15 *Let $1 \le j \le N$.*

(a) If $sgnQ = 0$, then $quot_j$ and $quotM_j$ are $(3j - 1)$-bit vectors,

$$quot_j = 8^{j-1} Q_j,$$

and

$$quotM_j = quot_j - 1.$$

(b) If $sgnQ = 1$, then $2^{65} - 2^{3j-1} < quot_j < 2^{65}$, $2^{65} - 2^{3j-1} < quotM_j < 2^{65}$,

$$quot_j = 2^{65} - 8^{j-1}Q_j,$$

and

$$quotM_j = quot_j - 1.$$

Proof The proof for $sgnQ = 0$ is the same as that of Lemma 18.7. Suppose $sgnQ = 1$ and assume that the lemma holds for some $j < N$. If $-q_j \geq 0$, then by Lemma 2.14,

$$quot_{j+1} = (8quot_j)[64 : 0] + (-q_j) \bmod 8 = 8quot_j[61 : 0] - q_j.$$

By Lemmas 2.3, 2.10, and 2.6,

$$quot_j = quot_j[64 : 0] = 2^{62}quot_j[64 : 62] + quot_j[61 : 0] = 2^{62} \cdot 7 + quot_j[61 : 0]$$

and hence,

$$\begin{aligned} quot_{j+1} &= 8(quot_j - 2^{62} \cdot 7) - q_j \\ &= 8(2^{65} - 8^{j-1}Q_j) - 2^{65} \cdot 7 - q_j \\ &= 2^{65} - 8^j(Q_j + 8^{-j}q_j) \\ &= 2^{65} - 8^j Q_{j+1}. \end{aligned}$$

On the other hand, if $-q_j < 0$, then similarly,

$$\begin{aligned} quot_{j+1} &= (8quotM_j)[64 : 0] + (-q_j) \bmod 8 \\ &= 8quotM_j[61 : 0] + 8 - q_j \\ &= 8quotM_j - 2^{65} \cdot 7 + 8 - q_j \\ &= 8quot_j - 2^{65} \cdot 7 - q_j \\ &= 2^{65} - 8^j Q_{j+1}. \end{aligned}$$

In either case, it is clear that $quot_{j+1}$ is a 65-bit vector, and since $quot_j > 2^{65} - 2^{3j-1}$,

$$\begin{aligned} quot_{j+1} &= 8quot_j - 2^{65} \cdot 7 - q_j \\ &\geq 8(2^{65} - 2^{3j-1} + 1) - 2^{65} \cdot 7 - q_j \end{aligned}$$

$$= 2^{65} - 2^{3(j+1)-1} + 8 - q_j$$
$$> 2^{65} - 2^{3(j+1)-1}.$$

The computation of $quotM_{j+1}$ is similar.   □

Another difference is that in *fdiv8*, the increment of the final partial quotient is determined by the data format, whereas here it is determined by the parameter K:

Lemma 19.16 *With the exception of the case* $sgnQ = 1$, $\Delta = 1$ *and* $Q_2 = \frac{1}{2}$,

$$quotP_N = quot_N + 2^K.$$

Proof For $N > 2$, the computation is similar to that of *fdiv8* and the proof is similar to that of Lemma 18.8 (a). In the case $N = 2$, the computation is performed directly as

$$quotP_2 = (quot_2 + 2^K) \bmod 2^{65},$$

and the lemma holds unless $quot_2 \geq 2^{65} - 2^K$, where $quot_2$ is related to Q_2 according to Lemma 19.15. This can occur only if $sgnQ = 1$ and $8Q_2 \leq 2^K$. Since $Q_2 = q_1 + 8^{-1}q_2$, where $q_1 \in \{1, 2\}$ and $K \in \{0, 1, 2\}$, this implies $q_1 = 1$, $q_2 = -4$, $Q_2 = \frac{1}{2}$, $K = 2$, and $\Delta = 1$.   □

19.4 Post-processing

In this section, in referring to the final partial quotient Q_N and remainder R_N and the associated variable values $quot_N$, etc., we shall omit the subscript N.

We have the following error bound for the final quotient:

Lemma 19.17 $\left| Q - \frac{x}{d} \right| \leq \frac{4}{7} \cdot 8^{1-N}$.

Proof This is an immediate consequence of (10.4) and Lemma 18.6 (c) with $j = N$.   □

The sign of the final remainder is given by the most significant bit of *rem*:

Lemma 19.18 $rem[70] = 1 \Leftrightarrow R < 0$.

Proof See the proof of Lemma 18.11.   □

Note that the local variables *quotSigned*, *quotMSigned*, and *quotPSigned* are defined by converting *quot*, *quotM*, and *quotP*, respectively, to signed integers. Thus, the values of these variables are related by

$$quotSigned = si(quot, 65),$$

$$quotMSigned = si(quotM, 65),$$

and

$$quotPSigned = si(quotP, 65).$$

As a matter of convenience, we define

$$\tilde{Q} = \begin{cases} Q & \text{if } sgnQ = 0 \\ -Q & \text{if } sgnQ = 1. \end{cases}$$

Lemma 19.19

(a) $quotSigned = 8^{N-1}\tilde{Q}$;
(b) $quotMSigned = 8^{N-1}\tilde{Q} - 1$;
(c) If $Q \neq \frac{1}{2}$, then $quotPSigned = 8^{N-1}\tilde{Q} + 2^K$.

Proof In light of Lemmas 19.15 and 19.16, since $K \leq 2$, it will suffice to show that $8^{N-1}Q + 4 < 2^{64}$. By Lemma 19.17,

$$8^{N-1}Q \leq 8^{N-1}\frac{x}{d} + \frac{4}{7}.$$

Since $N \leq 21$ implies

$$8^{N-1}Q + 4 \leq 8^{N-1}\frac{x}{d} + \frac{4}{7} + 4 < 8^{20} \cdot 2 + \frac{4}{7} + 4 = 2^{61} + \frac{4}{7} + 4 < 2^{64},$$

we may assume that $N = 22$, which implies

$$\Delta = expo(\hat{A}) - expo(\hat{B}) \in \{61, 62\}.$$

If $\Delta = 62$, then $expo(\hat{A}) = 63$ and $expo(\hat{B}) = 1$, which implies $|\hat{A}| < 2^{64}$ and (since B is not a power of 2) $|\hat{B}| = 3$, and by Lemma 19.14,

$$\frac{x}{d} = 2^{-62}\left|\frac{\hat{A}}{\hat{B}}\right| < 2^{-62}\frac{2^{64}}{3} = \frac{4}{3}.$$

If $\Delta = 61$, then either $expo(\hat{A}) = 62$ and $expo(\hat{B}) = 1$ or $expo(\hat{A}) = 63$ and $expo(\hat{B}) = 2$. In the first case, $|\hat{A}| < 2^{63}$, $|\hat{B}| = 3$, and again

$$\frac{x}{d} = 2^{-61}\left|\frac{\hat{A}}{\hat{B}}\right| < 2^{-61}\frac{2^{63}}{3} = \frac{4}{3}.$$

In the second case, $|\hat{A}| < 2^{64}$, $|\hat{B}| \geq 5$, and

$$\frac{x}{d} = 2^{-61} \left| \frac{\hat{A}}{\hat{B}} \right| < 2^{-61} \frac{2^{64}}{5} = \frac{8}{5}.$$

Thus, in all cases,

$$8^{N-1} Q + 4 \leq 8^{N-1} \frac{x}{d} + \frac{4}{7} + 4 < 8^{21} \cdot \frac{8}{5} + \frac{4}{7} + 4 = 2^{64} \cdot \frac{4}{5} + \frac{4}{7} + 4 < 2^{64}. \quad \square$$

The integer quotient is related to the final partial quotient Q and remainder R as follows:

Lemma 19.20

(a) If $\frac{A}{B} > 0$, then $I = \begin{cases} \lfloor 2^{\Delta} \tilde{Q} \rfloor - 1 & \text{if } 2^{\Delta} Q \in \mathbb{Z} \text{ and } R < 0 \\ \lfloor 2^{\Delta} \tilde{Q} \rfloor & \text{if } 2^{\Delta} Q \notin \mathbb{Z} \text{ or } R \geq 0. \end{cases}$

(b) If $\frac{A}{B} < 0$, then $I = \begin{cases} \lfloor 2^{\Delta} \tilde{Q} \rfloor & \text{if } 2^{\Delta} Q \in \mathbb{Z} \text{ and } R \geq 0 \\ \lfloor 2^{\Delta} \tilde{Q} \rfloor + 1 & \text{if } 2^{\Delta} Q \notin \mathbb{Z} \text{ or } R < 0. \end{cases}$

Proof By Eqs. (19.9) and (10.3), $2^K (2^{\Delta} \tilde{Q}) = 8^{N-1} \tilde{Q} \in \mathbb{Z}$, i.e., $2^{\Delta} \tilde{Q}$ is an integral multiple of 2^{-K}. By Lemma 19.17,

$$\left| \frac{A}{B} - 2^{\Delta} \tilde{Q} \right| = \left| 2^{\Delta} \frac{x}{d} - 2^{\Delta} Q \right| < 2^{\Delta} 8^{1-N} = 2^{\Delta - 3(N-1)} = 2^{-K}.$$

Thus, since $2^{\Delta} \tilde{Q} - \lfloor 2^{\Delta} \tilde{Q} \rfloor < 1$, $2^{\Delta} \tilde{Q} - \lfloor 2^{\Delta} \tilde{Q} \rfloor \leq 1 - 2^{-K}$ and

$$\left| \frac{A}{B} - \lfloor 2^{\Delta} \tilde{Q} \rfloor \right| \leq \left| \frac{A}{B} - 2^{\Delta} \tilde{Q} \right| + (2^{\Delta} \tilde{Q} - \lfloor 2^{\Delta} \tilde{Q} \rfloor) < 2^{-K} + (1 - 2^{-K}) = 1.$$

Suppose $\frac{A}{B} > 0$. Then $\left| \frac{A}{B} - \lfloor 2^{\Delta} Q \rfloor \right| < 1$, $I = \lfloor \frac{A}{B} \rfloor$, and we need only show that

$$\lfloor 2^{\Delta} Q \rfloor > \frac{A}{B} \Leftrightarrow 2^{\Delta} Q \in \mathbb{Z} \text{ and } R < 0.$$

But if $2^{\Delta} Q \in \mathbb{Z}$, then

$$R < 0 \Leftrightarrow \frac{x}{d} < Q \Leftrightarrow \frac{A}{B} < 2^{\Delta} Q \Leftrightarrow \frac{A}{B} < \lfloor 2^{\Delta} Q \rfloor$$

and if $2^{\Delta} Q \notin \mathbb{Z}$, then

$$\lfloor 2^{\Delta} Q \rfloor \leq 2^{\Delta} Q - 2^{-K} < \left(\frac{A}{B} + 2^{-k} \right) - 2^{-K} = \frac{A}{B}.$$

On the other hand, suppose $\frac{A}{B} < 0$. Then $\left| \frac{A}{B} - \lfloor -2^\Delta Q \rfloor \right| < 1$, $I = \lceil \frac{A}{B} \rceil$, and we need only show that

$$\lfloor -2^\Delta Q \rfloor \geq \frac{A}{B} \Leftrightarrow 2^\Delta Q \in \mathbb{Z} \text{ and } R \geq 0.$$

But if $2^\Delta Q \in \mathbb{Z}$, then

$$R \geq 0 \Leftrightarrow \frac{x}{d} \geq Q \Leftrightarrow \frac{A}{B} \leq -2^\Delta Q \Leftrightarrow \frac{A}{B} \leq \lfloor -2^\Delta Q \rfloor$$

and if $2^\Delta Q \notin \mathbb{Z}$, then

$$\lfloor -2^\Delta Q \rfloor \leq -2^\Delta Q - 2^{-K} < \left(\frac{A}{B} + 2^{-k} \right) - 2^{-K} = \frac{A}{B}. \qquad \square$$

Theorem 19.1 *Let*

$$D = idiv8(opa, opb, int32, sgnd),$$

where opa and opb are 64-bit vectors and int32 and sgnd are 1-bit vectors. Let A, B, and I be defined by Eqs. (19.1), (19.2), and (19.3). If $B = 0$, then $D = 0$; otherwise $D = I[63:0]$.

Proof We need only address the remaining computational case not covered by Lemma 19.8. As noted in Sect. 19.1, it suffices to show that $D = I[63:0]$.

First note that by Lemma 19.19,

$$2^{-K} quotSigned = 2^{-K}(8^{N-1}\tilde{Q}) = 2^{3(N-1)-K}\tilde{Q} = 2^\Delta \tilde{Q}$$

and therefore,

$$isLost = 0 \Leftrightarrow quotSigned[K-1:0] = 0 \Leftrightarrow 2^{-K} quotSigned = 2^\Delta \tilde{Q} \in \mathbb{Z}.$$

The following analysis mirrors the final conditional branching of *idiv8* and invokes Lemmas 19.18, 19.19, and 19.20:

First suppose $A/B < 0$ and either $2^\Delta Q \notin \mathbb{Z}$ or $R < 0$. We cannot have $Q_2 = \frac{1}{2}$ because in that case, $2^\Delta Q = 2^{\Delta-1} \in \mathbb{Z}$ and since $\frac{x}{d} > \frac{1}{2} = Q$, $R > 0$. Thus

$$I = \lfloor 2^\Delta \tilde{Q} \rfloor + 1 = \lfloor 2^\Delta \tilde{Q} + 1 \rfloor$$

and

$$D = quotP0$$
$$= \lfloor 2^{-K} quotPSigned \rfloor [63:0]$$

$$= \lfloor 2^{-K}(quotSigned + 2^K) \rfloor [63:0]$$
$$= \lfloor 2^{\Delta}\tilde{Q} + 1 \rfloor [63:0$$
$$= I[63:0].$$

If $A/B > 0$ and $2^{\Delta}Q \in \mathbb{Z}$ and $R < 0$, then

$$I = \lfloor 2^{\Delta}\tilde{Q} \rfloor - 1 = 2^{\Delta}\tilde{Q} - 1 = \lfloor 2^{\Delta}\tilde{Q} - 2^{-K} \rfloor$$

and

$$D = quotM0$$
$$= \lfloor 2^{-K} quotMSigned \rfloor [63:0]$$
$$= \lfloor 2^{-K}(quotSigned - 1) \rfloor [63:0]$$
$$= \lfloor 2^{\Delta}\tilde{Q} - 2^{-K} \rfloor [63:0]$$
$$= I[63:0].$$

In all other cases,

$$I = \lfloor 2^{\Delta}\tilde{Q} \rfloor$$

and

$$D = quot0 = \lfloor 2^{-K} quotSigned \rfloor [63:0] = \lfloor 2^{\Delta}\tilde{Q} \rfloor [63:0] = I[63:0]. \qquad \square$$

Chapter 20
Multi-precision Radix-4 SRT Square Root

The function `fsqrt4`, available at https://go.sn.pub/fsqrt4, performs double-, single-, and half-precision square root extraction. As noted in Chap. 18, it is derived from the same RTL module as the functions `fdiv8` and *idiv8*. The design shares hardware between the two floating-point operations for pre- and post-processing; therefore, the auxiliary functions `analyze`, `rounder`, and `final` are shared by `fdiv8` and `fsqrt4`.

The iterative phases, on the other hand, are implemented separately. In the design of this module, while the original radix-4 divider was replaced with a new radix-8 version, timing difficulties encountered in the development of the more complex square root operation resulted in the decision to retain the original square root design, an implementation of the minimally redundant radix-4 algorithm of Sect. 10.5. These difficulties were eventually overcome in a subsequent FPU design, leading to the algorithm of Sect. 10.6 and a unified radix-8 division and square root module. The selection of the hybrid design for this presentation was a pedagogical decision, based on its greater diversity of implementation techniques.

Notation The notational conventions established in Chap. 18 remain in force. Thus, for a variable that is assigned values within the main `for` loop, we shall use the subscript j to denote its value after $j - 1$ iterations of the loop, i.e., after j iterations of the algorithm. When the subscript is omitted from a loop variable, it is understood to be N, corresponding to the final value. We shall similarly abbreviate the final quotient Q_N and remainder R_N as Q and R.

20.1 Pre-processing

The input and output parameters of `fsqrt4` are the same as those of `fdiv8`, except that the second operand, `opb`, is not present. Once again, the initial phase of this function handles the trivial early termination cases, in which the operand is a zero, a

---

© The Author(s), under exclusive license to Springer Nature Switzerland AG 2022
D. M. Russinoff, *Formal Verification of Floating-Point Hardware Design*,
https://doi.org/10.1007/978-3-030-87181-9_20

NaN, an infinity, a negative value, or a power of 2. We shall focus on the remaining case, in which the operand is either a positive denormal that is not forced to 0 or a positive normal. The operand value A, precision p, exponent width e, exponent bias $bias$, and rounding mode $\mathcal{R}$ are defined as in Sect. 18.1.

The function *normalize* performs a mantissa shift in the case of a denormal operand and returns values that satisfy the following. Note that while tighter bounds on the root exponent $expQ$ could be achieved, those given are sufficient to preclude overflow and underflow:

Lemma 20.1

(a) $2^{52} \le siga < 2^{53}$ and $siga[52 - p : 0] = 0$;
(b) $A = 2^{expShft-bias-52} siga$;
(c) $expQ = \left\lfloor \frac{expShft+bias}{2} \right\rfloor$ and $0 < expQ < 2^e - 2$.

Proof This is easily proved by inspection of the definition of *normalize*.   □

We define

$$
x = \begin{cases} \frac{sig(A)}{4} & \text{if } expShft \text{ is odd} \\ \frac{sig(A)}{2} & \text{if } expShft \text{ is even.} \end{cases}
$$

Clearly $\frac{1}{4} \le x < 1$, as required by the algorithm of Sect. 10.5. Since we have defined x and q_j for $1 \le j \le N$, the definitions of the partial roots and remainders Q_j and R_j for $0 \le j \le N$ are given by Eqs. (10.15) and (10.16).

The algorithm computes an approximation of $\sqrt{x}$, which is related to the desired final result $\sqrt{A}$ according to the following:

Lemma 20.2 $A = 2^{2(expQ-bias+1)} x$.

Proof If $expShft$ is odd, then

$$
expQ = \frac{expShft + bias}{2} = \frac{expShft - bias}{2} + bias
$$

and

$$
A = (2^{expShft-bias+2})(2^{-54} siga) = 2^{2\left(\frac{expShft-bias}{2}+1\right)} x = 2^{2(expQ-bias+1)} x.
$$

If $expShft$ is even, then

$$
expQ = \frac{expShft + bias - 1}{2} = \frac{expShft - bias + 1}{2} + bias - 1
$$

and

$$A = (2^{expShft-bias+1})(2^{-53}siga) = 2^{2\left(\frac{expShft-bias+1}{2}\right)}x = 2^{2(expQ-bias+1)}x. \qquad \square$$

The remainder approximation A_j of Lemma 10.15 is defined by

$$A_j = \begin{cases} 4R_0 & \text{if } j = 0 \\ \frac{1}{8}si(RS7_j, 7) & \text{if } 1 \leq j \leq N. \end{cases}$$

The first iteration is performed by the function *firstIter*, the values of which satisfy the following:

Lemma 20.3

(a) q_1 is the greatest $k \in \{-2, \ldots, 2\}$ such that $m_k(8, 0) \leq A_0$;

(b) $root_1 = 2^{54}Q_1$, $rootM1_1 = root_1 - 2^{52}$, and $root_1[51:0] = rootM1_1[51:0] = 0$;

(c) $2^{55}R_1 = RP_1 - RN_1$ and $RP_1[52 - p : 0] = RN_1[52 - p : 0] = 0$.

Proof We shall consider the case in which *expShft* is odd and $siga[51] = 1$; the other three cases are similar:

(a) Since $q_1 = -1$, we must show that $m_{-1}(8, 0) \leq 4R_0 < m_0(8, 0)$, where $R_0 = x - 1$. But since $2^{52} + 2^{51} \leq siga < 2^{53}$ and $x = 2^{-54}siga$, $\frac{3}{8} \leq x < \frac{1}{2}$ and

$$m_{-1}(8, 0) = -\frac{5}{2} \leq 4(x - 1) < -2 < -1 = m_0(8, 0).$$

(b) $Q_1 = 1 + 4^{-1}(-1) = \frac{3}{4}$, $root_1 = 3 \cdot 2^{52} = 2^{54}Q_1$, and $rootM1_1 = 2^{53} = root_1 - 2^{52}$.

(c) $RP_1 = 2^3 siga + 2^{58} + 2^{57} = 2^{59} + 2^{57}(x - 1) = 2^{57}R_0$ and $RN_1 = 2^{53} + 2^{59} - 2^{56} = 2^{59} - 2^{53}7$. By Lemma 20.1 (a), $RP_1[52 - p : 0] = RN_1[52 - p : 0] = 0$, and by (10.16),

$$R_1 = 4R_0 - (-1)(2(1) + 4^{-1}(-1)) = 4R_0 - \frac{7}{4} = 2^{-55}(RP_1 - RN_1). \qquad \square$$

20.2   Iterative Phase

The remaining iterations are performed within the main `for` loop by the functions *nextDigit*, *nextRem*, and *nextRoot*. As in the case of radix-8 division, two iterations of the radix-4 algorithm are performed on each clock cycle. In this case, however, the resulting timing constraints do not require different approximations for the iterations within a cycle. This allows a simplification of the structure of the model: an iteration of the `for` loop of `fsqrt4` corresponds to a single iteration of the algorithm rather than a cycle. Since the first iteration of the algorithm is executed before the loop is entered and the loop variable j ranges from 1 to $N - 1$, the number of iterations of the algorithm is N.

A_j and R_j are related according to the comparison constants of Fig. 10.4 of Sect. 10.4 as specified by the following lemma:

Lemma 20.4 *The following conditions hold for all j, $1 \le j \le N$:*

(a) *If $j' = \min(j - 1, 2)$ and $i = 16\left(Q_{j'} - \frac{1}{2}\right)$, then q_j is the greatest $k \in \{-2, \ldots, 2\}$ such that $m_k(i, j - 1) \le A_{j-1}$;*

(b) *For all $k \in \{-1, \ldots, 2\}$,*

$$A_{j-1} < m_k(i, j - 1) \Rightarrow 4R_{j-1} < m_k(i, j - 1)$$

and

$$A_{j-1} \ge m_k(i, j - 1) \Rightarrow 4R_{j-1} > m_k(i, j - 1) - \frac{1}{32};$$

(c) *$\underline{B}(j) \le R_j \le \overline{B}(j)$;*

(d) *$\frac{1}{2} \le Q_j \le 1$;*

(e) *$root_j = 2^{54}Q_j$, $rootM1_j = root_j - 2^{54-2j}$, and $root_j[53 - 2j : 0] = rootM1_j[53 - 2j : 0] = 0$;*

(f) *$2^{55}R_j \in \mathbb{Z}$, $2^{55}R_j \equiv RP_j - RN_j$ (mod 2^{59}) and*

$$RP_j[52 - p : 0] = RN_j[52 - p : 0] = 0.$$

Proof We first consider the case $j = 1$. We have $Q_{j'} = Q_0 = 1$ and $i = 8$. Since $A_0 = 4R_0$, (b) holds trivially and (a), (e), and (f) correspond to Lemma 20.3 (a), (b), and (c); (c) and (d) then follow from Lemma 10.15.

Let $1 \le j < N$ and assume that the lemma holds for all ℓ, $1 \le \ell \le j$. We shall show that all claims hold for $j + 1$:

(a) First note that the value i_j computed by *fsqrt4* coincides with the value i defined above: If $j = 1$, then this is clear from the definition of *firstIter*, and if $j \ge 2$, then

$$i_j = i_2 = i_1 + q_2 = 16\left(Q_1 - \frac{1}{2}\right) + q_2 = 16\left(Q_1 + 4^{-2}q_2 - \frac{1}{2}\right) = 16\left(Q_2 - \frac{1}{2}\right).$$

Now consider the computation of *nextDigit*(RP_j, RN_j, i, j). Recall that $A_j = \frac{1}{8}si(RS7, 7)$. It is clear from the definition of *nextDigit* that q_{j+1} is the greatest k such that $8m_k(i, j) \le si(RS7, 7)$, or $m_k(i, j) \le A_j$.

(b) We must show that for all k,

$$A_j < m_k(i, j) \Rightarrow 4R_j < m_k(i, j)$$

and

$$A_j \geq m_k(i, j) \Rightarrow 4R_j > m_k(i, j) - \frac{1}{32}.$$

We shall invoke Lemma 2.37 with

$$X = RP4 - RN4 \equiv 4(RP_j - RN_j) \equiv 4(2^{55} R_j) = 2^{57} R_j \pmod{2^{59}}$$

and

$$Y = 2^{50} y,$$

where

$$y = RP4[58 : 50] - RN4[58 : 50].$$

Let $\bar{X} = X \bmod 2^{59}$, $\bar{Y} = Y \bmod 2^{59}$, and $\bar{y} = y \bmod 2^9$. By Lemmas 1.20 and 2.34,

$$si(\bar{Y}, 59) = si(2^{50} \bar{y}, 59) = 2^{50} si(\bar{y}, 9).$$

Since

$$|R_j| \leq \overline{B}(j) = 2 \cdot \frac{2}{3} Q_j + \left(\frac{2}{3}\right)^2 4^{-j} \leq \frac{4}{3} + \frac{1}{9} < 2 - 2^{-7},$$

$|2^{57} R_j| < 2^{58} - 2^{50}$, which implies $2^{57} R_j = si(\bar{X}, 59)$. Thus, the hypothesis of the lemma is satisfied, and we may conclude that

$$|2^{57} R_j - 2^{50} si(\bar{y}, 9)| = |si(\bar{X}, 59) - si(\bar{Y}, 59)$$
$$= |X - Y|$$
$$= |RP4[49 : 0] - RN4[49 : 0]|$$
$$< 2^{50}.$$

It is clear by a case analysis on $RP4[50]$ and $RN4[50]$ that $RS8[7 : 0] = \bar{y}[8 : 1]$, and it follows that $RS7 = \bar{y}[8 : 2]$. Consequently, for $m \in \mathbb{Z}$,

$$si(RS7, 7) \geq m \Leftrightarrow si(\bar{y}, 9) \geq 4m.$$

Thus

$$A_j < m_k(i, j) \Rightarrow si(RS7, 7) < 8m_k(i, j)$$
$$\Rightarrow si(\bar{y}, 9) < 32m_k(i, j)$$
$$\Rightarrow si(\bar{y}, 9) \leq 32m_k(i, j) - 1$$
$$\Rightarrow 2^{57} R_j < 2^{50}(32m_k(i, j) - 1) + 2^{50} = 2^{55} m_k(i, j)$$
$$\Rightarrow 4R_j < m_k(i, j).$$

Similarly

$$A_j \geq m_k(i, j) \Rightarrow si(RS7, 7) \geq 8m_k(i, j)$$
$$\Rightarrow si(\bar{y}, 9) \geq 32m_k(i, j)$$
$$\Rightarrow 2^{57} R_j > 2^{50} 32m_k(i, j) - 2^{50} = 2^{55} \left(m_k(i, j) - \frac{1}{32} \right)$$
$$\Rightarrow 4R_j > m_k(i, j) - \frac{1}{32}.$$

(c) This follows from Lemma 10.15, as does (d).

(e) This follows from Eq. (10.15), the definition of *nextRoot*, and induction.

(f) Consider the computation of $nextRem(RP_j, RN_j, root_j, rootM1_j, q_{j+1}, j, fmt)$. The case $q = 0$ is trivial:

$$RP_{j+1} - RN_{j+1} = RP4 - RN4 \equiv 4RP_j - 4RN_j \equiv 2^{57} R_j = 2^{55} R_{j+1} \pmod{2^{59}}.$$

In the remaining case, *RP4* and *RN4* are combined with the vector D by a 3:2 compressor. As a notational convenience, for a 59-bit vector V, let

$$V' = V[58 : 53 - p].$$

Note that if $V[52 - p : 0] = 0$, then $V' = 2^{p-53} V$.
   If $q < 0$, then

$$D = 4|q_{j+1}|rootM1_j + 2^{53-2j}|q_{j+1}|(8 - |q_{j+1}|)$$
$$= -4q_{j+1}2^{54}(Q_j - 4^{-j}) - 2^{53-2j}q_{j+1}(8 + q_{j+1})$$
$$= -2^{55}q_{j+1}(2Q_j + 4^{-(j+1)}q_{j+1}).$$

and since $D[52 - p : 0] = 0$

$$D' = 2^{p-53} D = -2^{p+2}q_{j+1}(2Q_j + 4^{-(j+1)}q_{j+1}).$$

If $q > 0$, then

$$Dcomp = 4q_{j+1}root_j + 2^{53-2j}q_{j+1}^2$$

$$= 4q_{j+1}2^{54}Q_j + 2^{53-2j}q_{j+1}^2$$

$$= 2^{55}q_{j+1}(2Q_j + 4^{-(j+1)}q_{j+1}),$$

$$Dcomp' = 2^{p-53}Dcomp = 2^{p+2}q_{j+1}(2Q_j + 4^{-(j+1)}q_{j+1}),$$

and

$$D' = 2^{p+6} - Dcomp' - 1 = 2^{p+6} - 2^{p+2}q_{j+1}(2Q_j + 4^{-(j+1)}q_{j+1}) - 1.$$

In both cases

$$D' \equiv -2^{p+2}q_{j+1}(2Q_j + 4^{-(j+1)}q_{j+1}) - \epsilon \pmod{2^{p+6}},$$

where

$$\epsilon = \begin{cases} 1 \text{ if } q_{j+1} > 0 \\ 0 \text{ if } q_{j+1} < 0. \end{cases}$$

Note that since $RP4 - RN4 \equiv 2^{57}R_j \pmod{2^{59}}$,

$$RP4' - RN4' \equiv 2^{p+4}R_j \pmod{2^{p+6}}.$$

Now according to the definitions of sum and car,

$$\sim sum' = RP4' \ ^\wedge \ \sim RN4' \ ^\wedge \ D'$$

and

$$car' = RP4' \ \& \ \sim RN4' \ | \ (RP4' \ | \ \sim RN4') \ \& \ D'),$$

and by Lemma 8.7,

$$2car' - sum' \equiv 2car' + \sim sum' + 1$$

$$\equiv RP4' + \sim RN4' + D' + 1$$

$$\equiv RP4' - RN4' + D'$$

$$\equiv 2^{p+4}R_j - 2^{p+2}q_{j+1}(2Q_j + 4^{-(j+1)}q_{j+1}) - \epsilon$$

$$= 2^{p+2}R_{j+1} - \epsilon \pmod{2^{p+6}}.$$

Thus

$$RP_{j+1} - RN_{j+1} = nextRP - nextRN$$

$$\equiv 2^{53-P}(2car' + \epsilon - sum')$$

$$\equiv 2^{55}R_{j+1} \pmod{2^{59}}. \qquad \square$$

On the final iteration, incremented versions of *root* and *rootM1* are computed, to be used by the rounder:

Lemma 20.5 *Let inc* $= \begin{cases} 2^{56-2N} & \textit{if fmt} = SP \\ 2^{55-2N} & \textit{if fmt} \neq SP. \end{cases}$

(a) $rootP = root_N + inc;$
(b) $rootM1P = rootM1_N + inc.$

Proof We refer to the function *incRoot*. The proof is based on Lemma 20.4 (e) and the recurrence relation

$$Q_j = Q_{j-1} + 4^{-j}q_j,$$

which follows from (10.15). Since *qLast*, *rootLast*, and *rootM1Last* are computed during iteration $N - 1$ after q_{N-1} is derived, we have *qLast* $= q_{N-1}$, *rootLast* $=$ $root_{N-2}$, and *rootM1Last* $= rootM1_{N-2}$.
Case 1: $lsbIs2 = 1.$
In this case, *fmt* $= SP$. The computation of *rootP* depends on q_N. First suppose $q_N \geq 0$. If $q_{N-1} \geq -1$, then

$$rootP = rootLast + 2^{base+2}(q_{N-1} + 1) + 2^{base}q_N$$

$$= root_{N-2} + 2^{56-2N}(q_{N-1} + 1) + 2^{54-2N}q_N$$

$$= 2^{54}(Q_{N-2} + 4^{-(N-1)}q_{N-1} + 4^{-N}q_N) + 2^{56-2N}$$

$$= 2^{54}Q_N + inc$$

$$= root_N + inc,$$

and if $q_{N-1} < -1$, then

$$rootP = rootM1_{N-2} + 2^{56-2N}((q_{N-1} + 1) \bmod 4) + 2^{54-2N}q_N$$

$$= 2^{54}Q_{N-2} - 2^{54-2(N-2)} + 2^{56-2N}(q_{N-1} + 1 + 4) + 2^{54-2N}q_N$$

$$= 2^{54}(Q_{N-2} + 4^{-(N-1)}q_{N-1} + 4^{-N}q_N) + 2^{56-2N}$$

$$= root_N + inc,$$

On the other hand, if $q_N < 0$, then

$$
\begin{aligned}
rootP &= root_{N-1} + 2^{54-2N}(q_N \bmod 4) \\
&= 2^{54}Q_{N-1} + 2^{54-2N}(q_N + 4) \\
&= 2^{54}(Q_{N-1} + 4^{-N}q_N) + 2^{56-2N} \\
&= 2^{54}Q_N + inc.
\end{aligned}
$$

The expression for $rootM1P$ is derived similarly.
Case 2: lsbIs2 $= 0$.
   In this case, $fmt \neq SP$.
Subcase 2.1: $q_N = 2$.
   If $q_{N-1} \geq -1$, then

$$
\begin{aligned}
rootP &= root_{N-2} + 2^{56-2N}(q_{N-1} + 1) \\
&= 2^{54}(Q_{N-2} + 2^{2-2N}q_{N-1} + 2^{-2N}(2 + 2)) \\
&= 2^{54}(Q_{N-2} + 4^{-(N-1)}q_{N-1} + 4^{-N}q_N) + 2^{55-2N} \\
&= 2^{54}Q_N + inc,
\end{aligned}
$$

and if $q_{N-1} < -1$, then

$$
\begin{aligned}
rootP &= rootM1_{N-2} + 2^{56-2N}((q_{N-1} + 1) \bmod 4) \\
&= 2^{54}Q_{N-2} - 2^{54-2(N-2)} + 2^{56-2N}(q_{N-1} + 1 + 4) \\
&= 2^{54}Q_{N-2} + 2^{56-2N}(q_{N-1} + 1) \\
&= 2^{54}Q_{N-2} + 2^{56-2N}q_{N-1} + 2^{54-2N}(q_N + 2) \\
&= 2^{54}(Q_{N-2} + 4^{-(N-1)}q_{N-1} + 4^{-N}q_N) + 2^{55-2N} \\
&= 2^{54}Q_N + inc.
\end{aligned}
$$

We also have

$$
\begin{aligned}
rootM1P &= root_{N-1} + 2^{54-2N}3 \\
&= 2^{54}Q_{N-1} + 2^{54-2N}(q_N - 1 + 2) \\
&= 2^{54}(Q_{N-1} + 4^{-N}q_N) - 2^{54-2N} + 2^{55-2N} \\
&= rootM1_N + inc.
\end{aligned}
$$

Subcase 2.2: $q_N < 2$.

$$
\begin{aligned}
rootP &= root_{N-1} + 2^{54-2N}(q_N + 2) \\
&= 2^{54}(Q_{N-1} + 4^{-N}q_N) + 2^{55-2N} \\
&= 2^{54}Q_N + inc.
\end{aligned}
$$

If $q_N \geq -1$, then

$$
\begin{aligned}
rootM1P &= root_{N-1} + 2^{54-2N}(q_N + 1) \\
&= 2^{54}Q_{N-1} + 2^{54-2N}(q_N - 1 + 2) \\
&= 2^{54}(Q_{N-1} + 4^{-N}q_N) - 2^{54-2N} + 2^{55-2N} \\
&= rootM1_N + inc,
\end{aligned}
$$

and if $q_N < -1$, then

$$
\begin{aligned}
rootM1P &= rootM1_{N-1} + 2^{54-2N}((q_N + 1) \bmod 4) \\
&= 2^{54}Q_{N-1} - 2^{54-2(N-1)} + 2^{54-2N}(q_N + 4 - 1 + 2) \\
&= 2^{54}(Q_{N-1} + 4^{-N}q_N) - 2^{54-2N} + 2^{55-2N} \\
&= rootM1_N + inc.
\end{aligned}
$$

$\square$

20.3 Post-processing and Rounding

In comparison to division, post-processing of the square root is simplified by the absence of underflow and overflow (Lemma 20.1 (c)) but has the minor complication that the result may round up to a power of 2.

On the other hand, the formal proof of correctness is significantly complicated by the limitations of the ACL2 logic, in which the square root function cannot be explicitly defined. Thus, the ACL2 specification of this operation, instead of referring directly to the desired rounded result $\mathcal{R}(\sqrt{A}, p)$, is expressed in terms of $\mathcal{R}(\sqrt[(p+2)]{A}, p)$, where $\sqrt[(p+2)]{A}$ is the $(p + 2)$-bit approximation of the square root discussed in Chap. 7.

However, in order to establish the required bound on the approximation error of the final root Q, it will be necessary to base our analysis instead on the value $\sqrt[(2N+1)]{A}$, with $(2N+1)$-bit accuracy, where $2N+1$ is either $p+2$ or $p+3$ depending on the data format. This is justified by Lemma 7.16, which ensures that

$$
\mathcal{R}(\sqrt[(p+2)]{A}, p) = \mathcal{R}(\sqrt[(2N+1)]{A}, p).
$$

The rounder actually produces a rounding of $\sqrt[(2N+1)]{x}$, which is related to the desired result as follows:

Lemma 20.6 $^{(2N+1)}\!\sqrt{A} = 2^{expQ-bias+1}\ ^{(2N+1)}\!\sqrt{x}.$

Proof This follows from Lemmas 20.2 and 7.15. □

Lemma 20.7

(a) $(Q - \frac{2}{3}4^{-N})^2 \le x \le (Q + \frac{2}{3}4^{-N})^2$;
(b) $|\ ^{(2N+1)}\!\sqrt{x} - Q| < 2^{-2N}$.

Proof (a) is a consequence of Lemmas 20.4 (c) and 10.10. It follows that

$$(Q - 4^{-N})^2 < x < (Q + 4^{-N})^2,$$

and (b) will follow from Lemma 7.18 once we show that $Q - 4^{-N}$ and $Q + 4^{-N}$ are both $2N$-exact.

Since $Q - 4^{-N} < 1$, $expo(Q - 4^{-N}) \le -1$ and since $4^N Q \in \mathbb{Z}$,

$$2^{2N-1-expo(Q-4^{-N})}(Q - 4^{-N}) = 2^{-1-expo(Q-4^{-N})}4^{-N}(Q - 4^{-N}) \in \mathbb{Z},$$

i.e., $Q - 4^{-N}$ is $2N$-exact. To draw the same conclusion about $Q + 4^{-N}$, it will suffice to show that $Q + 4^{-N} \le 1$.

Since $x < 1$ and x is p-exact, we have

$$(Q - 4^{-N})^2 < x \le 1 - 2^{-p} < (1 - 2^{-p-1})^2.$$

Thus, $Q - 4^{-N} < 1 - 2^{-p-1}$ and $Q < 1 - 2^{-p-1} + 2^{-2N} \le 1$. It follows that $Q \le 1 - 4^{-N}$, i.e., $Q + 4^{-N} \le 1$. □

Lemma 20.8 $^{(2N+1)}\!\sqrt{x} < 1 - 2^{-p-1}$ and $Q \le 1 - 2^{-p-1}.$

Proof As we have noted, $x < (1 - 2^{-p-1})^2$. The bound on $^{(2N+1)}\!\sqrt{x}$ follows from Lemma 7.18. To establish the bound on Q, first suppose $fmt \ne SP$. Then $2N = p+1$ and by Lemma 20.7,

$$Q < \ ^{(2N+1)}\!\sqrt{x} + 2^{-p-1} < 1,$$

and hence $Q \le 1 - 2^{-2N} = 1 - 2^{-p-1}$. On the other hand, if $fmt = SP$, then $2N = p + 2$ and

$$Q < \ ^{(2N+1)}\!\sqrt{x} + 2^{-p-2} < 1 - 2^{-p-1} + 2^{-p-2} = 1 - 2^{-p-2},$$

which again implies $Q \le 1 - 2^{-p-1}$. □

Once again, the rounder selects one of two values that are returned by *computeQ*:

Lemma 20.9

(a) $Qtrunc \equiv \lfloor 2^{p+1}\ ^{(2N+1)}\!\sqrt{x} \rfloor \pmod{2^p}$;
(b) $Qinc \equiv \lfloor 2^{p+1}\ ^{(2N+1)}\!\sqrt{x} \rfloor + 2 \pmod{2^p}$.

Proof The proof is similar to that of Lemma 18.12. It follows from Lemma 20.4 that $rem = 2^{55} R \bmod 2^{59}$, and therefore, by (10.16) and Lemma 7.18,

$$remSign = 0 \Leftrightarrow R \geq 0 \Leftrightarrow {}^{(2N+1)}\!\!\sqrt{x} \geq Q$$

and

$$remZero = 1 \Leftrightarrow R = 0 \Leftrightarrow {}^{(2N+1)}\!\!\sqrt{x} = Q.$$

By Eq. (10.15), $2^{2N} Q = 4^N Q \in \mathbb{Z}$. By Lemma 20.4 (e),

$$rootShft = root[54 : 54 - 2N] = 2^{2N-54} root = 2^{2N} Q$$

and similarly

$$rootM1Shft = 2^{2N-54} rootM1 = 2^{2N} Q - 1.$$

By Lemma 20.5,

$$rootPShft = 2^{2N-54} rootP = 2^{2N-54}(2^{54} Q + inc) = 2^{2N} Q + 2^{lsbIs2+1}$$

and similarly

$$rootM1PShft = 2^{2N} Q + 2^{lsbIs2+1} - 1.$$

By Lemma 20.7, $|2^{2N} Q - 2^{2N} \, {}^{(2N+1)}\!\!\sqrt{x}| < 1$. Suppose $R \geq 0$. Then

$$2^{2N} Q \leq 2^{2N} \, {}^{(2N+1)}\!\!\sqrt{x} < 2^{2N} Q + 1,$$

i.e.,

$$\left\lfloor 2^{2N} \, {}^{(2N+1)}\!\!\sqrt{x} \right\rfloor = 2^{2N} Q.$$

Referring to the definition of *computeQ*, we have

$$rootLo = rootShft = 2^{2N} Q = \left\lfloor 2^{2N} \, {}^{(2N+1)}\!\!\sqrt{x} \right\rfloor$$

and

$$rootLoP = rootPShft = 2^{2N} Q + 2^{lsbIs2+1} = \left\lfloor 2^{2N} \, {}^{(2N+1)}\!\!\sqrt{x} \right\rfloor + 2^{lsbIs2+1}.$$

But if $R < 0$, then we have the same expressions for *rootLo* and *rootLoP*:

$$2^{2N} Q - 1 < 2^{2N} \sqrt[(2N+1)]{x} < 2^{2N} Q,$$

$$\left\lfloor 2^{2N} \sqrt[(2N+1)]{x} \right\rfloor = 2^{2N} Q - 1,$$

$$rootLo = rootM1Shft = -2^{2N} Q - 1 = \left\lfloor 2^{2N} \sqrt[(2N+1)]{x} \right\rfloor$$

and

$$rootLoP = rootM1PShft = 2^{2N} Q + 2^{lsbIs2+1} - 1 = \left\lfloor 2^{2N} \sqrt[(2N+1)]{x} \right\rfloor + 2^{lsbIs2+1}.$$

Suppose $lsbIs2 = 0$. Then $2N = p + 1$

$$Qtrunc \equiv rootLo = \left\lfloor 2^{p+1} \sqrt[(2N+1)]{x} \right\rfloor \pmod{2^p}$$

and

$$Qinc \equiv rootLoP = \left\lfloor 2^{p+1} \sqrt[(2N+1)]{x} \right\rfloor + 2 \pmod{2^p}$$

On the other hand, if $lsbIs2 = 1$, then $2N = p + 2$,

$$Qtrunc \equiv \left\lfloor \frac{rootLo}{2} \right\rfloor = \left\lfloor \frac{\left\lfloor 2^{p+2} \sqrt[(2N+1)]{x} \right\rfloor}{2} \right\rfloor = \left\lfloor 2^{p+1} \sqrt[(2N+1)]{x} \right\rfloor \pmod{2^p},$$

and

$$Qinc \equiv \left\lfloor \frac{rootLoP}{2} \right\rfloor = \left\lfloor \frac{\left\lfloor 2^{p+2} \sqrt[(2N+1)]{x} \right\rfloor + 4}{2} \right\rfloor = \left\lfloor 2^{p+1} \sqrt[(2N+1)]{x} \right\rfloor + 2 \pmod{2^p} \qquad \square$$

The function *computeQ* also returns an indication of inexactness:

Lemma 20.10 $stk = 0 \Leftrightarrow 2^{p+1} \sqrt[(2N+1)]{x} \in Z.$

Proof As observed above, $remZero = 1 \Leftrightarrow \sqrt[(2N+1)]{x} = Q$. If $remSign = 0$, then since

$$\left\lfloor 2^{2N} \sqrt[(2N+1)]{x} \right\rfloor = 2^{2N} Q,$$

this condition is equivalent to $2^{2N} \sqrt[(2N+1)]{x} \in Z$. On the other hand, if $remSign = 1$, then $R < 0$, which implies $remZero = 0$, and since

$$2^{2N} Q - 1 < 2^{2N} \sqrt[(2N+1)]{x} < 2^{2N} Q,$$

$2^{2N} \sqrt[(2N+1)]{x} \notin \mathbb{Z}$. Thus, in general,

$$remZero = 1 \Leftrightarrow 2^{2N} \sqrt[(2N+1)]{x} \in \mathbb{Z}.$$

If $lsbIs2 = 0$, then $2N = p + 1$ and

$$stk = 0 \Leftrightarrow remSign = 1 \Leftrightarrow 2^{2N} \sqrt[(2N+1)]{x} = 2^{p+1} \sqrt[(2N+1)]{x} \in \mathbb{Z}.$$

On the other hand, if $lsbIs2 = 1$, then $2N = p + 2$,

$$rootLo \equiv \left\lfloor 2^{p+2} \sqrt[(2N+1)]{x} \right\rfloor \pmod{2^{p+2}}$$

and

$$stk = 0 \Leftrightarrow remZero = 1 \text{ and } quotLo \text{ is even}$$
$$\Leftrightarrow 2^{p+2} \sqrt[(2N+1)]{x} \in \mathbb{Z} \text{ and } 2^{p+2} \sqrt[(2N+1)]{x} \text{ is even}$$
$$\Leftrightarrow 2^{p+1} \sqrt[(2N+1)]{x} \in \mathbb{Z}. \qquad \square$$

Lemma 20.11

(a) $\sqrt[(2N+1)]{x}$ is p-exact $\Leftrightarrow inx = 0$;
(b) $\mathcal{R}\left(2^{p+1} \sqrt[(2N+1)]{x}, p\right) \equiv 2Qrnd[p-2:0] \pmod{2^p}$.

Proof The proof may be derived from that of Lemma 18.14 simply by replacing $2^p \frac{x}{d}$ with $2^{p+1} \sqrt[(2N+1)]{x}$. $\qquad \square$

It is not difficult to show that $\sqrt[(2N+1)]{x}$ rounds up to 1 only if the rounding mode is RUP and x has the maximum value $1 - 2^{-p}$. For timing reasons, however, the implementation instead makes this determination by observing that the sequence of root digits satisfies conditions that imply that the final root has its maximum value, $1 - 2^{-p-1}$. In this event, the variable $expInc$ is set, causing $expQ$ to be incremented:

Lemma 20.12 *The following are equivalent:*

(a) $\mathcal{R}(\sqrt[(2N+1)]{x}, p) = 1$;
(b) $\mathcal{R} = RUP$ and $x = 1 - 2^{-p}$;
(c) $\mathcal{R} = RUP$ and $Q = 1 - 2^{-p-1}$;
(d) $expInc = 1$.

Proof Since $\sqrt[(2N+1)]{x} < 1 - 2^{-p-1}$, $\mathcal{R}(\sqrt[(2N+1)]{x}, p) < 1$ unless $\mathcal{R} = RUP$.

If $x = 1 - 2^{-p}$, then $x > (1 - 2^{-p})^2$ and by Lemma 7.18, $\sqrt[(2N+1)]{x} > 1 - 2^{-p}$, which implies $RUP(\sqrt[(2N+1)]{x}, p) = 1$. Conversely, if $x \neq 1 - 2^{-p}$, then since x is p-exact, $x \leq 1 - 2^{1-p} < (1 - 2^{-p})^2$, which implies $\sqrt[(2N+1)]{x} < 1 - 2^{-p}$ and $RUP(\sqrt[(2N+1)]{x}, p) \leq 1 - 2^{-p}$.

If $Q = 1 - 2^{-p-1}$, then $\sqrt[(2N+1)]{x} > 1 - 2^{-p-1} - 2^{-2N} \geq 1 - 2^{-p}$ and $RUP(\sqrt[(2N+1)]{x}, p) = 1$. Otherwise, $Q < 1 - 2^{-p-1}$, which implies $Q \leq 1 - 2^{-p-1} - 2^{-2N}$, and by Lemma 20.7(a),

$$x \leq \left(Q + \frac{2}{3} \cdot 4^{-N}\right)^2 \leq \left(1 - 2^{-p-1} - \frac{1}{3} \cdot 2^{-2N}\right)^2 < 1 - 2^{-p}.$$

Regarding (d), it is clear that $expInc = 1$ iff the following conditions hold: opa is normal, $\mathcal{R} = RUP$, and

$$q_j = \begin{cases} 0 \text{ if } j < N \\ -1 \text{ if } j = N \text{ and } fmt \neq SP \\ -2 \text{ if } j = N \text{ and } fmt = SP. \end{cases}$$

Suppose $expInc = 1$. Then by (10.15) and the definition of N,

$$Q = \begin{cases} 1 - 4^{-N} = 1 - 2^{2N} = 1 - 2^{-p-1} & \text{if } fmt \neq SP \\ 1 - 2 \cdot 4^{-N} = 1 - 2^{1-2N} = 1 - 2^{-p-1} & \text{if } fmt = SP. \end{cases}$$

Conversely, suppose $RUP(\sqrt[(2N+1)]{x}, p) = 1$. Then opa must be normal, for if opa were denormal, then x would be $(p - 1)$-exact, contradicting $x = 1 - 2^{-p}$.

We shall show by induction on j that $q_j = 0$ for $1 \leq j < N$. Let $0 \leq j < N - 1$ and assume that $q_1 = \ldots q_j = 0$. Then $Q_j = 1$

$$R_j = 4^j(x - Q_j^2) = 4^j(1 - 2^{-p} - 1) = -2^{2j-p},$$

and

$$4R_j = -2^{2j+2-p} \geq -2^{2(N-2)+2-p} = -2^{2N-(p+2)} \geq -1 = m_0(8, j).$$

By Lemma 18.6(a), we also have $A_j \geq m_0(8, j)$, and consequently $q_{j+1} \geq 0$. But since $Q_{j+1} \leq 1$, this implies $q_{j+1} = 0$.

Thus, $Q_{N-1} = 1$ and $Q = 1 + 4^{-N}q_N = 1 - 2^{-p-1}$, which implies

$$q_N = -2^{2N-p-1} = \begin{cases} -1 \text{ if } fmt \neq SP \\ -2 \text{ if } fmt = SP. \end{cases} \qquad \square$$

We have an analog of Lemma 18.16, but since underflow cannot occur, we need no analog of Lemma 18.17:

Lemma 20.13

(a) $\mathcal{R}(\sqrt[(2N+1)]{A}, p) = \sqrt[(2N+1)]{A} \Leftrightarrow inx = 0;$
(b) $\mathcal{R}(\sqrt[(2N+1)]{A}, p) = 2^{expRnd-bias-(p-1)}(2^{p-1} + Qrnd[p-2:0]).$

Proof

(a) This follows from Lemmas 20.6 and 20.11 (a):

$$\mathcal{R}(\sqrt[2N+1]{A}, p) = \sqrt[2N+1]{A} \Leftrightarrow \sqrt[2N+1]{A} \text{ is } p\text{-exact}$$

$$\Leftrightarrow \sqrt[2N+1]{x} \text{ is } p\text{-exact}$$

$$\Leftrightarrow inx = 0.$$

(b) First suppose $\mathcal{R}(\sqrt[2N+1]{x}, p) < 1$. Then

$$expo(\mathcal{R}(2^{p+1} \sqrt[2N+1]{x}, p)) = expo(2^{p+1} \sqrt[2N+1]{x}) = p + 1 - 1 = p.$$

Therefore, by Lemma 20.11 (b),

$$\mathcal{R}(2^{p+1} \sqrt[2N+1]{x}, p) = 2^p + \mathcal{R}(2^{p+1} \sqrt[2N+1]{x}, p) \bmod 2^p = 2^p + 2Qrnd[p-2 : 0]$$

and by Lemmas 20.6 and 20.12,

$$\mathcal{R}(\sqrt[2N+1]{A}, p) = 2^{expQ-bias+1-(p+1)}\mathcal{R}(2^{p+1} \sqrt[2N+1]{x}, p)$$

$$= 2^{expRnd-bias+1-(p+1)}\mathcal{R}(2^{p+1} \sqrt[2N+1]{x}, p)$$

$$= 2^{expRnd-bias-(p-1)}(2^{p-1} + Qrnd[p-2 : 0]).$$

In the remaining special case, $\mathcal{R}(\sqrt[2N+1]{x}, p) = 1$ and by Lemma 20.11 (b),

$$2Qrnd[p-2 : 0] = \mathcal{R}(2^{p+1} \sqrt[2N+1]{x}, p) \bmod 2^p = 2^{p+1} \bmod 2^p = 0.$$

According to Lemma 20.12, $expInc = 1$, and therefore

$$\sqrt[2N+1]{A} = 2^{expQ-bias+1} \sqrt[2N+1]{x}$$

$$= 2^{(expRnd-1)-bias+1} \sqrt[2N+1]{x}$$

$$= 2^{expRnd-bias} \sqrt[2N+1]{x}$$

and

$$\mathcal{R}(\sqrt[2N+1]{A}, p) = 2^{expRnd-bias}\mathcal{R}(\sqrt[2N+1]{x}, p)$$

$$= 2^{expRnd-bias-(p-1)}(2^{p-1} + Qrnd[p-2 : 0]). \qquad \square$$

The statement and proof of our correctness theorem are quite similar to those of Theorem 18.1, although simplified by the absence of overflow and underflow:

Theorem 20.1 *Let*

$$\langle D_{spec}, R_{spec} \rangle = arm\text{-}sqrt\text{-}spec(opa, R_{in}, fmt),$$

where opa is a 64-bit vector, R_{in} is a 32-bit vector, and fmt $\in$ {DP, SP, HP}, and let

$$\langle D, flags \rangle = fsqrt4(opa, fmt, R_{in}[24], R_{in}[25], R_{in}[23 : 22]).$$

Then $D = D_{spec}$ and $R_{in} \mid flags = R_{spec}$.

Chapter 21
Multi-precision Radix-2 SRT Division

The division and square root module described in the preceding three chapters is a component of a "big core" CPU, for which execution speed is a high priority. In a smaller core, speed is typically sacrificed in order to reduce area and power consumption, and a lower radix is more suitable. The subject of this chapter is a floating-point divider that was designed for a small core, based on the radix-2 instantiation of the SRT algorithm of Sect. 10.1. It performs scalar double-precision and both scalar and vector single- and half-precision operations. The RAC pseudocode for this module is available online at https://go.sn.pub/fdiv2.

While generally simpler than higher-radix dividers, this radix-2 design employs sophisticated techniques to address the inherent difficulties associated with the relatively weak remainder bound that attends the maximally redundant case, as discussed at the end of Sect. 10.1. This results in subtle complications in the analysis, as seen in Sects. 21.3 and 21.4.

21.1 Overview

The input parameters of the top-level function `fdiv2` are as follows:

- `ui64 opa, opb`: Encodings of the dividend(s) and divisor(s), respectively.
- `ui2 fmt`: A 2-bit encoding of a floating-point format (DP = 3, SP = 2, HP = 1).
- `bool vec`: A boolean indication of a vector operation.
- `bool fz, dn`: The FZ and DN fields of the Floating-Point Status and Control Register (FPSCR).
- `ui2 rmode`: The RC field of the FPSCR, a 2-bit encoding of an IEEE rounding mode.

Two results are returned:

- `ui64 D`: The data result.
- `ui8 flags`: The cumulative exception flags, FPSCR[7:0].

© The Author(s), under exclusive license to Springer Nature Switzerland AG 2022
D. M. Russinoff, *Formal Verification of Floating-Point Hardware Design*,
https://doi.org/10.1007/978-3-030-87181-9_21

A DP operation must be scalar; a SP or HP operation may be scalar or vector. A vector operation performs two (resp., four) independent computations in the SP (resp., HP) case, in distinct data "lanes". While the RTL executes these lanes concurrently, the simpler RAC model, which need not reflect any of the timing aspects of the RTL, executes them sequentially. The function *fdiv2* simply extracts the appropriate segments of the operands and calls the auxiliary function *fdivLane*— once for a scalar operation, twice for SP vector, and four times for HP vector—and combines the results. We shall focus on *fdivLane*. Its parameters are the same as those of *fdiv2*, although for a SP or HP operation, only the low 32 or 16 bits of the data parameters are used.

The precision p, the exponent width and bias e and $bias$, the rounding mode $\mathcal{R}$, and the numerical values A and B of the operands are defined as in Sect. 18.1, but here we define the dividend and divisor as $x = sig(A)$ and $d = sig(B)$.

Since one fractional quotient bit is generated on each radix-2 iteration, the number of iterations required to produce the p-bit quotient and guard bit, allowing for a quotient less than 1, is $p + 2$. The RTL executes the first iteration before entering the iterative phase, in which three iterations are executed in each cycle. Thus, the total number of iterations is $N = 3C + 1$, where the number of iterative cycles is

$$
C = \begin{cases} 4 \text{ if } fmt = \text{HP} \\ 9 \text{ if } fmt = \text{SP} \\ 18 \text{ if } fmt = \text{DP.} \end{cases}
$$

Note that $N = p + 2$ as minimally required except for the SP case, in which $N = 28 = p + 4$, i.e., 2 more bits are generated than needed.

Following the RTL, the first iteration of the algorithm is executed before entering the `for` loop of *fdivLane*. Each iteration of the loop executes only one iteration of the algorithm. The loop variable j, therefore, ranges from 2 to N. For the variables that are updated iteratively, we shall use j as a subscript to indicate their values after j iterations of the algorithm. These include the following:

(1) q_j: the jth quotient digit, an element of $\{1, 0, -1\}$. Now that these have been defined along with x and d, the partial quotient and remainder, Q_j and R_j, are also defined, according to Eqs. (10.3) and (10.4).
(2) $quot_j$ and $quotM1_j$: 55-bit vectors representing the partial quotient and decremented partial quotient with one implicit leading integer bit. Thus, $quot_j = 2^{54}Q_j$ and $quotM1_j = 2^{54}(Q_j - 2^{1-j})$.
(3) RS_j and RC_j: 57-bit vectors that form a redundant carry-save representation of the partial remainder with 2 implicit leading integer bits. Thus

$$
2^{55}R_j \equiv RS_j + RC_j \pmod{2^{57}}, \tag{21.1}
$$

With regard to (2) and (3), for scalar operations the RTL assigns these values to corresponding signals of the same widths on each iteration. For vector operations, however, these signals are partitioned into smaller segments to accommodate the values computed in each lane. This presents a challenge for the equivalence check, which depends on bit-accurate mappings between essential signals and RAC variables. This is addressed in the model by limiting the representation of the partial remainder to the upper bits of the component vectors. More precisely, the width of the segments used for each partial remainder is

$$w = \begin{cases} 57 \text{ if } vec = 1 \\ 27 \text{ if } vec = 0 \text{ and } fmt = SP \\ 14 \text{ if } vec = 0 \text{ and } fmt = HP. \end{cases}$$

Therefore, the function $nextRem$, which updates the remainder, is designed to ensure that $RS_j[56 - w : 0] = RC_j[56 - w : 0] = 0$ for vector operations. Note that this requires passing an additional parameter that determines w.

As discussed in Sect. 10.1, we aim to establish the invariant

$$d \le R_j < d. \tag{21.2}$$

As usual, this relation is preserved by the selection of the quotient digit q_{j+1} based on an approximation A_j of R_j, in this case derived from the 3 most significant bits of RS_j and RC_j. As also noted in Sect. 10.1, however, the relation (21.1) is not sufficient to prove the required accuracy of A_j as was done in the proof of Lemma 18.6 to establish the analogous bounds on R_j. For this reason, the definition of $nextRem$ is designed to ensure that a stronger version of (21.1) holds:

$$2^{55} R_j = si(RS_j, 57) + si(RC_j, 57). \tag{21.3}$$

Note that (21.1) follows from (21.3) and the identity $si(r, n) \equiv r \pmod{2^n}$.

Accordingly, for $j \ge 1$, we define the remainder approximation as

$$A_j = \frac{1}{2} \left(si(RS_j[56 : 54], 3) + si(RC_j[56 : 54], 3) \right). \tag{21.4}$$

Under this definition, (21.3) yields the following bounds on the approximation error:

Lemma 21.1 If $2^{55} R_j = si(RS_j, 57) + si(RC_j, 57)$, then $A_j \le R_j < A_j + 1$.

Proof By Lemma 2.35,

$$si(RS_j, 57) = 2^{54} si(RS_j[56 : 54], 3) + RS_j \bmod 2^{54}$$

and

$$si(RC_j, 57) = 2^{54} si(RC_j[56 : 54], 3) + RC_j \bmod 2^{54}.$$

Adding these equations, we have

$$2^{55} R_j = 2^{55} A_j + RS_j \bmod 2^{54} + RC_j \bmod 2^{54}.$$

Since $0 \le RS_j \bmod 2^{54} < 2^{54}$ and $0 \le RC_j \bmod 2^{54} < 2^{54}$, this yields

$$0 \le 2^{55}(R_j - A_j) < 2^{55}$$

and the lemma follows.   □

We shall use this result in an inductive proof of the invariance of both (21.2) and (21.3).

21.2   Pre-processing

The initial phase of *fdivLane* handles the exceptional cases in which at least one operand is a zero, a NaN, or an infinity and identifies the case of division by a power of 2 (i.e., $d = 1$). As usual, we omit the analysis of these simple special cases. In the remaining computational case, the operands are normalized if necessary and the significands x and d are assigned to the 57-bit variables $x57$ and $d57$, each of which has two implicit leading integer bits.

Lemma 21.2

(a) $x57 = 2^{55}x$;
(b) $d57 = 2^{55}d$;
(c) $\left| \frac{A}{B} \right| = 2^{expDiff-bias} \left(\frac{x}{d} \right)$.

Proof The function *normalize* also appeared in the design of Chap. 18. See the proof of Lemma 18.2.   □

The invariants (21.2) and (21.3) hold after the first iteration:

Lemma 21.3

(a) $-d \le R_1 < d$;
(b) $2^{55} R_1 = si(RS_1, 57) + si(RC_1, 57)$.

Proof To establish (a), note that $q_1 = 1$, and therefore

$$R_1 = 2R_0 - q_1 d = x - d,$$

and

$$-d < R_1 < 2 - d < 2d - d = d.$$

For the proof of (b), let

$$p' = \begin{cases} p & \text{if } vec = 1 \\ 53 & \text{if } vec = 0. \end{cases}$$

Then

$$RS_1 = x57 + 2^{55-p'}$$

and

$$RC_1 = 2^{55-p}(\sim d57[56 : 55 - p'])$$
$$= 2^{55-p}(2^{p+2} - d57[56 : 55 - p'] - 1)$$
$$= 2^{55-p}(2^{p+2} - 2^{p'-55}d57 - 1)$$
$$= 2^{57} - d57 - 2^{55-p'}.$$

Since $RS_1 \geq 2^{56}$ and $RC_1 < 2^{56}$,

$$si(RS_1, 57) + si(RC_1, 57) = (RS_1 - 2^{57}) + RC_1$$
$$= x57 - d57$$
$$= 2^{55}(x - d)$$
$$= 2^{55}R_1. \qquad \square$$

We have the following initial values of the partial quotient and decremented partial quotient:

Lemma 21.4 $quot_1 = 2^{54}Q_1$ and $quotM1_1 = 2^{54}(Q_1 - 1)$.

Proof By (10.3), $Q_1 = q_1 = 1$. Thus

$$2^{54}Q_1 = 2^{54} = quot_1$$

and

$$2^{54}(Q_1 - 1) = 0 = quotM1_1. \qquad \square$$

21.3 Iterative Phase

The following result is readily derived from the definitions and Lemma 21.4 by induction. The proof is similar to that of Lemma 18.7 and is omitted here:

Lemma 21.5 *For* $1 \leq j \leq N$, $quot_j = 2^{54} Q_j$ *and* $quotM1_j = 2^{54}(Q_j - 2^{1-j})$.

The rest of this section is devoted to a proof of the invariance of (21.2) and (21.3). Our analysis begins with the following characterization of the quotient digits:

Lemma 21.6 *For* $1 \leq j < N$,

$$q_{j+1} = \begin{cases} +1 \ \text{if } A_j \geq 0 \\ \ \ 0 \ \text{if } A_j = -\frac{1}{2} \\ -1 \ \text{if } A_j \leq -1. \end{cases}$$

Proof We refer to the value of the signed integer $R4$ computed by *nextDigit*:

$$R4 = si((si(RS[56:54], 3) + si(RC[56:54], 3)) \bmod 16, 4)$$

$$= si(2A_j \bmod 16, 4).$$

It follows from Lemma 2.31 that $-8 = -4 + -4 \leq 2A_j < 4 + 4 = 8$. By Lemma 2.32, $R4 = 2A_j$, and the lemma follows. □

The invariance of (21.2) follows from Lemmas 21.1 and 21.6:

Lemma 21.7 *Let* $j \geq 1$ *and assume that* $-d \leq R_j < d$ *and*

$$2^{55} R_j = si(RS_j, 57) + si(RC_j, 57).$$

Then $-d \leq R_{j+1} < d$.

Proof We have $R_{j+1} = 2R_j - q_{j+1}d$, where $-2d \leq 2R_j < 2d$. It is clear from (21.4) that $2A_j \in Z$ and by Lemma 21.1, $2A_j \leq 2R_j < 2A_j + 2$.
Case 1: $q_{j+1} = +1$.
$R_{j+1} = 2R_j - d < 2d - d = d$ and $A_j \geq 0$. Thus, $2R_j \geq 2A_j \geq 0$ and $R_{j+1} \geq 0 - d = -d$.
Case 2: $q_{j+1} = 0$.
$R_{j+1} = 2R_j$ and $A_j = -\frac{1}{2}$. Thus, $-d \leq -1 \leq 2R_j < -1 + 2 = 1 \leq d$ and $-d \leq R_{j+1} < d$.
Case 3: $q_{j+1} = -1$.
$R_{j+1} = 2R_j + d \geq -2d + d = -d$ and $A_j \leq -1$. Thus, $2R_j < 2A_j + 2 \leq -2 + 2 = 0$ and $R_{j+1} < 0 + d = d$. □

The proof of invariance of (21.3) is based on the function *nextRem58* below, a 58-bit extension of *nextRem*. Note that this function is not called by *fdiv2* but is defined here solely for the purpose of this proof.

```
<ui58, ui58> nextRem58(ui58 sum0, ui58 car0, ui58 d58, int q, ui2 fmt) {
  ui58 sumIn = sum0 << 1, carIn = car0 << 1, dIn = 0, sumOut = 0, carOut = 0;
  if (q == 0) {
    sumOut = sumIn ^ carIn;
```

```
    carOut[57:1] = sumIn[55:0] & carIn[55:0];
  }
  else  {
    if (q == 1) {
      dIn = ~d58;
      switch (fmt) {
      case SP: dIn[29:0] = 0; break;
      case HP: dIn[42:0] = 0;
      }
    }
    else  {
      dIn = d58;
    }
    sumOut = sumIn ^ carIn ^ dIn;
    carOut[57:1] = sumIn[56:0] & carIn[56:0] | sumIn[56:0] & dIn[56:0] |
                   carIn[56:0] & dIn[56:0];
    if (q == 1) {
      switch (fmt) {
      case DP: carOut[0] = 1; break;
      case SP: carOut[30] = 1; break;
      case HP: carOut[43] = 1;
      }
    }
  }
  if (sumOut[56] != carOut[56]) {
    sumOut[56] = 0;
    carOut[56] = 1;
    sumOut[57] = 0;
    carOut[57] = 1;
  }
  else if (q == 1) {
    sumOut[56] = 0;
    carOut[56] = 0;
    sumOut[57] = 0;
    carOut[57] = 0;
  }
  else if (q == -1) {
    sumOut[56] = 1;
    carOut[56] = 1;
    sumOut[57] = 1;
    carOut[57] = 1,
  }
  return <sumOut, carOut>;
}
```

We apply this function to the arguments RS_j, RC_j, $d57_{j+1}$, q_{j+1}, and *fmtRem*, where $1 \leq j < N$. Let S and C be the returned values, and let S' and C' be the values of the local variables *sumOut* and *carOut* before the adjustment of bits 56 and 57.

We note two simple but critical properties of S and C:

Lemma 21.8 $RS_{j+1} = S[56 : 0]$ *and* $RC_{j+1} = C[56 : 0]$.

Proof The computations are precisely the same.   □

Lemma 21.9 $S[57] = S[56]$ *and* $C[57] = C[56]$.

Proof This obviously holds if $q_{j+1} = \pm 1$. For $q_{j+1} = 0$, we need only show $S'[56] \neq C'[56]$. In this case, $2A_j = -1$, which implies

$$RS_j[56:54] + RC_j[56:54] \bmod 8 = 7,$$

and by Lemma 8.4, $RS_j[56:54] \wedge RC_j[56:54] = 7$. In particular,

$$S'[56] = RS_j[55] \wedge RC_j[55] = 1$$

and

$$C'[56] = RS_j[54] \And RC_j[54] = 0. \qquad \square$$

A third critical property of S and C is given by Corollary 21.12. It is a consequence of the next two lemmas, which pertain to S' and C'.

Lemma 21.10 *If* $2^{55}R_j \equiv RS_j + RC_j \pmod{2^{57}}$, *then*

$$2^{55}R_{j+1} \equiv S' + C' \pmod{2^{58}}.$$

Proof We note first that the hypothesis implies

$$2^{56}R_j \equiv 2(RS_j + RC_j) \pmod{2^{58}}.$$

We also note that the invariant

$$RS_j[56 - w : 0] = RC_j[56 - w : 0] = 0$$

is easily established by induction and inspection of the definition of *nextRem*.

The proof is a case analysis based on the recurrence relation of Lemma 10.1, which in the present context yields $R_{j+1} = 2R_j - q_{j+1}d$.

Case 1: $q_{j+1} = 0$.
  $S' = 2RS_j \wedge 2RC_j$, $C' = 4RS_j \And 4RC_j \bmod 2^{58}$, and by Lemma 8.2,

$$S' + C' \equiv 2(RS_j + RC_j) \equiv 2^{56}R_j = 2^{55}R_{j+1} \pmod{2^{58}}.$$

Case 2: $q_{j+1} = -1$.
  We have *dIn* $= d58$

$$S' = 2RS_j \wedge 2RC_j \wedge d58,$$

$$C' = (4RS_j \And 4RC_j \mid 4RS_j \And 2d58 \mid 4RC_j \And 2d58) \bmod 2^{58},$$

and by Lemma 8.7,

$$S' + C' \equiv 2(RS_j + RC_j) + d58$$
$$\equiv 2^{56}R_j + 2^{55}d$$
$$\equiv 2^{55}R_{j+1} \pmod{2^{58}}.$$

Case 3: $q_{j+1} = +1$.
In this case,

$$S' = 2RS_j \text{ ^ } 2RC_j \text{ ^ } dIn$$

and

$$C' = ((4RS_j \text{ \& } 4RC_j \mid 4RS_j \text{ \& } 2dIn \mid 4RC_j \text{ \& } 2dIn) + 2^{53-p}) \bmod 2^{58},$$

where

$$dIn = 2^{53-p}(\sim d58[57 : 53 - p])$$
$$= 2^{53-p}(2^{p+5} - d58[57 : 53 - p] - 1)$$
$$= 2^{58} - d58 - 2^{53-p},$$

and by Lemma 8.7,

$$S' + C' \equiv 2(RS_j + RC_j) + dIn + 2^{53-p}$$
$$\equiv 2(RS_j + RC_j) + (2^{58} - d58 - 2^{53-p}) + 2^{53-p}$$
$$\equiv 2(RS_j + RC_j) - d58$$
$$\equiv 2^{56}R_j - 2^{55}d$$
$$\equiv 2^{55}R_{j+1} \pmod{2^{58}}. \qquad \square$$

The next lemma allows us to replace S' and C' with S and C in Lemma 21.10. Its proof requires some computation and offers little intuition.

Lemma 21.11 *If* $2^{55}R_j = si(RS_j, 57) + si(RC_j, 57)$ *and* $|R_j| < 2$, *then*

$$S + C \equiv S' + C' \pmod{2^{58}}.$$

Proof It is clear that $S[55 : 0] = S'[55 : 0]$ and $C[55 : 0] = C'[55 : 0]$, so we need only show that

$$(S + C)[57 : 56] = (S' + C')[57 : 56]. \tag{21.5}$$

Let $g = \lfloor 2^{-56}(S[55:0] + C[55:0]) \rfloor$. By Lemma 2.9,

$$(S + C)[57:56] = (S[57:56] + C[57:56] + g)[1:0]$$

and

$$(S' + C')[57:56] = (S'[57:56] + C'[57:56] + g)[1:0].$$

We have the following dependencies:

- A_j is determined by $RS_j[56:54]$ and $RC_j[56:54]$;
- q_{j+1} is determined by A_j;
- $d58[57:55] = 1$;
- $S'[57:56]$ and $C'[57:56]$ are determined by $RS_j[56:54]$, $RC_j[56:54]$, $d58[57:55]$, and q_{j+1};
- $S[57:56]$ and $C[57:56]$ are determined by $S'[56]$, $C'[56]$, and q_{j+1}.

Thus, A_j, $(S+C)[57:56]$, and $(S'+C')[57:56]$ are all determined by $RS_j[56:54]$, $RC_j[56:54]$, and g. There are $8 \cdot 8 \cdot 2 = 128$ possible combinations of values to be considered. By hypothesis and Lemma 21.1, we need only examine those cases in which $-3 < A_j < 4$. It may be shown by exhaustive computation that in all such cases, (21.5) holds.   □

Corollary 21.12 *If* $2^{55}R_j = si(RS_j, 57) + si(RC_j, 57)$ *and* $|R_j| < 2$, *then*

$$2^{55}R_{j+1} \equiv S + C \pmod{2^{58}}.$$

Proof This is a trivial consequence of the preceding two lemmas.   □

We can now establish the invariance of (21.3):

Lemma 21.13 *If* $2^{55}R_j = si(RS_j, 57) + si(RC_j, 57)$, $|R_j| < 2$, *and* $|R_{j+1}| < 2$, *then*

$$2^{55}R_{j+1} = si(RS_{j+1}, 57) + si(RC_{j+1}, 57).$$

Proof By Corollary 21.12,

$$2^{55}R_{j+1} \equiv S + C \equiv si(S, 58) + si(C, 58) \pmod{2^{58}}.$$

By Lemmas 21.8, 21.9, and 2.36,

$$si(S, 58) = si(sextend(S[56:0], 58), 57) = si(S[56:0], 57) = si(RS_{j+1}, 57)$$

and similarly

$$si(C, 58) = si(RC_{j+1}, 57).$$

Thus

$$2^{55} R_{j+1} \equiv si(RS_{j+1}, 57) + si(RC_{j+1}, 57) \pmod{258}.$$

But since $|R_{j+1}| < 2$, $|2^{55} R_{j+1}| < 2^{56}$, and by Lemma 2.31,

$$-2^{57} \leq si(RS_{j+1}, 57) + si(RC_{j+1}, 57) < 2^{57}.$$

Thus

$$|2^{55} R_{j+1} - (si(RS_{j+1}, 57) + si(RC_{j+1}, 57))| < 2^{58}$$

and the lemma follows. □

We have proved the following:

Lemma 21.14 *For* $1 \leq j \leq N$,

(a) $-d \leq R_j < d$;
(b) $2^{55} R_j = si(RS_j, 57) + si(RC_j, 57)$.

Proof The base case is Lemma 21.3. The inductive step follows easily from previous results: (a) is the conclusion of Lemma 21.7, and (b) follows from Lemma 21.13 and (a). □

21.4 Post-processing

Upon exit from the `for` loop of *fdivLane*, a truncation of the quotient is derived from the final partial quotient and remainder, along with an indication of inexactness. These computations are the subject of this section. The remaining code, which performs rounding, identifies exceptions, and computes the outputs, is quite similar to the designs of preceding chapters and will not be addressed here.

In referring to the final values of *RS*, *RC*, *R*, and *Q*, we may drop the subscript *N* without ambiguity. The sum of *RS* and *RC* is assigned to the 57-bit signed integer variable *RFinal*. Thus

$$RFinal = si((RS + RC) \bmod 2^{57}, 57).$$

This value is related to *R* as follows:

Lemma 21.15 $R = 2^{-55} RFinal$.

Proof It follows from Lemma 21.14 that

$$-2^{56} < 2^{55} R < 2^{56}$$

and

$$2^{55}R \equiv RS + RC \pmod{2^{57}}.$$

Thus, by Lemma 2.32,

$$si((RS + RC) \bmod 2^{57}, 57) = si(2^{55}R \bmod 2^{57}, 57) = 2^{55}R. \qquad \square$$

As usual, the sign of the final remainder is required to determine whether the final quotient is to be incremented as well as inexactness of the result. In this radix-2 case, since the possibility that $R = -d$ cannot be ignored, that condition must be identified as well.

Lemma 21.16

(a) $R < 0 \Leftrightarrow Rsign = 1$;
(b) $R = 0 \Leftrightarrow Rzero = 1$;
(c) $R = -d \Leftrightarrow RplusDis0 = 1$;

Proof (a) and (b) follow from Lemma 21.15. For the proof of (c), note that by Lemmas 8.7 and 21.14(b),

$$RplusDS + RplusDC \equiv RS + RC + d57 \equiv 2^{55}(R + d) \pmod{2^{57}}.$$

But by Lemma 21.14(a), $0 \le 2^{55}(R + d) < 2^{57}$, and hence

$$2^{55}(R + d) = RplusDS + RplusDC \bmod 2^{57}.$$

Thus, by Lemma 8.8,

$$R = -d \Leftrightarrow 2^{55}(R + d) = 0$$

$$\Leftrightarrow RplusDS + RplusDC \bmod 2^{57} = 0$$

$$\Leftrightarrow \left(2(RplusDS \mid RplusDC) - (RplusDS \,\hat{}\, RplusDC)\right) \bmod 2^{57} = 0$$

$$\Leftrightarrow \left(2(RplusDS \mid RplusDC)\right) \bmod 2^{57} = RplusDS \,\hat{}\, RplusDC$$

$$\Leftrightarrow RplusDor = RplusDxor$$

$$\Leftrightarrow RplusDis0 = 1. \qquad \square$$

The vector $Qtrunc$ is a shifted truncation of the quotient:

Lemma 21.17

(a) $Qtrunc = 2^{55-N} \lfloor 2^{N-1} \frac{x}{d} \rfloor$;
(b) $Qtrunc = 2^{54} \frac{x}{d} \Leftrightarrow stk0 = 0.$

Proof It follows from Lemma 21.5 and the definition (10.3) that

$$2^{N-1}Q = 2^{N-55}quot \in \mathbb{Z},$$

and consequently $2^{N-55}quotM1 \in \mathbb{Z}$ and $2^{N-55}Qtrunc \in \mathbb{Z}$. Thus

$$Qtrunc[54:55-N] = \lfloor 2^{N-55}Qtrunc \rfloor = 2^{N-55}Qtrunc.$$

Combining (10.4) with Lemma 21.14(a) and dividing by d yields

$$-1 \le 2^{N-1}\frac{x}{d} - 2^{N-1}Q < 1,$$

or

$$-1 \le 2^{N-1}\frac{x}{d} - 2^{N-55}quot < 1.$$

Furthermore

$$2^{N-1}\frac{x}{d} - 2^{N-55}quot = 0 \Leftrightarrow R = 0 \Leftrightarrow Rzero = 1,$$

$$2^{N-1}\frac{x}{d} - 2^{N-55}quot < 0 \Leftrightarrow R < 0 \Leftrightarrow RSign = 1,$$

and

$$2^{N-1}\frac{x}{d} - 2^{N-55}quot = -1 \Leftrightarrow R = -d \Leftrightarrow RplusDis0 = 1.$$

Suppose $R \ge 0$. Then $Qtrunc = quot$ and

$$0 \le 2^{N-1}\frac{x}{d} - 2^{N-55}Qtrunc < 1,$$

which implies (a). Since $RplusDis0 = 0$,

$$Qtrunc = 2^{54}\frac{x}{d} \Leftrightarrow 2^{N-55}Qtrunc = 2^{N-1}\frac{x}{d} \Leftrightarrow R = 0 \Leftrightarrow Rzero = 1 \Leftrightarrow stk0 = 0.$$

On the other hand, suppose $R < 0$. Then $Qtrunc = quotM1$,

$$-1 \le 2^{N-1}\frac{x}{d} - 2^{N-55}quot < 0, \tag{21.6}$$

$$2^{N-55}Qtrunc = 2^{N-55}quotM1 = 2^{N-55}(quot - 2^{55-N}) = 2^{N-55}quot - 1,$$

and adding 1 to (21.6), we have once again $0 \le 2^{N-1}\frac{x}{d} - 2^{N-55}Qtrunc < 1$. Since $Rzero = 0$,

$$Qtrunc = 2^{54}\frac{x}{d} \Leftrightarrow 2^{N-55}Qtrunc = 2^{N-1}\frac{x}{d}$$

$$\Leftrightarrow 2^{N-55}quot - 1 = 2^{N-1}\frac{x}{d}$$

$$\Leftrightarrow 2^{N-1}\frac{x}{d} - 2^{N-55}quot = -1$$

$$\Leftrightarrow RplusDis0 = 1$$

$$\Leftrightarrow stk0 = 0. \qquad\qquad \square$$

Lemma 21.18

(a) If $\left|\frac{A}{B}\right| \ge spn(fmt)$, then $expQ > 0$, $expo(Qshft) = p$

$$Qshft = \left\lfloor 2^{p+bias-expQ}\left|\frac{A}{B}\right| \right\rfloor$$

and

$$Qshft = 2^{p+bias-expQ}\left|\frac{A}{B}\right| \Leftrightarrow stk = 0;$$

(b) If $\left|\frac{A}{B}\right| < spn(fmt)$, then $expQ \le 0$, $expo(Qshft) < p$

$$Qshft = \left\lfloor 2^{p+bias-1}\left|\frac{A}{B}\right| \right\rfloor,$$

and

$$Qshft = 2^{p+bias-1}\left|\frac{A}{B}\right| \Leftrightarrow stk = 0.$$

Proof
Case 1: $expDiff > 0$ and $Qtrunc[54] = 1$.
   In this case, $expQ = expDiff > 0$. Since $Qtrunc = 2^{55-N}\left\lfloor 2^{N-1}\frac{x}{d} \right\rfloor \ge 2^{54}$, $\frac{x}{d} \ge 1$. Thus

$$\left|\frac{A}{B}\right| = 2^{expDiff-bias}\frac{x}{d} \ge 2^{1-bias} = spn(fmt).$$

$Qshft$ is produced by right-shifting $Qtrunc$ by $54 - p$ bits, and hence

$$expo(Qshft) = expo(Qtrunc) - (54 - p) = 54 - (54 - p) = p.$$

By Lemma 1.2,

$$Qshft = \left\lfloor 2^{p-54} Qtrunc \right\rfloor$$

$$= \left\lfloor 2^{p-54} \left(2^{55-N} \left\lfloor 2^{N-1} \frac{x}{d} \right\rfloor \right) \right\rfloor$$

$$= \left\lfloor 2^{p+1-N} \left\lfloor 2^{N-1} \frac{x}{d} \right\rfloor \right\rfloor$$

$$= \left\lfloor 2^p \frac{x}{d} \right\rfloor$$

$$= \left\lfloor 2^{p+bias-expDiff} \left| \frac{A}{B} \right| \right\rfloor$$

$$= \left\lfloor 2^{p+bias-expQ} \left| \frac{A}{B} \right| \right\rfloor,$$

and

$$Qshft[= 2^{p+bias-1} \left| \frac{A}{B} \right| \Leftrightarrow 2^{N-1} \frac{x}{d} \in \mathbb{Z} \text{ and } 2^{p-54} Qtrunc \in \mathbb{Z}$$

$$\Leftrightarrow stk0 = Qtrunc[53 - p : 0] = 0$$

$$\Leftrightarrow stk = 0.$$

Case 2: expDiff ≤ 0.
  In this case, $expQ = expDiff \leq 0$

$$\left| \frac{A}{B} \right| = 2^{expDiff-bias} \frac{x}{d} < 2^{1-bias} = spn(fmt),$$

and *Qtrunc* is right-shifted twice, first by $1 - expQ$ bits and then by $54 - p$. Thus

$$expo(Qshft) = expo(Qtrunc) - (1 - expQ) - (54 - p) \leq 54 - 1 - (54 - p) = p - 1,$$

$$Qshft = \left\lfloor 2^{p-54} \left\lfloor 2^{expQ-1} Qtrunc \right\rfloor \right\rfloor$$

$$= \left\lfloor 2^{p-54} \left\lfloor 2^{expQ-1} \left(2^{55-N} \left\lfloor 2^{N-1} \frac{x}{d} \right\rfloor \right) \right\rfloor \right\rfloor$$

$$= \left\lfloor 2^{expQ-1+p} \frac{x}{d} \right\rfloor$$

$$= \left\lfloor 2^{bias+p-1} \left| \frac{A}{B} \right| \right\rfloor,$$

and

$$Qshft = 2^{bias+p-1}\left|\frac{A}{B}\right| \Leftrightarrow 2^{N-1}\frac{x}{d} \in \mathbb{Z} \text{ and } 2^{(p-54)+(expQ-1)}Qtrunc \in \mathbb{Z}$$

$$\Leftrightarrow stk0 = Qtrunc[54 - p - expQ] = 0$$

$$\Leftrightarrow stk = 0.$$

Case 3: $expDiff > 1$ and $Qtrunc[54] = 0$.
   We have $expQ = expDiff - 1 > 0$ and

$$\left|\frac{A}{B}\right| = 2^{expDiff-bias}\frac{x}{d} \geq 2^{1-bias}\frac{1}{2} = 2^{1-bias} = spn(fmt).$$

The proof is completed as in Case 1, but with $Qtrunc$ and $expDiff$ replaced by $2Qtrunc$ and $expDiff - 1$.

Case 4: $expDiff = 1$ and $Qtrunc[54] = 0$.
   In this case, $expQ = expDiff - 1 = 0$. Since $Qtrunc = 2^{55-N}\left\lfloor 2^{N-1}\frac{x}{d}\right\rfloor < 2^{54}$, $\frac{x}{d} < 1$ and

$$\left|\frac{A}{B}\right| = 2^{expDiff-bias}\frac{x}{d} < 2^{1-bias} = spn(fmt).$$

$Qtrunc$ is right-shifted by $54 - p$. Thus

$$expo(Qshft) = expo(Qtrunc) - (54 - p) = 53 - (54 - p) = p - 1.$$

The proof is completed as in Case 1, but with $expQ$ replaced by 1.   □

As in previous chapters, we define the rounding mode

$$\mathcal{R}' = \begin{cases} RDN & \text{if } \mathcal{R} = RUP \text{ and } sign = 1 \\ RUP & \text{if } \mathcal{R} = RDN \text{ and } sign = 1 \\ \mathcal{R} & \text{otherwise.} \end{cases}$$

Lemma 21.19

(a) If $\left|\frac{A}{B}\right| \geq spn(fmt)$, then

$$Qrnd = 2^{p-1+bias-expQ}\mathcal{R}'\left(\left|\frac{\hat{A}}{\hat{B}}\right|, p\right)$$

and

$$\mathcal{R}'\left(\left|\frac{\hat{A}}{\hat{B}}\right|, p\right) = \left|\frac{\hat{A}}{\hat{B}}\right| \Leftrightarrow inx = 0;$$

(b) If $\left|\frac{A}{B}\right| < spn(fmt)$, then

$$Qrnd = 2^{p-2+bias}drnd\left(\left|\frac{\hat{A}}{\hat{B}}\right|, \mathcal{R}', fmt\right)$$

and

$$drnd\left(\left|\frac{\hat{A}}{\hat{B}}\right|, \mathcal{R}', fmt\right) = \left|\frac{\hat{A}}{\hat{B}}\right| \Leftrightarrow inx = 0.$$

Proof We invoke Lemma 6.97 with

$$z = \begin{cases} 2^{p+bias-expQ}\left|\frac{\hat{A}}{\hat{B}}\right| & \text{if } \left|\frac{\hat{A}}{\hat{B}}\right| \geq spn(fmt) \\ 2^{p+bias-1}\left|\frac{\hat{A}}{\hat{B}}\right| & \text{if } \left|\frac{\hat{A}}{\hat{B}}\right| < spn(fmt) \end{cases}$$

and $n = expo(Qshft)$. Unless $n < 1$, the lemma yields

$$Qrnd = \frac{1}{2}\mathcal{R}'(z, n),$$

If $\left|\frac{A}{B}\right| \geq spn(fmt)$, then $n = p$ and this reduces to the first claim.

If $\left|\frac{A}{B}\right| < spn(fmt)$ and $n \geq 1$, then

$$n = expo\left(\frac{\hat{A}}{\hat{B}}\right) + p + bias - 1 = expo\left(\frac{\hat{A}}{\hat{B}}\right) + p - expo(spn(fmt))$$

and according to Definition 6.9, we have

$$Qrnd = \frac{1}{2}\mathcal{R}'\left(2^{p+bias-1}\left|\frac{\hat{A}}{\hat{B}}\right|, n\right) = 2^{p+bias-2}drnd\left(\left|\frac{\hat{A}}{\hat{B}}\right|, \mathcal{R}', fmt\right).$$

In the remaining case, $n = 0$, $Qshft = 1$, and the same result follows from Lemma 6.101.

In all cases, the rounded result is exact iff $Qshft$ is exact and $Qshft[0] = 0$, or equivalently, $inx = 0$. $\qquad\square$

The outputs are computed by the function *final*, which is essentially the same as the corresponding function of Chap. 18, as is the proof of the following:

Theorem 21.1 *Let*

$$\langle D_{spec}, R_{spec} \rangle = arm\text{-}binary\text{-}spec(DIV, opa, opb, R_{in}, fmt),$$

where fmt $\in$ *{DP, SP, HP}, opa and opb are 64-bit vectors, and R_{in} is a 32-bit vector, and let*

$$\langle D, flags \rangle = fdiv2(opa, opb, fmt, R_{in}[24], R_{in}[25], R_{in}[23:22]).$$

Then $D = D_{spec}$ *and* $R_{in} \mid flags = R_{spec}$.

Chapter 22
Fused Multiply-Add of a Graphics Processor

The defining characteristic of a graphics processing architecture is the coordination of hundreds or thousands of identical multipurpose arithmetic units, which may be executed in parallel to perform the same operations on large sets of data. Though originally designed for efficient image rendering, such hardware has been used in a variety of scientific applications that are naturally accelerated by massive data parallelism. From the perspective of arithmetic circuit design, the primary implication of this scheme is the prioritization of area and power conservation over execution speed.

The subject of this chapter is a component of an Arm graphics processing unit that performs single-precision addition, multiplication, and a variety of fused multiply-add and scaling operations. This module, the pseudocode version of which is available online at https://go.sn.pub/fma32, is replicated 128 times per core and accounts for 13% of the core's area and 7% of its power consumption. Most of the choices underlying its design are driven by these considerations. Note that area and power are sometimes conflicting objectives: in this case, the SP and HP adders were separated in order to lower power at the expense of an area penalty.

Since the multiplier is not a time-critical component of this design, unlike the DP multiplier of Chap. 16, the 24×24 bit vector multiplication is treated as a primitive to be implemented in synthesis. On the other hand, in contrast to the adder of Chap. 17, this design uses a custom-coded adder. When left to a synthesis tool, the selection of an adder from a design library is generally based on prevailing timing constraints, under the assumption that its inputs are all available at the same time and its outputs are simultaneously generated. In the present case, this assumption is both false and costly. The most significant bit of the sum is the sign bit that determines whether the remaining bits are to be complemented. Consequently, this bit is needed several gate delays before the rest of the sum. In a trial synthesis run, this subtlety was not addressed, and the result was an expensive Kogge-Stone adder that produced the top bit as required but wasted hardware computing the full sum earlier than needed. This was replaced by a hand-coded Han-Carlson adder with more suitable

© The Author(s), under exclusive license to Springer Nature Switzerland AG 2022   409
D. M. Russinoff, *Formal Verification of Floating-Point Hardware Design*,
https://doi.org/10.1007/978-3-030-87181-9_22

timing properties, resulting in an overall 10% reduction in the area of the FMA unit. The implication of such a decision for formal verification at the RTL level is the requirement of analysis of the previously unexplored design discipline of parallel prefix adders (Sect. 8.2).

Another feature of this module that stems from area considerations is its leading zero anticipator. In contrast to the LZA of Chap. 17, which is based on an assumed ordering of the operands, this design uses the LZA of Lemma 8.28, which produces a correct result under either ordering. As discussed in Sect. 8.3, it is more complicated and slower but reduces overall area because the simpler design must be replicated to account for the two possible orderings.

22.1 Overview

The inputs of the top-level function `fma32` are as follows:

- `ui32 a, b, c`: Single-precision encodings of the operands;
- `bool scaleOp`: A boolean indication of a scaled operation;
- `si32 d`: The scale factor as a signed integer, valid only for a scaled operation;
- `ui3 rMode`: A 3-bit encoding of a rounding mode. As displayed in Table 22.1, in addition to the usual IEEE modes, the modes *RNA* and *RTO* are also implemented; the values 6 and 7 are disallowed.

Table 22.1 Rounding control

Encoding	Rounding mode
000	*RUP*
001	*RDN*
010	*RTZ*
011	*RNE*
100	*RNA*
101	*RTO*

For the purpose of this analysis, we shall assume that a, b, and c are all numerical (normal, denormal, or zero); the trivial case of a NaN or infinity operand is disallowed. Let A, B, and C denote their decoded values, and let $P = AB$. We also ignore the trivial case $P = C = 0$. The objective is to compute the result of rounding the value

$$S = \begin{cases} 2^d(P + C) & \text{if } scaleOp = 1 \\ P + C & \text{if } scaleOp = 0 \end{cases}$$

according to the rounding mode $\mathcal{R}$ represented by *rMode*. The sole value returned is a single-precision encoding of this result; the module does not record exceptions.

The function *fma32* calls three auxiliary functions in succession: the multiplier, *fmul32*; a rescaling module, *scale128*; and the adder, *fadd32*. The multiplier returns an unnormalized representation of the product, consisting of a sign bit, a 9-bit exponent with a bias of 255, and a 48-bit significand field. In the event of a denormal addend c and under certain additional conditions, the operands are rescaled, shifting c into the normal range, as described in Sect. 22.3. The results are then passed to the adder.

22.2 Multiplication

The following lemma refers to the values computed by *fmul32*:

Lemma 22.1 *If $P = 0$, then $pExp = pMant = 0$, and otherwise,*

(a) $4 \leq pExp \leq 510$;
(b) If $pMant < 2^{23}$, then $pExp = 4$;
(c) $P = (-1)^{pSign} 2^{pExp-255-47} pMant$.

Proof The zero case is trivial. In the remaining case, it is clear that $P < 0$ iff $pSign = 1$ and that

$$|A| = 2^{aExp-127-23} aMant$$

and

$$|B| = 2^{bExp-127-23} bMant,$$

where $1 \leq aExp \leq 254$ and $1 \leq bExp \leq 254$. Thus

$$4 \leq pExp = aExp + bExp + 2 \leq 254 + 254 + 2 = 510$$

and

$$\begin{aligned} |P| &= |AB| \\ &= 2^{aExp-127-23} aMant \cdot 2^{bExp-127-23} bMant \\ &= 2^{pExp-255-47} aMant bMant \\ &= 2^{pExp-255-47} pMant. \end{aligned}$$

Finally, if $pMant < 2^{23}$, then $aMant < 2^{23}$ and $bMant < 2^{23}$, which implies $aExp - bExp = 1$ and $pExp = 4$. $\square$

22.3   Rescaling

The purpose of rescaling is to avoid the following situation. In the event of a denormal addend, for a certain range of values of the product exponent, the addition may produce massive cancellation resulting in two full mantissa widths of leading zeroes (or ones). If this were to occur in a scaled operation with a large positive scaling factor, then an additional mantissa width of significant bits of the sum would have to be retained to be shifted into the final result. Under these circumstances, the resulting necessity of widening the adder is obviated by rescaling: the addend is effectively multiplied by 2^{128} to shift it into the normal range, and the product exponent and scale factor are adjusted accordingly to preserve the sum.

The global inputs $scaleOp$, a, b, c, and d are passed to $scale128$ along with the product exponent $pExp$. Let c', $pExp'$, and $scale$ denote the outputs. Under the conditions described above, these are generated by adjustments of c, $pExp$, and d; otherwise, the inputs are simply passed through, except that $scale$ is coerced to 0 for an unscaled operation, or into the range $-512 \leq scale \leq 511$ for a scaled operation. We define the adjusted values

$$\tilde{C} = decode(c', SP),$$

$$\tilde{P} = (-1)^{pSign} 2^{pExp'-302} pMant,$$

and

$$\tilde{S} = 2^{scale}(\tilde{P} + \tilde{C}).$$

Lemma 22.2   *Assume that either $scaleOp = 0$ or $-512 \leq d \leq 511$.*
If $scaleOp = 1$, $P \neq 0$, C is denormal, $64 \leq a[30:23] + b[30:23] < 256$, and $d \geq 16$, then

$$\tilde{P} = 2^{128}P \text{ and } \tilde{C} = 2^{128}C;$$

otherwise

$$\tilde{P} = P \text{ and } \tilde{C} = C.$$

Proof   We shall derive the value of $\tilde{C}$ in the first case; the other three claims are trivial.

Let $s = c[31]$, $e = c[30:23]$, and $m = c[22:0]$. It is clear that $e = 0$ and

$$C = ddecode(c, SP) = (-1)^s 2^{-149} m.$$

Let c' be the transformation of c returned by $scale128$ and let $s' = c'[31]$, $e' = c'[30:23]$, and $m' = c'[22:0]$. It is also clear that $e' > 0$ and

$$\tilde{C} = (-1)^{s'} 2^{e'-150}(2^{23} + m').$$

Now $s' = s$, $e' = 129 - clz$, where $clz = 23 - expo(m)$, and therefore

$$expo(2^{clz}m) = clz + expo(m) = 23$$

and

$$m' = (2^{clz}m) \bmod 2^{23} = 2^{clz}m - 2^{23}.$$

Thus

$$\tilde{C} = (-1)^{s} 2^{(129-clz)-150}(2^{23} + (2^{clz}m - 2^{23})) = (-1)^{s} 2^{-21}m = 2^{128}C. \qquad \square$$

Corollary 22.3 $\tilde{S} = 0 \Leftrightarrow S = 0$ and $\tilde{S} > 0 \Leftrightarrow S > 0$.

Proof $\tilde{S} = 0 \Leftrightarrow \tilde{P} + \tilde{C} = 0 \Leftrightarrow P + C = 0 \Leftrightarrow S = 0$; the other equivalence is similar. $\qquad \square$

Lemma 22.4 *Assume $S \neq 0$.*

(a) *If $scaleOp = 1$ and $d > 511$, then $|S| > 2 \cdot lpn(SP)$ and $|\tilde{S}| > 2 \cdot lpn(SP)$;*
(b) *If $scaleOp = 1$ and $d < -512$, then $|S| < \frac{1}{2}spd(SP)$ and $|\tilde{S}| < \frac{1}{2}spd(SP)$;*
(c) *If $scaleOp = 0$ or $-512 \leq d \leq 511$, then $S = \tilde{S}$.*

Proof
(a) Since A, B, and C are all multiples of $spd(SP) = 2^{-149}$, $P + C$ is a multiple of 2^{-298} and

$$|S| = 2^{d}|P + C| = 2^{512}2^{-298} = 2^{214} > 2 \cdot lpn(SP).$$

Furthermore, $scale = 511$ and

$$|\tilde{S}| = 2^{scale}|\tilde{P} + \tilde{C}| \geq 2^{511}|P + C| \geq 2^{511}2^{-298} = 2^{213} > 2 \cdot lpn(SP).$$

(b) Since A, B, and C are all bounded by $lpn(SP) = 2^{127}$, $|P + C| < 2^{255}$ and

$$0 < |S| = 2^{d}|P + C| < 2^{-512}2^{255} = 2^{-257} < \frac{1}{2}spd(SP).$$

Furthermore, $scale = -512$ and

$$0 < |\tilde{S}| = 2^{scale}|\tilde{P} + \tilde{C}| = 2^{-512}|P + C| < \frac{1}{2}spd(SP).$$

(c) If $P \neq 0$, C is denormal, $66 \leq aExp + bExp < 258$, and $d \geq 16$, then $scale = d - 128$, and by Lemma 22.2,

$$\tilde{S} = 2^{scale}(\tilde{P} + \tilde{C}) = 2^{d-128}(2^{128}P + 2^{128}C) = 2^d(P + C) = S.$$

In the remaining case, $scale = d$ and

$$\tilde{S} = 2^{scale}(\tilde{P} + \tilde{C}) = 2^d(P + C) + S. \qquad \square$$

The purpose of the shift performed by $scale128$ is to ensure the following condition, which is used in the proof of Lemma 22.16:

Lemma 22.5 *If* $0 < cMant < 2^{47}$ *and* $67 \leq pExp \leq 257$, *then scale* < 16.

Proof The hypothesis implies that c has not been normalized by $scale128$, and therefore

$$pExp = aExp + bExp + 2,$$

which implies $64 \leq a[30:23] + b[30:23] < 256$. It follows that $scale < 16$, for otherwise c would be normalized. $\qquad \square$

22.4 Addition

Unless otherwise noted, the following lemmas refer to the parameters and local variables of *fadd32*. Lemmas 22.6–22.10 are self-evident and stated without proof:

Lemma 22.6

(a) *If* $pMant = 0$, *then* $pExp = 0$;
(b) *If* $pMant > 0$, *then* $4 \leq pExp \leq 510$;
(c) *If* $0 < pMant < 2^{23}$, *then* $pExp = 4$.

Lemma 22.7

(a) $2^{-24}cMant \in \mathbb{Z}$;
(b) *If* $cMant < 2^{47}$, *then* $cExp = 128$;
(c) *If* $cMant > 0$, *then* $128 \leq cExp \leq 382$ *and* $cMant \geq 2^{25}$.

We define the following parameters:

- $e_L = \begin{cases} cExp \text{ if } cLarger = 1 \\ pExp \text{ if } cLarger = 0 \end{cases}$

- $e_S = \begin{cases} pExp \text{ if } cLarger = 1 \\ cExp \text{ if } cLarger = 0 \end{cases}$

- $m_L = \begin{cases} cMant & \text{if } cLarger = 1 \\ pMant & \text{if } cLarger = 0s \end{cases}$

- $m_S = \begin{cases} pMant & \text{if } cLarger = 1 \\ cMant & \text{if } cLarger = 0 \end{cases}$

- $\Delta = e_L - e_S$

- $\Phi = \begin{cases} 4(m_L + 2^{-\Delta}m_S) & \text{if } sub = 0 \\ 4(m_L - 2^{-\Delta}m_S) & \text{if } sub = 1 \end{cases}$

Lemma 22.8

(a) $add1 = 2^{51} + \lfloor 2^{2-\Delta}m_S \rfloor$ and $2^{2-\Delta}m_S \in \mathbb{Z} \Leftrightarrow inx = 0$;

(b) $add2 = \begin{cases} 2^{51} + 4m_L & \text{if } sub = 0 \\ 2^{51} - 4m_L - 1 & \text{if } sub = 1. \end{cases}$

Lemma 22.9

(a) If $sub = 0$, then $2^{52} < add1 + add2 < 2^{52} + 2^{51}$;

(b) If $sub = 1$, then $2^{51} + 2^{50} < add1 + add2 < 2^{52} + 2^{50}$.

Lemma 22.10

(a) $inx = 0 \Leftrightarrow \Phi \in \mathbb{Z}$;

(b) If $inx = 1$ and $cLarger = 0$, then $\Delta \geq 27$;

(c) If $inx = 1$ and $cLarger = 1$, then $\Delta \geq 3$.

Lemma 22.11 $|\tilde{S}| = 2^{scale + e_L - 304}\Phi$.

Proof If $sub = 1$, then

$$\begin{aligned} |\tilde{S}| &= 2^{scale}(\tilde{P} + \tilde{C}) \\ &= 2^{scale}\lfloor 2^{e_L - 302}m_L - 2^{e_S - 302}m_S \rfloor \\ &= 2^{scale + e_L - 304}|4(m_L - 2^{-\Delta}m_S)| \\ &= 2^{scale + e_L - 304}\Phi. \end{aligned}$$

The case $sub = 0$ is similar.   □

Lemma 22.12 Assume $\tilde{S} \neq 0$.

(a) If $sub = 0$, then $\tilde{S} > 0 \Leftrightarrow signLarger = 0$;

(b) If $sub = 1$ and $m_L > 2^{-\Delta}m_S$, then $\tilde{S} > 0 \Leftrightarrow signLarger = 0$;

(c) If $sub = 1$ and $m_L < 2^{-\Delta}m_S$, then $\tilde{S} > 0 \Leftrightarrow signLarger = 1$.

Proof If $sub = 0$, then $signLarger = cSign = pSign$ and (a) follows trivially from Lemma 22.4.

Suppose $sub = 1$ and $cLarger = 1$. We have $signLarger = cSign$, $e_L = cExp$, $e_S = pExp$, $m_L = cMant$, and $m_S = pMant$. By Lemma 22.4,

$$\tilde{S} = 2^{scale}(\tilde{C} + \tilde{P})$$
$$= 2^{scale}(-1)^{cSign}(2^{e_L-302}m_L - 2^{e_S-302}m_S)$$
$$= (-1)^{signLarger}2^{scale+e_L-302}(m_L - 2^{-\Delta}m_S)$$

and (b) and (c) follow. The case $signLarger = 0$ is similar.   □

The next two lemmas refer to the function *computeSum*.

Lemma 22.13

(a) $gOut = 1 \Leftrightarrow add1 + add2 \geq 2^{52}$;

(b) $pOut = 1 \Leftrightarrow add1 + add2 = 2^{52} - 1$;

(c) $sum = \begin{cases} (add1 + add2 + inc)[50 : 0] & \text{if } gOut[51] = 1 \\ \sim(add1 + add2 + inc)[50 : 0] & \text{if } gOut[51] = 0. \end{cases}$

Proof The Han-Carlson adder *HC64* is easily shown to be an implementation of Definition 8.8. The lemma follows from Lemmas 8.25, 8.11, and 8.9.   □

The case of "severe cancellation" is handled separately:

Lemma 22.14 *Assume severe* $= 1$.

(a) *If* $inx = 0$, *then* $\Phi = \tilde{S} = S = 0$;

(b) *If* $inx = 1$, *then* $0 < \Phi < 1$, $\tilde{S} \neq 0$, *and* $\tilde{S} > 0 \Leftrightarrow resSign = 0$.

Proof By Lemma 22.13, $add1 + add2 = 2^{52} - 1$ and $gOut[51] = 0$, which implies $sub = 1$ and $toggleSign = 1$. By Lemma 22.8,

$$2^{52} - 1 = add1 + add2$$
$$= (2^{51} - 4m_L - 1) + (2^{51} + \lfloor 2^{2-\Delta}m_S \rfloor)$$
$$= 2^{52} - (4m_L - \lfloor 2^{2-\Delta}m_S \rfloor) - 1,$$

which implies $4m_L - \lfloor 2^{2-\Delta}m_S \rfloor = 0$.

If $inx = 0$, then $2^{-\Delta}m_S \in \mathbb{Z}$ and

$$\Phi = |4m_L - 2^{2-\Delta}m_S| = |4m_L - \lfloor 2^{2-\Delta}m_S \rfloor| = 0.$$

Suppose $inx = 1$. Since $2^{-\Delta}m_S \notin \mathbb{Z}$, $4m_L = \lfloor 2^{2-\Delta}m_S \rfloor < 2^{2-\Delta}m_S$,

$$\Phi = 2^{2-\Delta}m_S - 4m_L > 0,$$

and

$$\lfloor \Phi \rfloor = \lfloor 2^{2-\Delta}m_S - 4m_L \rfloor = \lfloor \lfloor 2^{2-\Delta}m_S \rfloor - 4m_L \rfloor = 0,$$

which implies $\Phi < 1$. By Lemma 22.11, $\tilde{S} \neq 0$. By Lemma 22.12 and the definition of *sign*,

$$\tilde{S} > 0 \Leftrightarrow signLarger = 1 \Leftrightarrow sign = 0. \qquad \square$$

Lemma 22.15 *If severe* $= 0$*, then*

(a) $\Phi > 0$*;*
(b) $sum = \lfloor \Phi \rfloor$*;*
(c) $\tilde{S} > 0 \Leftrightarrow resSign = 0$.

Proof We consider the following cases:
Case 1: $sub = 0$.

Since $add1 + add2 > 2^{52}$, $gOut[51] = 1$, and since $inc = 0$, by Lemmas 22.8, 22.9, and 22.13,

$$
\begin{aligned}
sum &= (add1 + add2)[50:0] \\
&= add1 + add2 - 2^{52} \\
&= 4m_L + \lfloor 2^{2-\Delta} m_S \rfloor \\
&= \lfloor 4m_L + 2^{2-\Delta} m_S \rfloor \\
&= \lfloor \Phi \rfloor,
\end{aligned}
$$

and $\Phi \geq add1 + add2 - 2^{52} > 0$.

Since $sign = signLarger$, (c) follows from Lemma 22.12.
Case 2: $sub = 1, gOut[51] = 1$.

In this case, $inc = 1$, $toggleSign = 0$, and $sign \neq signLarger$. By Lemmas 22.8, 22.9, and 22.13,

$$
\begin{aligned}
sum &= (add1 + add2 + 1) \bmod 2^{50} \\
&= add1 + add2 + 1 - 2^{52} \\
&= (2^{51} + \lfloor 2^{2-\Delta} m_S \rfloor) + (2^{51} - 4m_L - 1) + 1 - 2^{52} \\
&= \lfloor 2^{2-\Delta} m_S \rfloor - 4m_L \\
&= \lfloor 2^{2-\Delta} m_S - 4m_L \rfloor \\
&= \lfloor \Phi \rfloor,
\end{aligned}
$$

and $\Phi \geq sum = add1 + add2 + 1 - 2^{52} > 0$.
By Lemma 22.12, $\tilde{S} > 0 \Leftrightarrow signLarger = 1 \Leftrightarrow sign = 0$.

Case 3: $sub = 1$, $gOut[51] = 0$.

We have $toggleSign = 1$, $sign = signLarger$

$$(add1 + add2 + inc)[50:0] = (add1 + add2 + inc) \bmod 2^{51}$$
$$= add1 + add2 + inc - 2^{51},$$

and

$$sum = \sim(add1 + add2 + inc)[50:0]$$
$$= 2^{51} - (add1 + add2 + inc - 2^{51}) - 1$$
$$= 2^{52} - (add1 + add2 + inc + 1)$$
$$= 2^{52} - ((2^{51} + \lfloor 2^{2-\Delta} m_S \rfloor) + (2^{51} - 4m_L - 1) + inc + 1)$$
$$= 4m_L - \lfloor 2^{2-\Delta} m_S \rfloor - inc.$$

If $inx = 1$, then $inc = 1$, $\Phi \notin \mathbb{Z}$, $2^{2-\Delta} m_S \notin \mathbb{Z}$, and

$$sum = 4m_L - \lfloor 2^{2-\Delta} m_S \rfloor - 1$$
$$= 4m_L + \lfloor -2^{2-\Delta} m_S \rfloor$$
$$= \lfloor 4m_L - 2^{2-\Delta} m_S \rfloor$$
$$= \lfloor \Phi \rfloor.$$

If $inx = 0$, then $inc = 0$, $\Phi \in \mathbb{Z}$

$$sum = 4m_L - \lfloor 2^{2-\Delta} m_S \rfloor$$
$$= 4m_L - 2^{2-\Delta} m_S$$
$$= \lfloor 4m_L - 2^{2-\Delta} m_S \rfloor$$
$$= \lfloor \Phi \rfloor,$$

and we also have

$$\Phi = sum = 2^{52} - (add1 + add2 + 1) > 2^{52} - ((2^{52} - 1) + 1) = 0.$$

In either case, by Lemma 22.12, $\tilde{S} > 0 \Leftrightarrow signLarger = 0 \Leftrightarrow sign = 0$. □

Lemma 22.16 *If severe $= 0$, then*

(a) $2^{sumExp-1098} sum \leq |\tilde{S}| < 2^{sumExp-1098}(sum + 1)$;
(b) $2^{sumExp-1098} sum = |\tilde{S}| \Leftrightarrow inx = 0$.

Proof This follows from Lemmas 22.11 and 22.15 and the definition of *sumExp*.

□

Lemma 22.17 *Assume that either scale > 15 or $e_L > 128$.*

(a) If severe $= 1$, then inx $= 0$;
(b) If severe $= 0$, then sum ≥ 2;
(c) If severe $= 0$ and inx $= 1$, then sum $\geq 2^{24}$.

Proof Note that by Lemma 22.14, in the case of (a), we have $\Phi < 1$ and by Lemma 22.15, in cases (b) and (c), it suffices to show that the bounds are satisfied by Φ.

Case 1: cMant $= 0$.

In this case, $cLarger = 0$, $e_S = cMant = 0$, $\Phi = 4m_L = 4pMant \geq 4$, which implies $severe = 0$, and $2^{2-\Delta}m_S = 0 \in \mathbb{Z}$, which implies $inx = 0$.

Case 2: cMant > 0, sub $= 0$.

If $cLarger = 0$, then $e_L = pExp > cExp \geq 128$, and by Lemma 22.6, $m_L = pMant \geq 2^{23}$, and if $cLarger = 1$, then $m_L = cMant \geq 2^{25}$. In either case,

$$\Phi = 4m_L + 2^{2-\Delta}m_S \geq 4m_L \geq 2^{25}.$$

Case 3: cMant > 0, sub $= 1$, pMant $= 0$.

In this case, $cLarger = 1$ and $\Phi = 4cMant \geq 2^{27}$.

Case 4: cMant > 0, sub $= 1$, $0 < pMant < 2^{23}$.

Since $pExp = 4 < cExp$, $cLarger = 1$ and

$$\Phi = cMant - 2^{2-\Delta}pMant > 2^{27} - 2^2 2^{23} > 2^{26}.$$

Case 5: cMant > 0, pMant $\geq 2^{23}$, sub $= 1$, cLarger $= 0$.

If $inx = 1$, then by Lemma 22.10, $\Delta \geq 27$ and

$$\Phi = 4pMant - 2^{2-\Delta}cMant > 2^{25} - 2^{-25}2^{48} > 2^{24}.$$

Thus, we may assume $severe = inx = 0$. Since $\Phi \in \mathbb{Z}$ (Lemma 22.10) and $\Phi > 0$ (Lemma 22.15), we need only show $\Phi \neq 1$. But if $\Phi = 1$, then $\Delta \geq 26$, for otherwise

$$\frac{1}{2}\Phi = \frac{1}{2}(4pMant - 2^{2-\Delta}cMant) = 2pMant - 2^{25-\Delta}2^{-24}cMant \in \mathbb{Z},$$

and hence

$$\Phi = 4pMant - 2^{2-\Delta}cMant > 2^{25} - 2^{-24}2^{48} = 2^{24},$$

a contradiction.

Case 6: pMant $\geq 2^{23}$, sub $= 1$, cLarger $= 1$, cMant $\geq 2^{47}$.

If $inx = 1$, then by Lemma 22.10, $\Delta \geq 3$ and

$$\Phi = 4cMant - 2^{2-\Delta}pMant > 2^{49} - 2^{-2}2^{48} > 2^{48}.$$

As in Case 5, we may assume $severe = inx = 0$ and need only show $\Phi \neq 1$. But if $\Phi = 1$, then $\Delta \geq 2$, for otherwise

$$\frac{1}{2}\Phi = 2cMant - 2^{1-\Delta}pMant \in \mathbb{Z},$$

and hence

$$\Phi = 4cMant - 2^{2-\Delta}pMant > 2^{49} - 2^{48} = 2^{48}.$$

Case 7: $pMant \geq 2^{23}$, $sub = 1$, $cLarger = 1$, $cMant < 2^{47}$.

In this case, $e_L = cExp = 128$ and we must have $scale \geq 16$. By Lemma 22.5, either $pExp < 67$ or $pExp > 257$, but since $pExp \leq e_L = 128$, we have $pExp < 67$, $\Delta = 128 - pExp \geq 62$, and

$$\Phi = 4cMant - 2^{2-\Delta}pMant > 4 \cdot 2^{25} - 2^{60}2^{48} > 2^{24}. \qquad \square$$

Lemma 22.18 *If* $severe = 1$ *or* $sumExp < 896$, *then* $|\tilde{S}| < \frac{1}{2}spd(SP)$.

Proof If $severe = 1$, then by Lemmas 22.11, 22.15, and 22.17,

$$|\tilde{S}| = 2^{scale+e_L-304}\Phi < 2^{15+128-304} = 2^{-161} < 2^{-150} = \frac{1}{2}spd(SP).$$

If $severe = 0$ and $sumExp < 896$, then by Lemma 22.16

$$|\tilde{S}| < 2^{sumExp-1098}(sum+1) < 2^{896-1098}2^{51} = 2^{-152} < \frac{1}{2}spd(SP). \qquad \square$$

We define the following parameters:

- $clz' = \begin{cases} clz & \text{if } overShft = 0 \\ clz - 1 & \text{if } overShft = 1 \end{cases}$
- $sumShft' = 2^{clz'}sum$
- $\Psi = \begin{cases} sumShft[77:53] & \text{if } overShft = 1 \\ sumShft[76:52] & \text{if } overShft = 0 \end{cases}$
- $\Omega = sumExp - clz' + 1$

Lemma 22.19 *Assume* $severe = 0$ *and* $sumExp \geq 896$.

(a) If $expo(sum) < 972 - sumExp$, *then* $\Omega = 897$ *and*

$$clz' = clz = sumExp - 896 \leq 76 - expo(sum);$$

(b) If $expo(sum) \geq 972 - sumExp$, *then* $\Omega \geq 897$ *and*

$$clz' = 76 - expo(sum).$$

Proof The following properties the intermediate value w computed by *LZA52* are provided by Lemma 8.28:

(1) If $add1 + add2 > 2^{52}$ or $add1 + add2 < 2^{52} - 2$, then

$$expo(w) \leq expo(sum) \leq expo(w) + 1;$$

(2) If $add1 + add2 = 2^{52}$ or $add1 + add2 = 2^{52} - 2$, then $w = 0$ and $sum \leq 1$.

Also note that $972 - sumExp \leq 972 - 896 = 76$.

(a) In this case, $expo(w) \leq expo(vec) \leq 972 - sumExp$, which implies

$$expo(vec) = 972 - sumExp \geq expo(sum),$$

$$clz = 76 - expo(vec) \leq 76 - expo(sum),$$

and

$$expo(sumShft) = expo(sum) + clz \leq 76,$$

which implies $overShft = sumShft[77] = 0$ and $clz' = clz$.
By definition,

$$\Omega = sumExp - clz' + 1$$
$$= sumExp - (sumExp - 896) + 1$$
$$= 897.$$

(b) First note that if $sum < 2$, then by Lemma 22.19, $scale \leq 15$ and $e_L = 128$, which implies

$$sumExp \leq 15 + 128 + 794 = 937$$

and

$$972 - sumExp \geq 972 - 937 = 35 > expo(sum),$$

a contradiction. Therefore, $expo(sum) \in \{expo(w), expo(w) + 1\}$.
If $expo(sum) = expo(w)$, then $expo(vec) = expo(sum)$, $clz = 76 - expo(sum)$,

$$expo(sumShft) = expo(sum) + clz = 76,$$

$overShft = sumShft[77] = 0$, and $clz' = clz$.
If $expo(sum) = expo(w) + 1 = 972 - sumExp$, then once again $expo(vec) = expo(sum)$ and the same conclusions follow.

In the remaining case, $expo(sum) = expo(w) + 1 > 972 - sumExp$, which implies

$$expo(vec) = expo(w) = expo(sum) - 1,$$

$$clz = 76 - (expo(sum) - 1) = 77 - expo(sum),$$

$$expo(sumShft) = expo(sum) + clz = 77,$$

$overShft = 1$, and $clz' = clz - 1 = 76 - expo(sum)$.
   In this case,

$$
\begin{aligned}
\Omega &= sumExp - clz' + 1 \\
&\geq 972 - expo(sum) - clz' + 1 \\
&= 972 - (76 - clz') - clz' + 1 \\
&= 897.
\end{aligned}
$$

□

Lemma 22.20 *If* $severe = 0$ *and* $sumExp - clz' > 1149$, *then*

$$expo(\tilde{S}) > expo(lpn(SP)).$$

Proof By Lemmas 22.11 and 22.15 and the definition of $sumExp$,

$$
\begin{aligned}
expo(\tilde{S}) &= scale + e_L - 304 + expo(\Phi) \\
&= (sumExp - 794) - 304 + expo(sum) \\
&= sumExp - 1098 + expo(sum).
\end{aligned}
$$

By Lemma 22.19, since $clz' \neq sumExp - 896$, we must have $clz' = 76 - expo(sum)$
and

$$
\begin{aligned}
expo(\tilde{S}) &= sumExp - 1098 + (76 - clz') \\
&= (sumExp - clz') - 1022 \\
&\geq 1150 - 1022 \\
&= 128 \\
&= expo(lpn(SP)) + 1.
\end{aligned}
$$

□

Lemma 22.21 *If* $severe = 0$ *and* $896 \leq sumExp \leq 1149 + clz'$, *then*

$$897 \leq \Omega \leq 1150.$$

Proof The lower bound is given by Lemma 22.19; the upper bound follows from the definition of Ω:

$$\Omega = sumExp - clz' + 1 \le (1149 + clz') - clz' + 1 = 1150. \qquad \square$$

Lemma 22.22 *Assume severe = 0 and $896 \le sumExp \le 1149 + clz'$.*

(a) If $\Psi[24] = 0$, then $resExp = 0$ and $\Omega = 897$;
(b) If $\Psi[24] = 1$, then $resExp = \Omega - 896$ and $1 \le resExp \le 254$.

Proof
(a) In this case, $sumShft[76] = sumShft[77] = 0$. If $expo(sum) \ge 972 - sumExp$, then by Lemma 22.19,

$$expo(sumShft') = expo(sum) + clz' = 76,$$

contradicting $sumShft'[76] = \Psi[24] = 0$. Thus. $expo(sum) < 972 - sumExp$, which implies $\Omega = 897$. By direct computation, $resExp = 0$.
(b) Let $e = \Omega \bmod 256$. It is clear that

$$resExp = \begin{cases} e - 128 \text{ if } e[7] = 1 \\ e + 128 \text{ if } e[7] = 0. \end{cases}$$

Suppose $\Omega \ge 1024$. Since $\Omega \le 1150$, $e = \Omega - 1024 \le 126$, which implies $e[7] = 0$ and

$$resExp = e + 128 = \Omega - 1024 + 128 = \Omega - 896.$$

Similarly, if $\Omega < 1024$, $\Omega \ge 897$ implies $e - \Omega - 768 \ge 129$ and

$$resExp = e - 128 = \Omega - 768 - 128 = \Omega - 896.$$

The bounds on $resExp$ follow from Lemma 22.21. $\qquad \square$

Lemma 22.23 *Assume severe = 0 and $896 \le sumExp \le 1149 + clz'$.*

(a) $resMant = sumShft'[75 : 53] = \Psi[23 : 1]$;
(b) $grd = sumShft'[52] = \Psi[0]$;
(c) $stk = 0 \Leftrightarrow sumShft'[51 : 0] = 0$.

Lemma 22.24 *Assume severe = 0 and $896 \le sumExp \le 1149 + clz'$.*

(a) $2^{\Omega - 1047} \Psi \le |\tilde{S}| < 2^{\Omega - 1047} (\Psi + 1)$;
(b) $2^{\Omega - 1047} \Psi = |\tilde{S}| \Leftrightarrow stk = 0$.

Proof Since $expo(sumShft') = expo(sum) + clz' \le 76$,

$$sumShft' = 2^{52} sumShft'[76 : 52] + sumShft'[51 : 0] = 2^{52} \Psi + sumShft'[51 : 0],$$

which implies

$$2^{52}\Psi \leq sumShft' < 2^{52}(\Psi + 1),$$

with equality iff $sumShft'[51:0] = 0$. Multiplying by $2^{\Omega-1099}$ and observing that

$$2^{\Omega-1099}sumShft' = 2^{sumExp-clz'+1-1099}2^{clz'}sum = 2^{sumExp-1098}sum$$

yields

$$2^{\Omega-1047}\Psi \leq 2^{sumExp-1098}sum < 2^{\Omega-1047}(\Psi + 1)$$

with the same condition for equality. Combining this with Lemmas 22.16 and 22.23 (c), we have

$$2^{\Omega-1047}\Psi \leq |\tilde{S}| < 2^{\Omega-1047}(\Psi + 1).$$

with equality holding iff $stk = 0$.

It remains to show that $|\tilde{S}| < 2^{\Omega-1047}(\Psi + 1)$. Since $stk = 0$ implies $|\tilde{S}| = 2^{\Omega-1047}\Psi$, we may assume $stk = 1$. It will suffice to show that $clz' \leq 52$, for in this case,

$$sum < 2^{52-clz'}(\Psi + 1) \in \mathbb{Z}$$

implies

$$sum + 1 < 2^{52-clz'}(\Psi + 1)$$

and

$$\begin{aligned}
|\tilde{S}| &< 2^{sumExp-1098}(sum + 1) \\
&\leq 2^{sumExp-1098+52-clz'}(\Psi + 1) \\
&= 2^{\Omega-1047}(\Psi + 1).
\end{aligned}$$

If $expo(sum) \geq 24$, then $clz' \leq 76 - 24 = 52$. In the remaining case, according to Lemma 22.17, $scale \leq 15$ and $e_L = 128$, which implies

$$sumExp = scale + e_L + 794 \leq 15 + 128 + 794 = 937$$

and

$$972 - sumExp \geq 972 - 937 = 35 > expo(sum).$$

It then follows from Lemma 22.19 that

$$clz' = clz = sumExp - 896 \leq 937 - 896 = 41. \qquad \Box$$

Lemma 22.25 *Assume severe = 0 and $896 \leq sumExp \leq 1149 + clz'$.*

(a) $|\tilde{S}| < 2 \cdot lpn(SP)$;
(b) $|\tilde{S}| \geq spn(SP) \Leftrightarrow \Psi[24] = 1$;
(c) $|\tilde{S}| \geq spd(SP) \Leftrightarrow \Psi \geq 2$;
(d) $|\tilde{S}| \geq \frac{1}{2}spd(SP) \Leftrightarrow \Psi \geq 1$;
(e) $\Psi \geq 1 \Leftrightarrow \tilde{S} = S$;
(f) $\Psi = 0 \Leftrightarrow |\tilde{S}| < \frac{1}{2}spd(SP)$ and $0 < |S| < \frac{1}{2}spd(SP)$.

Proof
(a) By Lemmas 22.21 and 22.24,

$$|\tilde{S}| < 2^{\Omega-1047}(\Psi + 1) \leq 2^{1150-1047}2^{25} = 2^{128} = 2 \cdot lpn(SP),$$

(b) If $\Psi[24] = 1$, then by Lemmas 22.21, 22.22, and 22.24

$$|\tilde{S}| \geq 2^{\Omega-1047}\Psi \geq 2^{1150-1047}2^{25} = 2^{-126} = spn(SP);$$

otherwise,

$$|\tilde{S}| < 2^{\Omega-1047}(\Psi + 1) \leq 2^{897-1047}2^{24} = spn(SP).$$

(c) and (d) follow similarly.
(e) and (f) now follow from Lemma 22.4. $\qquad \Box$

Lemma 22.26 *If severe = 0, $896 \leq sumExp \leq 1149 + clz'$, and $\Psi \geq 2$, then*

$$\mathcal{R}(S, expo(\Psi)) = (-1)^{resSign}2^{\Omega-1046}(\Psi[24:1] + rndInc).$$

Proof Let

$$\mathcal{R}' = \begin{cases} RDN & \text{if } \mathcal{R} = RUP \text{ and } resSign = 1 \\ RUP & \text{if } \mathcal{R} = RDN \text{ and } resSign = 1 \\ \mathcal{R} & \text{otherwise.} \end{cases}$$

By Lemmas 6.82 and 22.15 (c), it suffices to show that

$$\mathcal{R}(|S|, expo(\Psi)) = 2^{\Omega-1046}(\Psi[24:1] + rndInc).$$

We shall invoke Lemma 6.97 with $\mathcal{R} = \mathcal{R}'$, $z = 2^{1047-\Omega}|S|$, and $n = expo(\Psi)$. By Lemma 22.24, $\lfloor z \rfloor = \Psi$ and $x \in \mathbb{Z}$ iff $stk = 0$. By Definition 6.2,

$$RTZ(\psi, n) = 2\lfloor 2^{-1}\Psi \rfloor = 2\Psi[24:1],$$

and by Definition 4.3,

$$fp^+(2\Psi[24:1], n) = 2\Psi[24:1] + 2.$$

By a straightforward case analysis based on Lemma 6.97 and the definition of *computeInc*,

$$\mathcal{R}'(z, n) = \begin{cases} 2\Psi[24:1] & \text{if } rndInc = 0 \\ 2\Psi[24:1] + 2 & \text{if } rndInc = 1. \end{cases}$$

Thus

$$\mathcal{R}'(|S|, n) = 2^{\Omega-1047}\mathcal{R}'(s, n) = 2^{\Omega-1046}(\Psi[24:1] + 1). \qquad \Box$$

Lemma 22.27 *Assume severe* $= 0$, $896 \le sumExp \le 1149 + clz'$, *and* $|S| \ge spn(SP)$. *Let* $R = \mathcal{R}(S, 24)$.

(a) *If* $|R| \le lpn(SP)$, *then* $resRnd = nencode(R, SP)$;
(b) *If* $|R| > lpn(SP)$, *then* $resRnd = iencode(resSign, SP)$ *and one of the following holds (i)* $\mathcal{R} = RNE$, *(ii)* $\mathcal{R} = RUP$ *and* $resSign = 0$, *or (iii)* $\mathcal{R} = RDN$ *and* $resSign = 1$.

Proof By Lemma 22.25, $\Psi[24] = 1$ and hence $expo(\Psi) = 24$. By Lemma 22.23, $\Psi[24:1] = 2^{23} + resMant$. By Lemmas 22.26 and 22.22,

$$R = (-1)^{resSign}2^{\Omega-1046}(\Psi[24:1] + resInc)$$
$$= (-1)^{resSign}2^{resExp-150}(2^{23} + resMant + resInc),$$

where $1 \le resExp \le 254$. Let $m = resRnd[22:0]$, $e = resRnd[30:23]$, and $s = resRnd[31] = resSign$.

Suppose $resMant + resInc < 2^{23}$. Then $m = resMant + resInc$, $e = resExp$, *resRnd* is a normal encoding with

$$ndecode(resRnd, SP) = (-1)^s 2^{e-150}(2^{23} + m) = R,$$

which implies $|R| \le lpn(SP)$, and by Lemma 5.2, $resRnd = nencode(R, SP)$.

Thus, we may assume $resMant + resInc = 2^{23}$, which implies $m = 0$, $e = resExp + 1$, and

$$R = (-1)^{resSign}2^{e-151}(2^{23} + 2^{23}) = (-1)^{resSign}2^{e-127}.$$

If $resExp = 254$, then $e = 255$, $resRnd = iencode(resSign, SP)$, and

$$|R| = 2^{128} = 2 \cdot lpn(SP).$$

In this case, we must have $resInc = 1$ and the claim regarding $\mathcal{R}$ follows.

On the other hand, if $resExp \leq 253$, then $e \leq 254$, $resRnd$ is a normal encoding with

$$ndecode(resRnd, SP) = (-1)^s 2^{e-150} 2^{23} = (-1)^s 2^{e-127} = R,$$

and the claim again follows from Lemma 5.2.   □

Lemma 22.28 *Assume* $severe = 0$, $896 \leq sumExp \leq 1149 + clz'$, *and* $|S| < spn(SP)$. *Let* $D = drnd(S, \mathcal{R}, SP)$.

(a) If $D = 0$, *then* $resRnd = zencode(resSign, SP)$;
(b) If $0 < |D| < spn(SP)$, *then* $resRnd = dencode(D, SP)$;
(c) If $|D| = spn(SP)$, *then* $resRnd = nencode(D, SP)$.

Proof We first consider the case $|\tilde{S}| < spd(SP)$. If $|\tilde{S}| \geq \frac{1}{2}spd(SP)$, then Lemma 22.25 implies $S = \tilde{S}$ and $\Psi = 1$, Lemma 22.23 implies $resMant = 0$ and $grd = 1$, and the lemma follows from Lemma 6.101. On the other hand, if $|\tilde{S}| < \frac{1}{2}spd(SP)$, then Lemma 22.25 implies $0 < |\tilde{S}|$, $0 < |S| < \frac{1}{2}spd(SP)$, and $\Psi = 0$, which implies $resMant = grd = 0$ and $stk = 1$, and (a) and (b) follow from Lemma 6.101 and Corollary 22.22.

Now suppose $|\tilde{S}| \geq spd(SP)$. By Lemma 22.25, $1 \leq expo(\Psi) \leq 24$. By Lemmas 22.24 and 22.22,

$$expo(S) = expo(\Psi) + \Omega - 1047 = expo(\Psi) - 150,$$

or

$$expo(\Psi) = expo(S) + 150 = expo(S) + 24 - expo(spn(SP)).$$

By Definition 6.9 and Lemmas 22.26, 22.22, and 22.23,

$$D = (-1)^{resSign} 2^{\Omega - 1046} (\Psi[24 : 1] + rndInc)$$
$$= (-1)^{resSign} 2^{-149} (resMant + rndInc)$$

and $resExp = 0$. Let $m = resRnd[22 : 0]$, $e = resRnd[30 : 23]$, and $s = resRnd[31] = resSign$.

Suppose $resMant + rndInc < 2^{23}$. Then $m = resMant + rndInc$, $e = resExp = 0$, $resRnd$ is a denormal encoding with

$$ddecode(resRnd, SP) = (-1)^s 2^{-149} m = D,$$

and by Lemma 5.8, $resRnd = dencode(D, SP)$.

In the remaining case, $resMant + rndInc = 2^{23}$, which implies $m = 0$, $e = resExp + 1 = 1$, and

$$D = (-1)^{resSign} 2^{-149} 2^{23} = (-1)^{resSign} 2^{-126}.$$

Thus, $|D| = spn(SP)$ and $resRnd = nencode(D, SP)$.   □

22.5 Final Result

The absence of NaN and infinity operands, control inputs (other than rounding mode), and exceptions allows a simpler interface specification than those of the preceding chapters:

Theorem 22.1 *Let*

$$resRnd = fma32(a, b, c, d, rMode, scaleOp),$$

where a, b, and c are numerical SP encodings with values A, B, and C, respectively, and d is a 32-bit signed integer. rMode is a 3-bit encoding of a rounding mode $\mathcal{R}$ as specified in Table. 22.1, and scaleOp is boolean. Let

$$S = \begin{cases} 2^d(AB + C) & \text{if } scaleOp = 1 \\ AB + C & \text{if } scaleOp = 0, \end{cases}$$

$$R = \mathcal{R}(S, 24),$$

and

$$D = drnd(S, \mathcal{R}, S).$$

(a) If $S = 0$, then

$$resRnd = zencode(\sigma, SP),$$

where $\sigma = c[31]$ if $c[31] = a[31] \,\hat{}\, b[31]$, and otherwise $\sigma = 1 \Leftrightarrow \mathcal{R} = RDN$;
(b) Suppose $|R| > lpn(SP)$. If (i) $\mathcal{R} = RDN$ and $R > 0$, (ii) $\mathcal{R} = RUP$ and $R < 0$, or (iii) $\mathcal{R} = RTZ$, then

$$resRnd = nencode(sgn(R) \cdot lpn(SP), SP),$$

and otherwise

$$resRnd = \begin{cases} iencode(0, SP) \text{ if } R > 0 \\ iencode(1, SP) \text{ if } R < 0; \end{cases}$$

(c) If $0 < |S| < spn(SP)$, then

$$decode(resRnd, SP) = D$$

and if $D = 0$, then $resRnd[31] = 1 \Leftrightarrow S < 0$;
(d) In the remaining case, $resRnd = nencode(R, SP)$.

Proof The trivial case $C = AB = 0$ may be readily verified by inspection. We consider the following remaining cases:
Case 1: $severe = 1$ and $inx = 0$.

By Corollary 22.3 and Lemma 22.14, $S = 0$. The theorem follows by inspection of the definitions.
Case 2: $severe = 1$ and $inx = 1$ or $severe = 0$ and $sumExp < 896$.

By Lemmas 22.4 and 22.18, $|S| = |\tilde{S}| < \frac{1}{2}spd(SP)$, and by Lemma 22.14, $S \neq 0$ and $S > 0 \Leftrightarrow resSign = 0$. In this case, the theorem follows from Lemma 6.101 and inspection of the definition of *fadd32*.
Case 3: $severe = 0$ and $sumExp - clz' > 1149$.

By Lemmas 22.4 and 22.20, $expo(S) > expo(lpn(SP))$. The theorem follows by inspection of the definitions.
Case 4: $severe = 0$, $896 \leq sumExp - clz' \leq 1149$, and $|S| \geq spn(SP)$.

The theorem follows from Lemma 22.27.
Case 5: $severe = 0$, $896 \leq sumExp - clz' \leq 1149$, and $|S| < spn(SP)$.

The theorem follows from Lemma 22.28.   □

Bibliography

1. S. Boldo and G. Melquiond. Emulation of a FMA and correctly rounded sums: Proved algorithms using rounding to odd. Technical Report HAL Id: 00080427, Inria, November 2010.
2. A. D. Booth. A signed binary multiplication technique. *Quarterly Journal of Mechanics and Applied Mathematics*, 4, 1951.
3. A. W. Burks, H. H. Goldstine, and J. von Neumann. Preliminary discussion of the logical design of an electronic computing instrument. In B. Randell, editor, *The Origins of Digital Computers*. Springer, Berlin, 1982.
4. J. Cocke and D. W. Sweeney. High speed arithmetic in a parallel device. Available at http://archive.computerhistory.org/resources/text/IBM/Stretch/pdfs/06-08/102632302.pdf, February 1957.
5. M. D. Ercegovac and T. Lang. *Division and Square Root Digit-Recurrence Algorithms and Implementations*. Kluwer Academic Publishers, 1994.
6. R. Gamboa. *Real-Valued Algorithms in ACL2*. PhD thesis, University of Texas at Austin, Department of Computer Sciences, May 1999.
7. J. Harrison. Formal verification of IA-64 division algorithms. In M. Aagaard and J. Harrison, editors, *Theorem Proving in Higher Order Logics: 13th International Conference*, volume 1869 of *Lecture Notes in Computer Science*. Springer-Verlag, 2000.
8. Institute of Electrical and Electronic Engineers. IEEE standard for floating point arithmetic, Std. 754-1985, 1985.
9. Institute of Electrical and Electronic Engineers. IEEE standard for floating point arithmetic, Std. 754-2008, 2008.
10. N. M. Josuttis. *The C++ Standard Library: A Tutorial and Reference*. Pearson Education, Inc., 2nd edition, 2012.
11. M. Kaufmann, P. Manolios, and J S. Moore, editors. *Computer-Aided Reasoning: ACL2 Case Studies*. Kluwer Academic Press, 2000.
12. M. Kaufmann, P. Manolios, and J S. Moore. *Computer-Aided Reasoning: An Approach*. Kluwer Academic Press, 2000.
13. M. Kaufmann and J S. Moore. ACL2 web site. http://www.cs.utexas.edu/users/moore/acl2/.
14. M. Kline. *Mathematical Thought from Ancient to Modern Times*. Oxford University Press, 1972.
15. I. Koren. *Computer Arithmetic Algorithms*. A. K. Peters, 2nd edition, 1993.
16. J. Levine. *Flex and Bison*. O'Reilly Media, 2009.
17. D. R. Lutz. *The Power of the Half-Adder Form*. PhD thesis, The Ohio State University, 1996.

© The Author(s), under exclusive license to Springer Nature Switzerland AG 2022
D. M. Russinoff, *Formal Verification of Floating-Point Hardware Design*,
https://doi.org/10.1007/978-3-030-87181-9

18. D. R. Lutz. Fused multiply-add architecture comprising separate early-normalizing multiply and add pipelines. In *20th IEEE Symposium on Computer Arithmetic*, 2011. Available at https://www.computer.org/csdl/proceedings/arith/2011/9457/00/05992117.pdf.

19. P. W. Markstein. Computation of elementary functions on the IBM RISC System/6000 processor. *IBM Journal of Research and Development*, 34(1), January 1990.

20. Mentor Graphics Corp. Algorithmic C datatypes. Available at https://www.mentor.com/hls-lp/downloads/ac-datatypes.

21. Mentor Graphics Corp. Sequential logic equivalence checker. https://www.mentor.com/products/fv/questa-slec.

22. R. Mizouni, Tahar S, and P. Curzon. Hybrid verification integerating HOL theorem proving with MDG model checking. *Microelectronics Journal*, 37(11), November 2006.

23. J S. Moore, T. Lynch, and M. Kaufmann. A mechanically checked proof of the correctness of the kernel of the $AMD5_K86$ floating point division algorithm. *IEEE Transactions on Computers*, 47(9), September 1998.

24. V. Oklobdzija. An algorithmic and novel design of a leading zero detector circuit: Comparison with logic synthesis. *IEEE Transactions on Very Large Scale Integration (VLSI) Systems*, 2(1):124–128, March 1994.

25. J. W. O'Leary and D. M. Russinoff. Modeling algorithms in SystemC and ACL2. In *ACL2 2014: 12th International Workshop on the ACL2 Theorem Prover and its Applications*, Vienna, July 2014. Available at http://www.russinoff.com/papers/masc.html.

26. J.-A. Pineiro, S. F. Oberman, J.-M. Muller, and J. D. Bruguera. High-speed function approximation using a minimax quadratic interpolator. *IEEE Transactions on Computers*, 54(3), March 2005. Available at http://perso.ens-lyon.fr/jean-michel.muller/QuadraticIEEETC0305.pdf.

27. S. Ray and R. Sumners. Combining theorem proving with model checking through predicate abstraction. *IEEE Design and Test Special Issue on Advances in Functional Validation through Hybrid Techniques*, 2007.

28. J. E. Robertson. A new class of digital division methods. *IRE Transactions on Electronic Computers*, EC-7, 1958.

29. D. M. Russinoff. A mechanically checked proof of IEEE compliance of the AMD-K7 floating point multiplication, division, and square root instructions. *London Mathematical Society Journal of Computation and Mathematics*, 1:148–200, December 1998. Available at http://www.russinoff.com/papers/k7-div-sqrt.html.

30. D. M. Russinoff. A mechanically checked proof of IEEE compliance of the AMD-K5 floating point square root microcode. *Formal Methods in System Design*, 14:75–125, 1999. Available at http://www.russinoff.com/papers/fsqrt.html.

31. D. M. Russinoff. A case study in formal verification of register-transfer logic with ACL2: the floating point adder of the AMD Athlon processor. In *Formal Methods in Computer-Aided Design*, 2000. Available at http://www.russinoff.com/papers/fadd.html.

32. D. M. Russinoff. A mechanically verified commercial SRT divider. In David S. Hardin, editor, *Design and Verification of Microprocessor Systems for High-Assurance Applications*, pages 23–63. Springer, 2010.

33. D. M. Russinoff. Computation and formal verification of SRT quotient and square root digit selection tables. *IEEE Transactions on Computers*, 62(5):900–913, May 2013. Available at http://www.russinoff.com/papers/srt8.html.

34. D. M. Russinoff and A. Flatau. *RTL verification: a floating-point multiplier*, chapter 13, pages 201–231. In Kaufmann et al. [11], 2000. Available at http://www.russinoff.com/papers/acl2.pdf.

35. C. H. Seger, R. B. Jones, J. W. O'Leary, T. Melham, M. D. Aagaard, C. Barrett, and D. Syme. An industrially effective environment for formal hardware verification. *IEEE Transactions on Computer-Aided Design of Integrated Circuits and Systems*, 24(9):1381–1405, September 2005.

36. Synopsys, Inc. Hector. http://www.synopsys.com/Tools/Verification/FunctionalVerification/
Pages/hector.aspx.

37. K. D. Tocher. Techniques of multiplication and division for automatic binary computers.
Quarterly Journal of Mechanics and Applied Mathematics, 2, 1958.

Index

© The Author(s), under exclusive license to Springer Nature Switzerland AG 2022
D. M. Russinoff, *Formal Verification of Floating-Point Hardware Design*,
https://doi.org/10.1007/978-3-030-87181-9

Printed in the United States
by Baker & Taylor Publisher Services

Printed in the United States
by Baker & Taylor Publisher Services